A GUIDE TO PETROLEUM NOMOGRAPHY

A GUIDE TO

PETROLEUM NOMOGRAPHY

PPC Books
Tulsa, OK

CONTENTS

PART I How to Make Nomograms

PART II Nomograms and Production

PART III Nomograms and Pipelines

PART IV Nomograms and Gas Processing

PART V NOMOGRAMS AND REFINING

The Petroleum Publishing Company
Tulsa, OK

ISBN: 0-87814-042-5

Printed in U. S. A.

Using these six standard nomogram forms, you can learn . . .

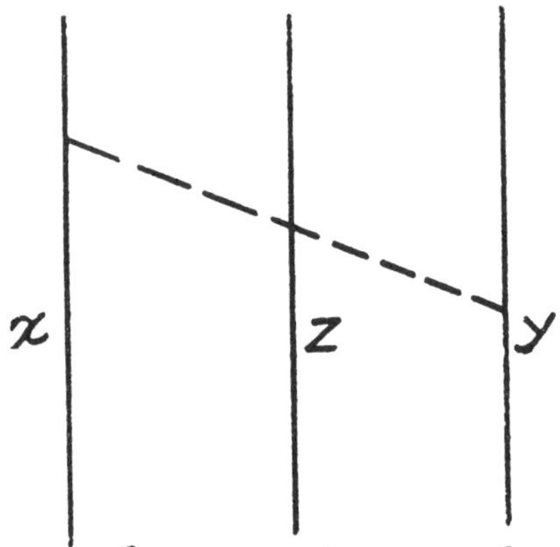

$f_1(x)+f_2(y)=f_3(Z)$

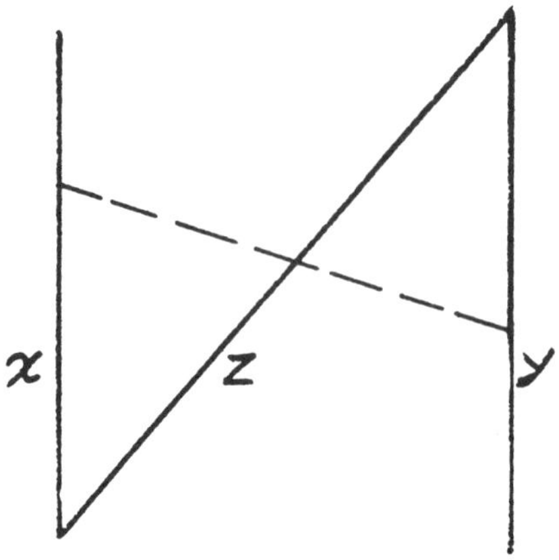

$f_1(x)=f_2(y)\cdot f_3(Z)$

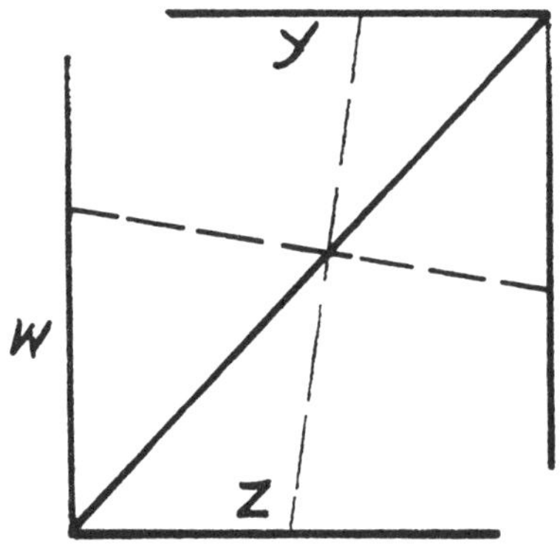

$f_1(w)/f_2(x)=f_3(y)/f_4(Z)$

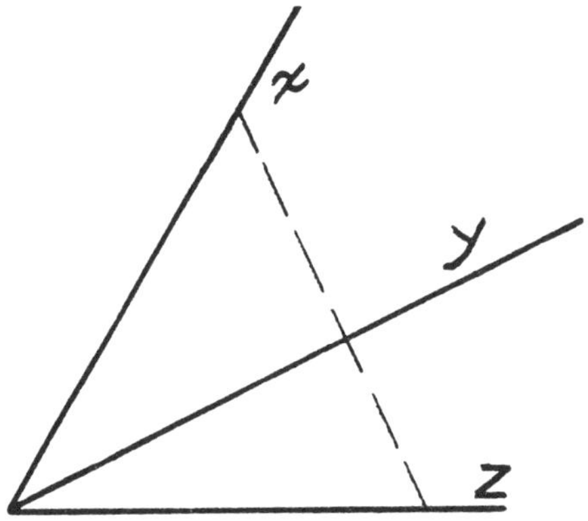

$1/f_1(x)+1/f_2(y)=1/f_3(Z)$

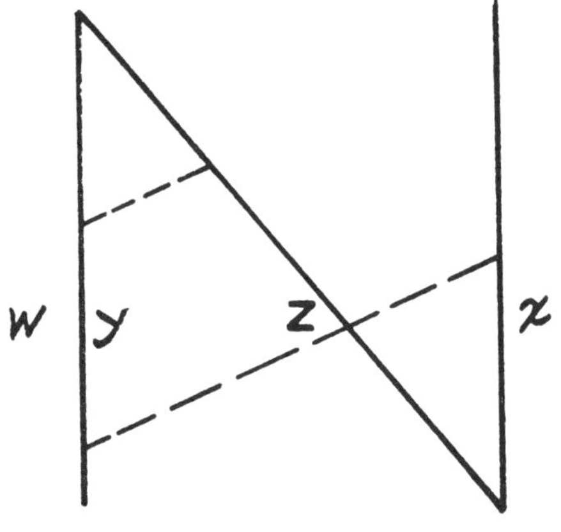

$f_1(w)+f_2(x)=f_3(y)/f_4(Z)$

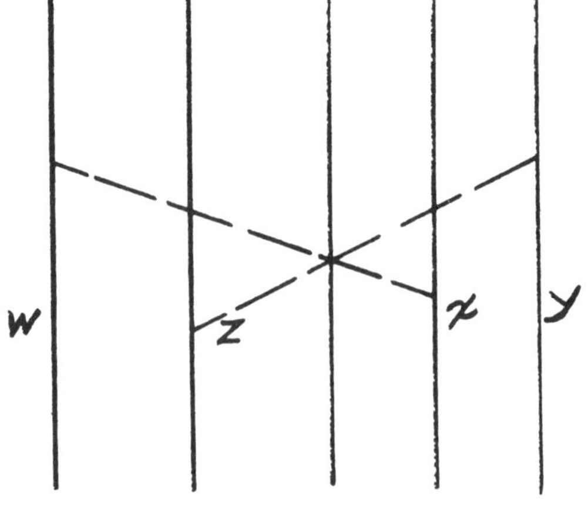

$f_1(w)+f_2(x)+f_3(y)=f_4(Z)$

. . . HOW TO MAKE NOMOGRAMS

Here are the fundamentals of nomograms and the basic charts and equations used in their construction.

by Stephen C. Lane

NOMOGRAMS, or alignment charts,* find wide practical use for repeated solutions of engineering problems. Though the majority of nomograms available today were constructed for solving design problems of various types, it is both possible and wise to use them in shop or field.

No matter where they are used, nomograms have a number of outstanding advantages. These include a saving in calculation time (the usual problem can be solved in a few seconds), elimination of the need for many reference books, reduction of errors, a saving in energy on the part of the user, and the possibility of assigning semi-skilled personnel to the solution of engineering problems.

Though many nomograms are directed at solving problems of an engineering nature, they are equally suitable for equations arising in management, production, purchasing, safety, transportation, economics, etc.

Construction methods . . . Nomograms may be constructed by a number of different methods—plane geometry, analytic geometry, and determinants. While none of these are difficult, the methods employing analytic geometry and determinants require a knowledge of a number of symbols and equations which are easily forgotten unless one makes frequent use of them. Fortunately, the method using plane geometry can be applied without many symbols and equations, and the results obtained are just as accurate as with the other two methods.

This series of articles will deal exclusively with charts constructed by means of plane geometry. The method lends itself to use in the design office, is easy to apply, and requires a minimum number of directions for constructing a given type of chart. If the reader is interested in the mathematical proof behind any particular chart form, he can refer to one or more of the texts to be listed in the bibliography at the end of the series.

Definitions . . . A nomogram is a chart consisting of three or more scales arranged so a line or lines joining points on two or more of the scales will cut another scale at the point that satisfies the relation between the variables used in constructing the nomogram. In simpler terms, the nomogram permits us to solve for an unknown in the equation on which the chart is based.

Here we have the key to constructing nomograms: **Every nomogram is based on an equation between three or more variables.** A variable, always assigned a letter like x, y, or z, is a quantity which may assume several values in an equation. Thus, in $x \pm y = z$, the variables x, y, and z can have an infinite number of values.

The function of a variable, commonly written as f(x), where x is the variable, is a mathematical expression containing the variable and any number of constants such that when a definite value is assigned to the variable, the value of the function can be found. Thus, if the variable is x, expressions like $1/x$, x^2, 2x, log x, etc., are functions of x because we can find their values by assigning numerical values to x for each. A constant has a fixed value in an equation at all times.

The concepts of a variable, function of a variable, and function of a constant provide the means for simplifying complex equations so they can be readily plotted in nomographic form. For as soon as you are familiar with the standard chart forms, it is a simple matter to rearrange most equations so they can be plotted in one or more forms.

Standard forms . . . At left are six standard nomogram forms and the types of equations for which each is used. Although the functional expressions do not indicate it, any number of constants may appear in these equations without altering the form of the nomogram. The only effect of a constant is a change in the scale on one of the axes. Later in this series, after we have learned to plot the six forms shown, we will be ready for more complex forms having curved scales of various types. We will also learn how to construct the useful intersection chart for a large number of different equations.

Tools . . . Next month we will construct our first two charts—one for the sum of two variables, the other for the difference of two variables. To be sure you are ready to follow the easy steps, secure the following: one 10-in., 30°-by-60° triangle; one 6-in., 45° triangle; one 6-in. engineer's scale; one T-square with a blade about 9 in. long; several sheets of 8½ by 11-in. white bond paper; and two sharp 2H pencils. That's all you need to start. As we proceed we'll add a few inexpensive items to speed our work.

Meanwhile, keep these facts in mind. To construct a nomogram we must: (1) have an equation containing three or more variables; (2) simplify this equation so it takes one of the standard forms at left; and (3) plot the variables on suitable scales. That's all there is to it.

*The terms **nomogram** and **alignment chart** are used interchangeably. In this series we will use nomogram because it is simpler.

Plotting the sum of two variables:

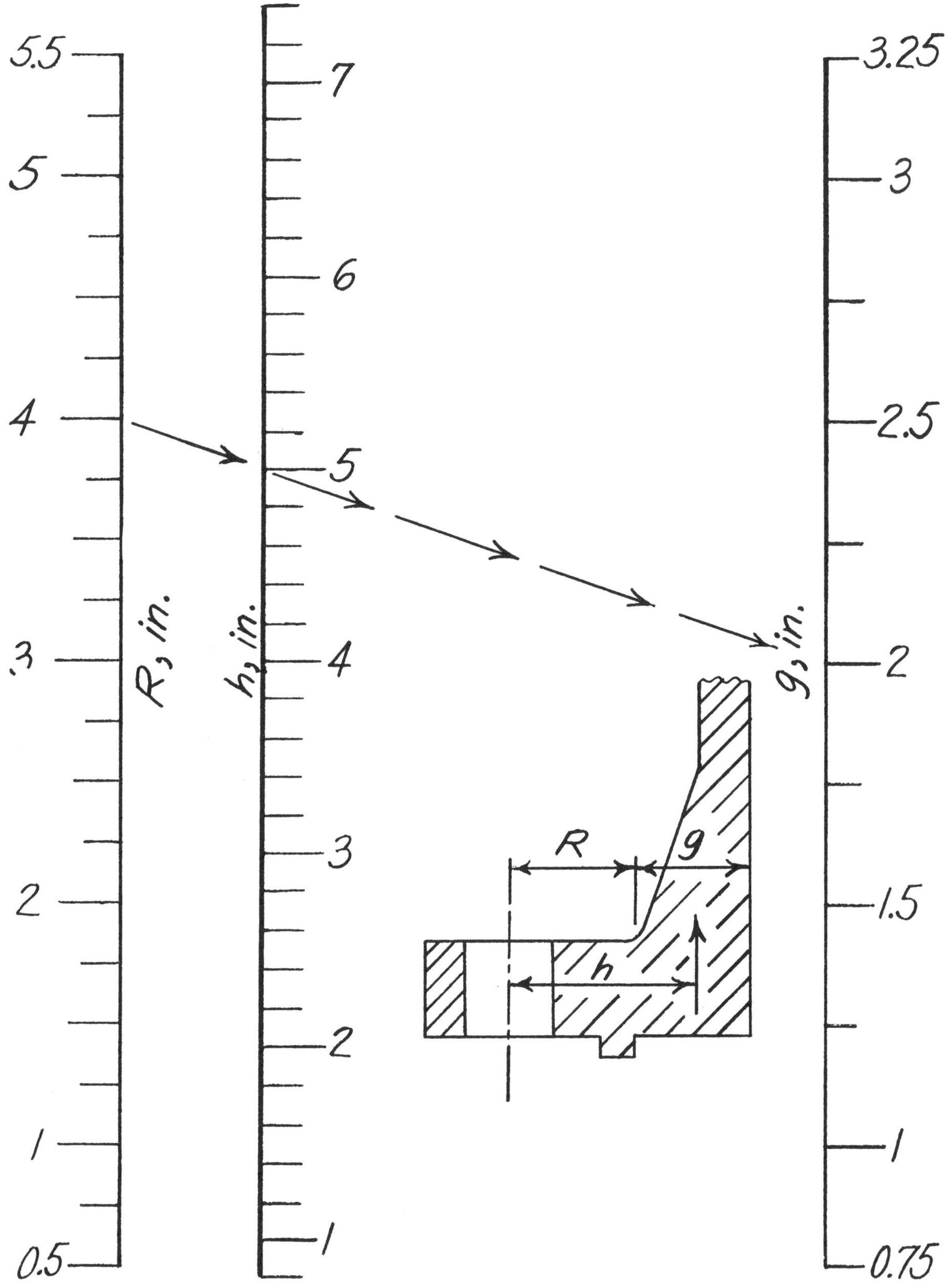

by Stephen C. Lane

HOW TO MAKE NOMOGRAMS

THE simplest nomogram we can construct consists of three parallel scales arranged so a straight line cutting all scales intersects each at a point satisfying the relation for which the chart was plotted. Likewise, the simplest equation we can write is that expressing the sum of two variables. Let's see how we can combine these two facts to construct our first nomogram.

Sum of two variables . . . While it is usually easier to add two numbers directly than to use a chart to find their sum, the methods for constructing a chart for the sum of two variables illustrate many important points useful for more complex equations. Also, the chart form used may be required where the sum of two variables is part of an expression involving multiplication, division, and other operations.

To make our illustrative construction more useful, we will choose an equation whose solution is often required during the design of flanges for refinery and petrochemical pressure vessels. Since it contains a constant, it is not too readily solved by inspection.

Example: Construct a nomogram for: h = R + 0.5g, where h = lever arm for total hydrostatic end force in a flanged vessel, in.; R = radial distance from the bolt circle to point of intersection of hub and back of flange, in.; g = thickness of hub at back of flange, in.

Solution: Use the following steps to plot this chart, and any other involving the sum of two variables:

1. Attach a piece of 8½-by-11-in. white bond paper to a drawing board or a suitable pad.

2. Draw two parallel vertical lines about 6 in. apart on the paper, Fig. 1.

3. Label the left line R, the right one g. (These correspond to the **x** and **y** scales in the general relation for the sum of two variables z = x + y).

4. Determine the **limits** of R and g, i.e., the maximum and minimum values normally encountered when applying the equation to refinery and petrochemical pressure vessels. Finding the correct limits for the variables in an equation is often the most difficult part of constructing a chart. This is true because normal experience does not too often lead us to a consideration of the smallest and largest values to be expected. Instead, we are usually more concerned with some average value. Determination of limits is so important in chart construction that we will deal with it later in a separate article. For the present, let's assume we know from experience that these variables, R and g, have upper and lower limits of 5.5 and 0.5; and 3.25 and 0.75, respectively.

5. Assume a convenient length for each scale. This involves inspection of the available area on the chart paper and a decision as to what is suitable for each scale. Since we know this chart will have only three scales, we can use the maximum length afforded by the paper size, or 10 in. Measure this distance accurately and mark the upper and lower points carefully, leaving about half an inch between the top and bottom of each scale and the edge of the paper.

6. Label the bottom points 0.5 and 0.75, respectively, Fig. 1. Label the top points 5.5 and 3.25. Note that we have set up the chart so both scales ascend in the same direction. The z scale, as we saw in Part I, also ascends in the same direction as the x and y scales.

7. We must now subdivide each scale uniformly. To do this, subtract the minimum value from the maximum value and divide by the scale length, in inches. For R, each inch on the scale = (5.5−0.5)/10 = 0.5; for g, (3.25−0.75)/(10)(2) = 0.125. (The expression for g is divided by 2, in addition to 10, because the constant 0.5 = ½, appears before it in the equation.) Thus, we will actually plot on the scale values which are half those corresponding to our numbered divisions. For example, at 0.75, we will be plotting 0.375 (= 0.75/2) but we will label the division 0.75. If this seems abstruse, think about it for a moment and you will see that we are providing, in the chart construction, a factor that automatically provides for the constant that will always appear in this equation when expressed in this form.

8. Lay off the R and g scales with a rule graduated in tenths of an inch. This permits a reading to two places beyond the decimal point and is sufficiently accurate for all normal flange calculations.

9. To locate the central scale, select two values each of R and g that give the same value for h when substituted in the equation h = R + 0.5g. These can be 2:2 and 1.5:3, because when inserted in the equation the value of h is 3.0 in each case. Draw a light line from R = 2 through g = 2, and another light line through R = 1.5 and g = 3. Through the point of intersection of these two lines draw a vertical line parallel to the R and g scales. This is the h stem and must now be marked with suitable scale values to satisfy the equation for which the nomogram is being plotted.

10. Determine another value of h greater or less than the point used in locating the h stem. Substitute R = 3 and g = 2, h = 4. Draw a light line through these points and mark where it intersects the h stem. Measure the distance between the points h = 3 and h = 4, and graduate the h stem to its upper and lower limits, using this distance. Subdivide this unity dimension as shown, Fig. 1. A uniformly divided central scale is another characteristic of this type of chart for the sum or difference of two variables.

11. Letter the main divisions of each scale and the units for each. The chart is now complete and ready for use. As a final check, set up a few values of R and g and solve the equation of the chart for h. Enter the chart with each set of values and compare the chart result with that obtained from the equation.

Difference of two variables . . . The same construction methods are used for a nomogram of the relation z = x — y except that the x or left-hand scale ascends from the bottom to the top of the page while the y or right-hand scale ascends from the top to the bottom of the page. The z scale is parallel to the x and y scales and ascends in the same direction as the x scale. All three scales are uniformly divided.

For either type of chart, the location of the central scale can be varied by increasing or reducing the overall length of the x or y scales, or both. For the chart plotted in this article, the off-center location was favored because it permits easier reading and leaves room for the flange diagram. The latter is useful for reference during solution of a problem.

Plotting the product of two variables:

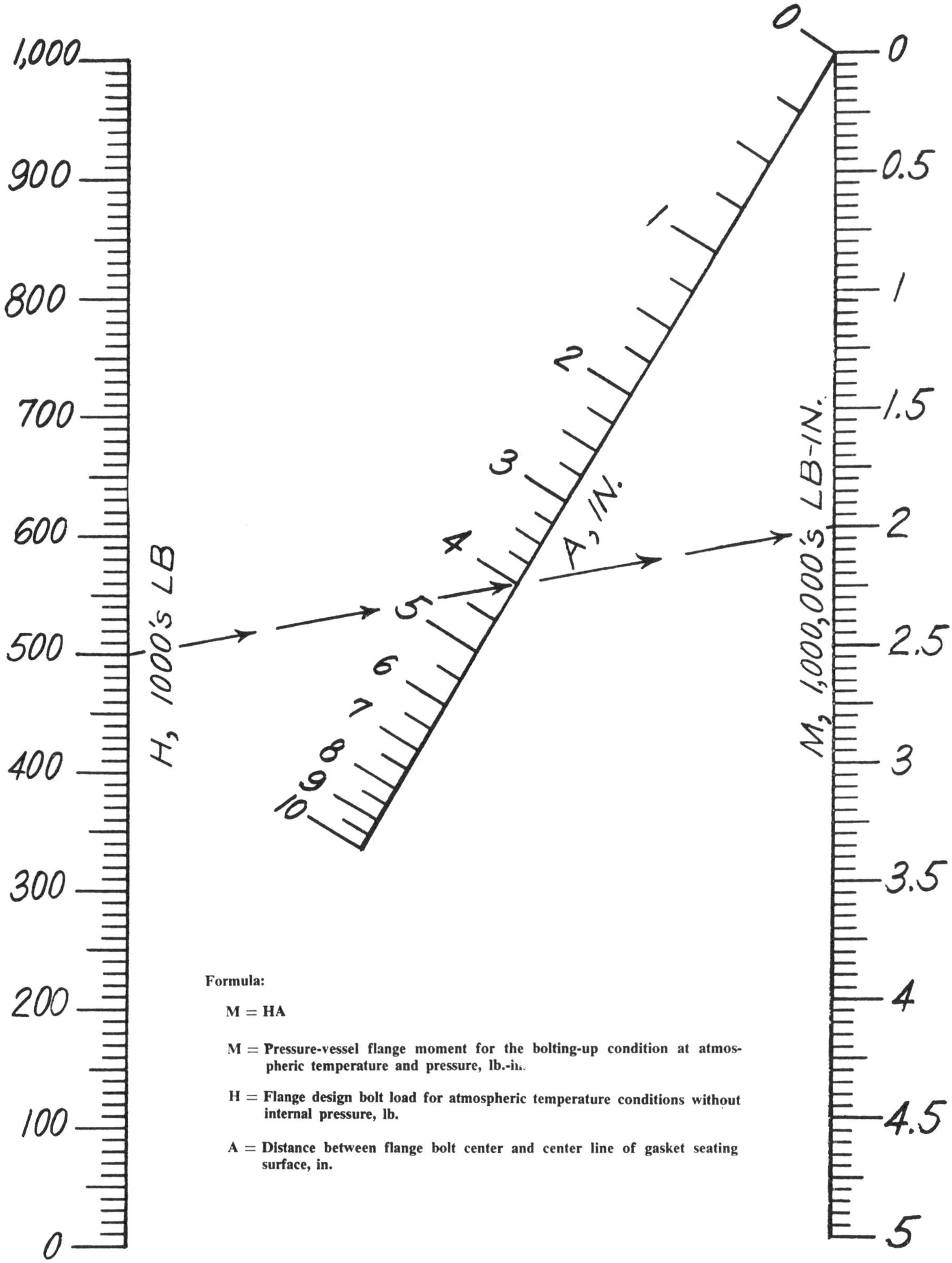

Formula:

M = HA

M = Pressure-vessel flange moment for the bolting-up condition at atmospheric temperature and pressure, lb.-in.

H = Flange design bolt load for atmospheric temperature conditions without internal pressure, lb.

A = Distance between flange bolt center and center line of gasket seating surface, in.

HOW TO MAKE NOMOGRAMS

by **Stephen C. Lane**

ONE of the most common operations in the solution of an equation is finding the product of two variables. Many equations consist of three variables, and possibly a constant, with two of the variables linked by a multiplication sign. This month we will learn how to construct the simplest chart for the product of two variables. Next month we will construct a somewhat more complex form for the product of two variables.

N-Chart construction . . . The N-chart derives its name from its resemblance to the letter N. Having two uniformly divided scales, it can be plotted in a few minutes with a minimum of computation for scale lengths and divisions. One of the major advantages of this form is its easy use after it has been plotted.

Example: Construct a chart for the relation M = HA, where M = flange moment for the bolting-up condition at atmospheric temperature and pressure, lb. in.; H = flange-design bolt load for atmospheric temperature conditions without internal pressure, lb.; A = distance between flange bolt center and centerline of gasket seating surface, in.

Solution: This equation is often met during design or checking of flanges for petroleum and petrochemical vessels. To solve it, we must know the values of H and A. The value of A can be obtained directly from a tabulation of flange dimensions, while that for H must be computed using standard equations available in flange-design handbooks.

1. Our first step in plotting an N chart for this or any other similar equation is to choose the upper and lower values for the two variables whose range is known—H and A. Experience shows that H varies from a low of less than 100,000 lb. to a high of somewhat more than 500,000 lb. To make our chart as useful as possible, let's assume a low of 0 for H, and a high of 1,000,000 lb. The lever arm A varies from something greater than zero to a maximum of about 5 in. or more in large flanges. Therefore, we'll use the limits of 0, 5 in., for A.

2. Attach a sheet of 8½ by 11 in. white bond paper to a drawing board or pad. Then, 1 in. from the right and left borders of the paper, draw two vertical, parallel lines about 10 in. long. Label the left line H and the right one M.

3. Since H varies from 0 to 1,000,000 lb., a stem length of 10 in. with 1-in. subdivisions (= 100,000 lb.), would probably be suitable for the size of chart we plan to construct. With a rule divided in tenths, lay off the H scale, Fig. 1, beginning at 0 and ascending upwards to 1,000,000.

4. Our righthand stem will contain the M scale and on it we will read the product of HA. Inspecting our equation, we can see that the maximum value of M occurs when H = 1,000,000 lb. and A = 5 in. At this point, M = (1,000,000 (5) = 5,000,000 lb. in. When H = 0 and A = 0, M = 0. Therefore, our M scale must range from 0 to 5,000,000 lb. in. With a 10-in.-long stem available, a total length of 10 in. appears most suitable because then 1 in. = 5,000,000/10 = 500,000 lb. in. Starting at the top of the M scale, lay off 10 in. with a tenths rule, ascending downwards to 5,000,000 lb. in., Fig. 1.

5. Draw a light diagonal line from the point M = 0 in the upper righthand corner to the point H = 0 in the lower lefthand corner. Be careful to draw this line accurately between the two points. We now have the skeleton of our N chart. As usually plotted, this form resembles the letter N written backwards.

6. The diagonal stem will contain the A scale and will vary from 0 to 5 in., for values of A. However, this scale will not be uniformly divided, as the H and M scales are. Nor can we arbitrarily assign scale divisions because this is a projective scale, i.e., its subdivisions are obtained by joint projection from the H and M scales.

Using the upper limit of the H scale, 1,000,000 lb., as one projection point, lay a straight edge from it through 1,000,000 lb. in. (= 1,000,000 × 1.0 in. arm) on the M scale. This is equivalent to solving for M when H = 1,-000,000 lb. and A = 1.0 in. Mark the point where the straight edge intersects the diagonal. This point corresponds to A = 1.0 in.

7. Do the same for A = 2, 3, 4 and 5 in. To take care of those rare cases where A exceeds 5 in., plot values through 10 in., using a value of H less than 1,000,000; say 500,000. Depending on how accurately you wish to read A, divide the distance between each unit into tenths or quarters, using the same method of projection as for the major divisions.

8. Letter the various scales as shown in Fig. 1, and the chart is finished. Make a few longhand calculations for M and check them on the chart. If the results are accurate, the chart is ready for use. Should you discover any errors, review the construction described above and check the accuracy of scale subdivisions, the diagonal, and the stems.

Chart characteristics . . . Note several facts about this chart. Values on the lefthand scale must increase from the bottom of the scale to the top. Just the opposite is true for the righthand scale. Both scales must increase from a zero starting point when the chart is constructed as shown here. While it is possible to construct this type of chart with either or both scales starting at a value other than zero, we are not now quite ready to discuss the methods to be used. This will be taken up in Part IV of this series.

The diagonal must always be drawn between the zero points of the two scales. Though the N chart, as shown here, is extremely useful, there are some equations for which another form is better for finding the product of two variables. This will be discussed in a future part of this series. Next month we will construct a chart for the dividend of two variables.

Plotting the dividend of two variables:

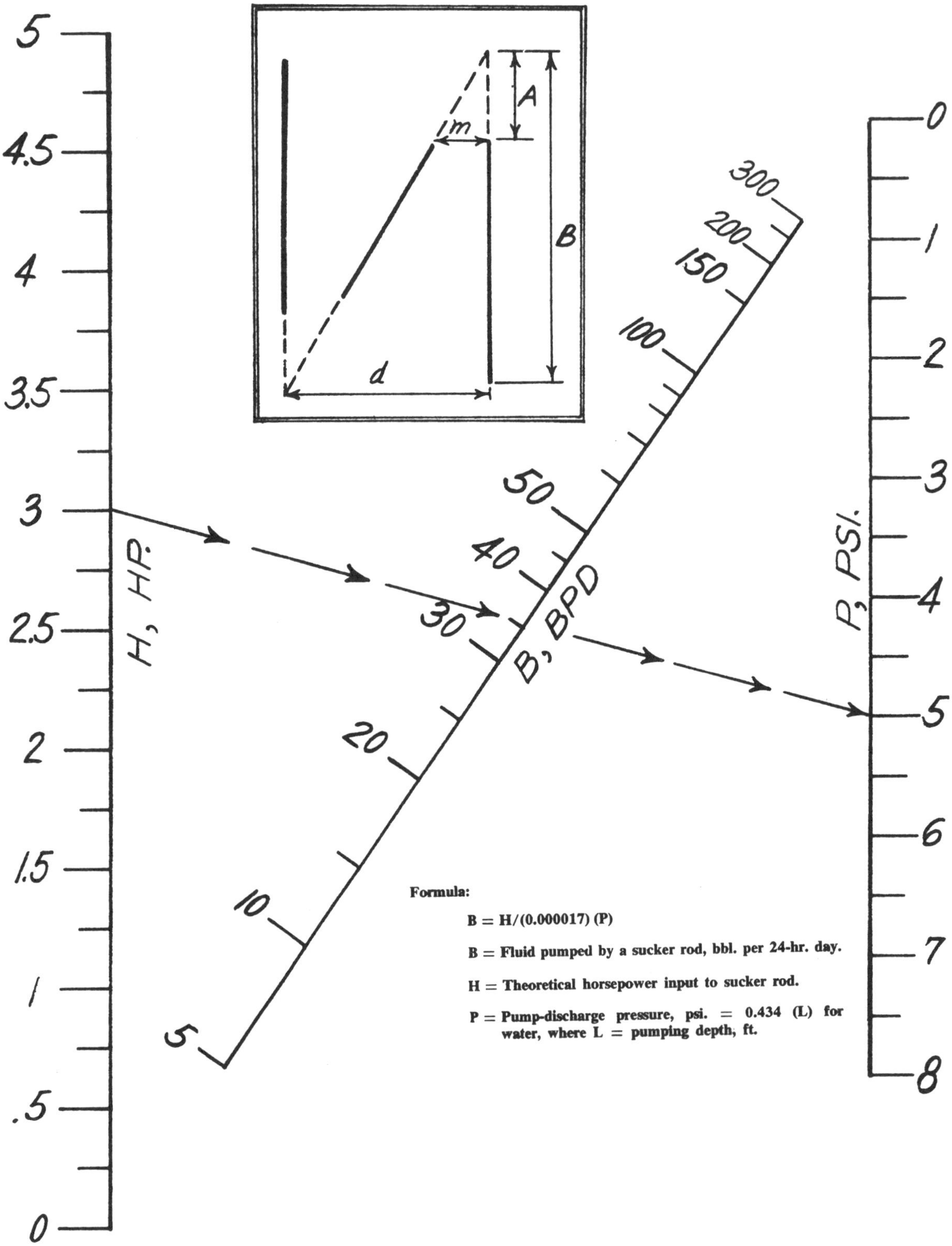

Formula:

B = H/(0.000017) (P)

B = Fluid pumped by a sucker rod, bbl. per 24-hr. day.

H = Theoretical horsepower input to sucker rod.

P = Pump-discharge pressure, psi. = 0.434 (L) for water, where L = pumping depth, ft.

HOW TO MAKE NOMOGRAMS

by **Stephen C. Lane**

NEXT to multiplication, division is probably the most common mathematical operation used in engineering equations. And since the division of one variable by another is usually a more difficult problem, mentally, than finding the product of two variables, we have greater reason to plot the type of chart involving division. Though the chart resembles that plotted in Part 3 for the product of two variables, there are certain important basic differences.

Example: Construct a chart for the relation B = H/(0.000017)(P), where B = fluid pumped by a sucker rod, bbl. per 24-hour day; H = theoretical horsepower input to the sucker rod; and P = pump discharge pressure, psi. —P = 0.434(L) for water, where L = pumping depth, ft. This equation is often used during preliminary analyses of sucker rods and surface equipment.

Solution: To solve this equation for B we must know the values of H and P. By plotting a chart applicable to water, we can secure results which are always on the safe side for oil. They can be multiplied by a suitable factor for salt water or oil.

1. In choosing our limits, studies show that the horsepower input, H, will vary from a low near 0 to a high of about 5.0 at a depth of 15,000 ft. This, of course, is only the theoretical power input. Actual input depends on the efficiency of the equipment, etc. The depth, L, varies from a low of 0 to an assumed high of 15,000 ft. in this case. Therefore, for water, P varies from 0 to 0.434(15,000) = 6,510 psi., say 8,000 psi.

2. Attach a sheet of 8½ by 11-in. white bond paper to a drawing board or pad. Then draw two vertical lines about 10-in. long, roughly 1 in. from the right and left-hand borders of the paper. Label the left line H and the right one P.

3. Since horsepower input varies from 0 to 5.0, a stem length of 10 in. gives a unit division of 1 in. = 5.0/10 = 0.50 hp. With a rule divided in tenths, lay off the H scale from 0 to 5.0 hp., as shown in Fig. 1. Note that the H scale ascends from the bottom of the paper.

4. The P scale will vary from 0 to 8,000 psi. A scale length of 8 in. provides a unit division of 1 in. = 8,000/8 = 1,000 psi. Starting near the top of the right-hand stem, lay off 8 in. with a tenths rule. To divide we always plot the right-hand scale of the N chart so its values ascend from the top of the page downwards, opposite to the left-hand scale. Label the subdivisions on both the P and H scales.

5. Draw a light diagonal from the point P = 0 in the upper right-hand corner through the point H = 0 in the lower left-hand corner. Be sure to draw this line accurately between the two points because it will contain the B or results scale. However, this scale will not be uniformly divided because it is a projective scale. That is, its subdivisions are obtained by joint projection from the H and P scales.

6. Since our equation contains a constant, 0.000017, in the denominator, we must take this into consideration when plotting the projective scale on the diagonal. Since this constant is less than unity, it has the same effect as multiplying the numerator by a number greater than unity —in this case 1/0.000017 = 58,700, closely. So we can rewrite our original equation as B = 58,700H/P.

Our task now is to find a series of factors which, when multiplied by 58,700, will give us products between 5 and 300 (the approximate lower and upper limits of the diagonal scale; they're established by solving the equation for B using low and high values for H and P). Use a slide rule and list the factors like this: 10, 0.00017; 50, 0.00085; 100, 0.0017; 150, 0.00255; 200, 0.0034; 250, 0.00425; 300, 0.0051.

With this list of factors we're ready to begin plotting values on our diagonal scale. But, before a specific value can be plotted, we must find points on the H and P scales such that H/P = the factor for the value in question. For H = 0.17 and P = 1,000, the factor = 0.00017, corresponding to a value of B = 10.

Using this procedure, plot the various values of B on the diagonal scale. To plot more values, use the same method.

7. Letter the various scales as shown in Fig. 1, and the chart is finished. Make a few longhand calculations for B and check them on the chart. If the results are accurate, the chart is ready for use. Should you discover any errors, review the construction described above and check the accuracy of the scale subdivisions, the diagonal, and the stems.

Chart characteristics . . . The remarks concerning the general construction features of an N-chart for the product of two variables apply equally well to the dividend, plotted here. Where the minimum value of either or both variables on the vertical scales is greater than zero, or must be greater than zero to allow plotting in the available area, proceed with the construction as described above until you are ready to draw the diagonal. Then set up the proportion A/B = m/d, as shown on the inset of Fig. 1.

Dimension A, in inches, is the distance between the first graduation on the scale and the zero point of the scale. This is easily found by extrapolation, using the unit scale length and the assigned value of the first graduation. The dimension B is found in a similar way. Measure d on the drawing, substitute in the proportion, and solve for m. The same method can be used at either end of the chart.

Plotting the product of two variables by log chart . . .

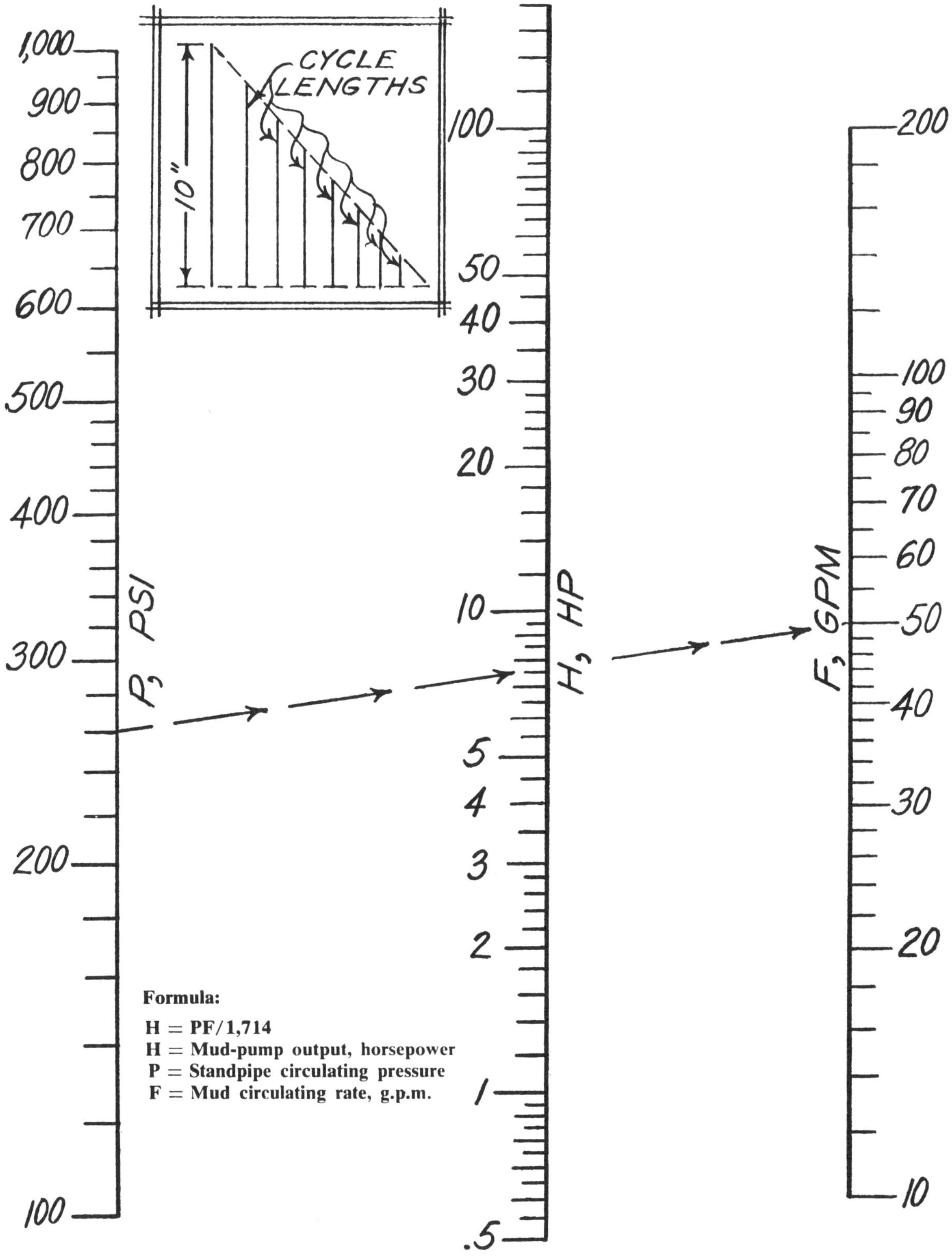

HOW TO MAKE NOMOGRAMS

by **Stephen C. Lane**

IN Parts 3 and 4, we plotted the product and dividend of two variables, respectively, using an N-chart in each case. The N-chart was introduced first for these two operations because its two uniformly divided scales are easier to understand and plot than a chart having nonuniformly divided scales.

Now, however, we are ready to study and plot a more general type of chart for the product of the two variables. All its scales are nonuniformly divided, but this should present no difficulty if the instructions are carefully followed.

Logarithmic chart . . . This type of chart for the product of two variables consists of three parallel scales, Fig. 1. The outer scales usually contain the two variables being multiplied together, while the central scale contains their product. However, it is possible to vary this arrangement if any advantages are gained thereby. Right now, we'll confine our attention to the standard forms used for plotting this type of chart.

Example . . . Construct a logarithmic-type chart for the equation H = PF/1,714, where H = mud-pump horsepower output; P = standpipe circulating pressure, psi; F = mud circulating rate, gallons per minute; and 1,714 = an empirically derived constant.

Solution . . . This equation gives the mud-pump horsepower for air-mud drilling rigs. A relatively recent innovation, aerated-mud drilling appreciably increases penetration rates and alleviates loss of circulation. By substituting the circulating pressure before and after injection, the above equation can be used to find the mud-pump horsepower output without and with air.

1. Our first step is to determine limits for P and F. Recent tests show that the stand-pipe circulating pressure for many installations varies from a low of 0 to a high of 1,000 psi. Since the values between 0 and 100 psi. are not of too much importance in practical work, we will assign limits of 100 and 1,000 psi. to P. Mud flow, F, appears to range between a minimum of 0 g.p.m. and a maximum of 200 g.p.m. So we'll assign lower and upper limits of 10 and 200 g.p.m. respectively, because the values between 0 and 10 g.p.m. are not too important.

2. Attach a sheet of 8½ by 11 in. white bond paper to a drawing board or pad. Then, 1 in. from the right and left borders of the paper, draw two vertical parallel lines, each about 10 in. long. Label the left line P and the right one F.

3. Since P varies from 100 to 1,000 psi, a 10-in. long logarithmic scale ranging between the same values would be satisfactory. Lay off this scale along the left-hand stem, Fig. 1, using the points from a 10-in. slide rule, or from some other plot of a logarithmic scale ranging from 1 to 10.

4. The variable F ranges from 10 to 200 g.p.m. To cover this range, we'll need 1.2 cycles of a logarithmic scale. A cycle 7 in. long will give us a total length of about 9.125 in. So, using the 7-in. cycle, plot the F scale, Fig. 1.

5. We must now determine the location of our central scale. The easiest way to do this is to find two different values of P and F giving the same value of H. The solution lines will then intersect at a point through which the central scale passes. And since, in this form, the central scale is parallel to the two outside scales, one point is enough for locating it.

Values satisfying the requirement for a common value of H are P = 100, F = 100, and P = 1,000, F = 10, because the product is 10,000 in both cases and the constant is the same. Draw light lines between these values. Through the point of intersection draw a perpendicular line parallel to the P and F scales. This is the H stem. Its values will ascend in the same direction as the P and F scales.

6. Next, determine values of P and F such that when substituted in the equation given for this chart, the value of H = 1.0. When P = 171.4 and F = 10, H = 1.0. Using these two values, draw a light line across the chart to intersect the H scale. Mark this point unity. Do the same for a point representing H = 10; values of P and F for this are P = 171.4, F = 100.

The distance between H = 1 and H= 10 is the cycle length for the H scale. Extend the scale above and below these points, as shown in Fig. 1. The exact upper and lower limits occur at the point found by drawing a line through the two maximum and the two minimum values on the P and F scales. However, it is usual practice to extend the central scale slightly beyond these points to make the chart easier to read.

7. Letter the various scales as shown in Fig. 1 and the chart is finished. Make a few longhand calculations for H and check the results on the chart. If they agree, the chart is ready for use. Should you discover any errors, check the construction, as given above.

Chart characteristics . . . All scales ascend in the same direction when using this type of chart for the product of two variables. Since a logarithmic scale can never be plotted for the value 0, the minimum scale value on this type of chart is always greater than 0.

Logarithmic scales of a number of different cycle lengths are usually required when constructing this type of chart. So it is wise to prepare a master log chart as shown in the inset, Fig. 1. All that is needed is one logarithmic scale ranging between 1 and 10. This may be taken from either a 10- or a 5-in. slide rule. Where charts are to be plotted on 8½ by 11 in. paper, the maximum cycle length need not exceed 10 in.

Next month we will construct a chart of this type for the dividend of two variables. Many charts in succeeding articles will be based on equations submitted by readers.

Plotting the dividend of two variables by logarithmic chart

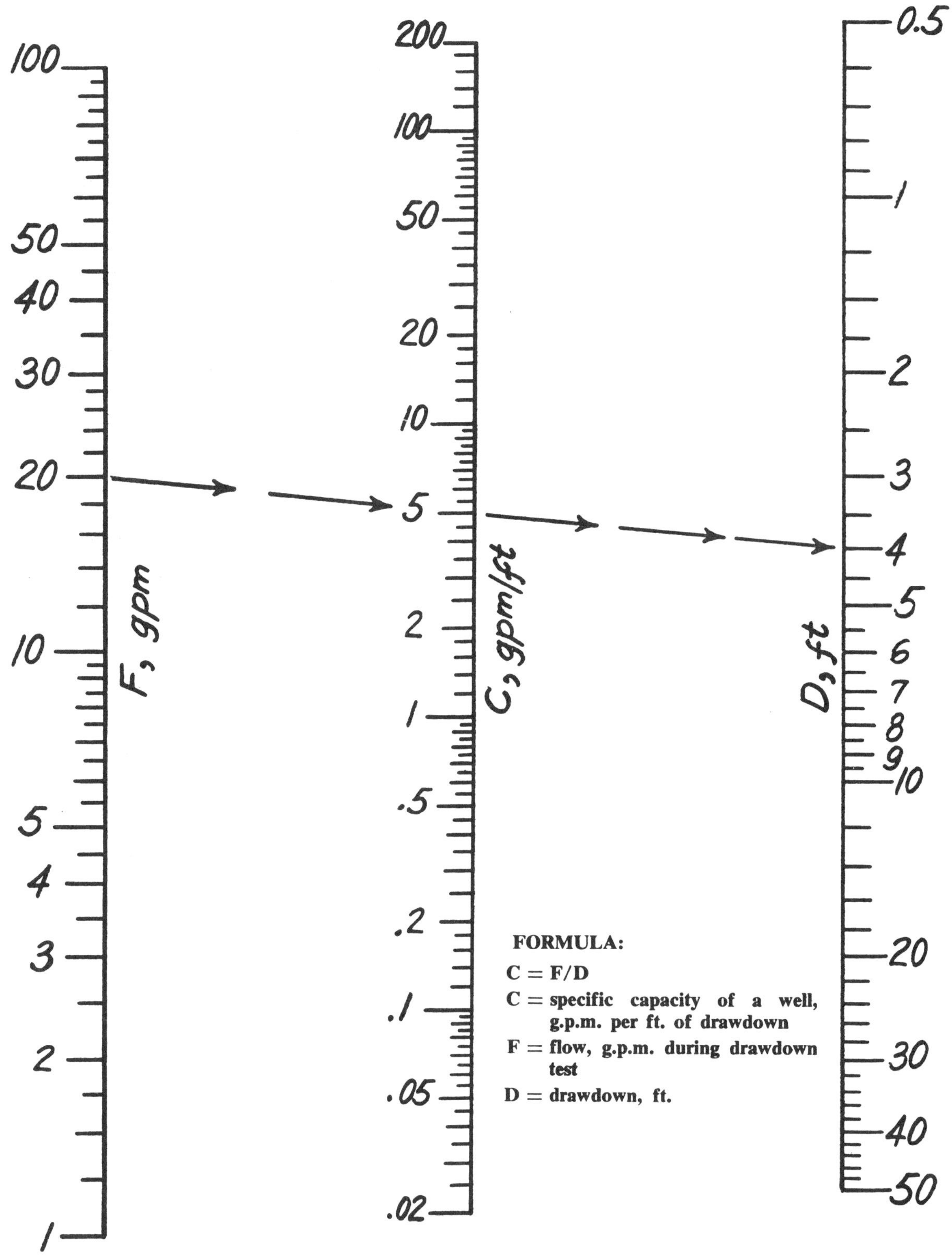

FORMULA:

C = F/D

C = specific capacity of a well, g.p.m. per ft. of drawdown

F = flow, g.p.m. during drawdown test

D = drawdown, ft.

HOW TO MAKE NOMOGRAMS

by **Stephen C. Lane**

IN Parts 3 and 4 we plotted the product and dividend of two variables, using an N-chart. The N-chart was introduced first for these two operations because its uniformly divided scales are easier to understand and plot than a chart having nonuniformly divided scales.

Now we are ready to study and plot a more general type of chart for the dividend of two variables All its scales are nonuniformly divided. But since the construction resembles that presented in Part 5 for the product of two variables, it should present no difficulty if the instructions are carefully followed.

Logarithmic chart . . . This type of chart for the dividend of two variables consists of three parallel scales, Fig. 1. The outer scales usually contain the numerator and denominator, while the central scale contains their dividend. It is possible, however, to vary this arrangement if any advantages can be gained. But for the present we'll confine our attention to the standard forms used for plotting this type of chart.

Example . . . Construct a logarithmic-type chart for the equation C = F/D, where C = specific capacity of a well, gallons per minute per foot of drawdown; F = flow, gallons per minute during drawdown test; D = drawdown, feet.

Solution . . . This equation gives the specific capacity of a well and is often used when measuring the performance of a new or existing well. It does not give a measure of the pump performance and should not be used for this purpose.

1. Our first step is to determine the limits for F and D. Past experience shows that a minimum of 1 g.p.m. and a maximum of 100 g.p.m. will cover the majority of small and medium-size installations. This range will also permit us to extend the limits of the chart should this be necessary. The way this can be done is explained later in this article. Drawdown, for the types of wells we have in mind, varies between a low of about 0.5 ft. and a maximum of about 50 ft. This range will make the chart suitable for a large number of calculations. Also, its included values are such that we will be able to extend the limits of the scale if necessary. When the ranges of the numerator and denominator scales can be extended at will, the chart permits solving the equation for which it was plotted—no matter how large or small the numerical values of the variables may be.

2. Attach a sheet of 8½ by 11 in. white bond paper to a drawing board or pad. Then, 1 in. from the right and left borders of the paper, draw two vertical parallel lines, each about 10 in. long. Label the left line F and the right one D.

3. Since F varies from 1 to 100 g.p.m., we will need two cycles of a logarithmic scale to cover this range. The first cycle will cover 1 to 10 g.p.m., and the second 10 to 100 g.p.m. With a total available length of 10 in., each cycle must be 5 in. long. (Note that when more than one cycle is used on a stem, each cycle must be the same length. For example, if four cycles were needed on the above stem, each would be 10/4 = 2.5 in. long, (or shorter). Using a 5-in. cycle, plot the F scale, Fig. 1.

4. A trial shows that a 5-in. cycle is also suitable for the D scale. To check the suitability of a given cycle length, simply lay off its range on the stem and see if it fits. It is advisable to leave a small amount of space at the top and bottom of each stem. Plot the D scale, Fig. 1.

5. We must now determine the location of our central scale. The easiest way to do this is to find two different valves of F and D giving the same value of C, just as we did last month for the product of two variables. The solution lines then intersect at a point through which the central scale passes. And since, in this form, the central scale is parallel to the two outside scales, one point is sufficient for locating it. Values satisfying the requirement for a common value of C are F = 10, D = 10, and F = 50, D = 50, because the dividend in each case is 1. Draw light lines between these values. Through the point of intersection draw a perpendicular line parallel to the F and D scales. This is the C stem and its values will ascend in the same direction as the F stem.

6. Next, determine values of F and D so that when they are substituted in the equation for this chart, the value of C = 10. This occurs when F = 50 and D = 5. Draw a light line between these values. The point where it intersects the C stem corresponds to C = 10. The distance between the points 1 and 10 on the C stem is the cycle length of the logarithmic scale on the C stem. Step this distance off and subdivide as shown in Fig. 1. (Note that by selecting the intersection points C = 1 and C = 10, we automatically saved ourselves some time because we obtained the cycle length while locating the C stem.)

7. Letter the various scales as shown in Fig. 1 and the chart is finished. Make a few longhand calculations for C and check the results on the chart. If they agree, the chart is ready for use. Should you discover any errors, check the construction as given above.

Chart characteristics . . . The two outside scales ascend in the opposite directions, while the central scale ascends in the same direction as the lefthand scale. Note that this is different from the product chart plotted in Part 5. Since a logarithmic scale can never be plotted for the value 0, the minimum scale value of this type of chart is always greater than 0.

To extend the limits of this chart, assume we have a problem in which F = 1,000, D = 10. To solve, move the decimal point one place to the left in the numerator to give a value of F = 100 and solve in the normal way. Then move the decimal point one place to the right in the result. Thus, by suitable adjustment of the decimal point we can use this chart for any value of F greater than 100 or less than 1. The same manipulation can be performed with values on the D scale.

Next month we will plot an expression involving the product and dividend of variables, based on an equation submitted by a reader of this series.

Plotting multiple operations by log arithmic chart

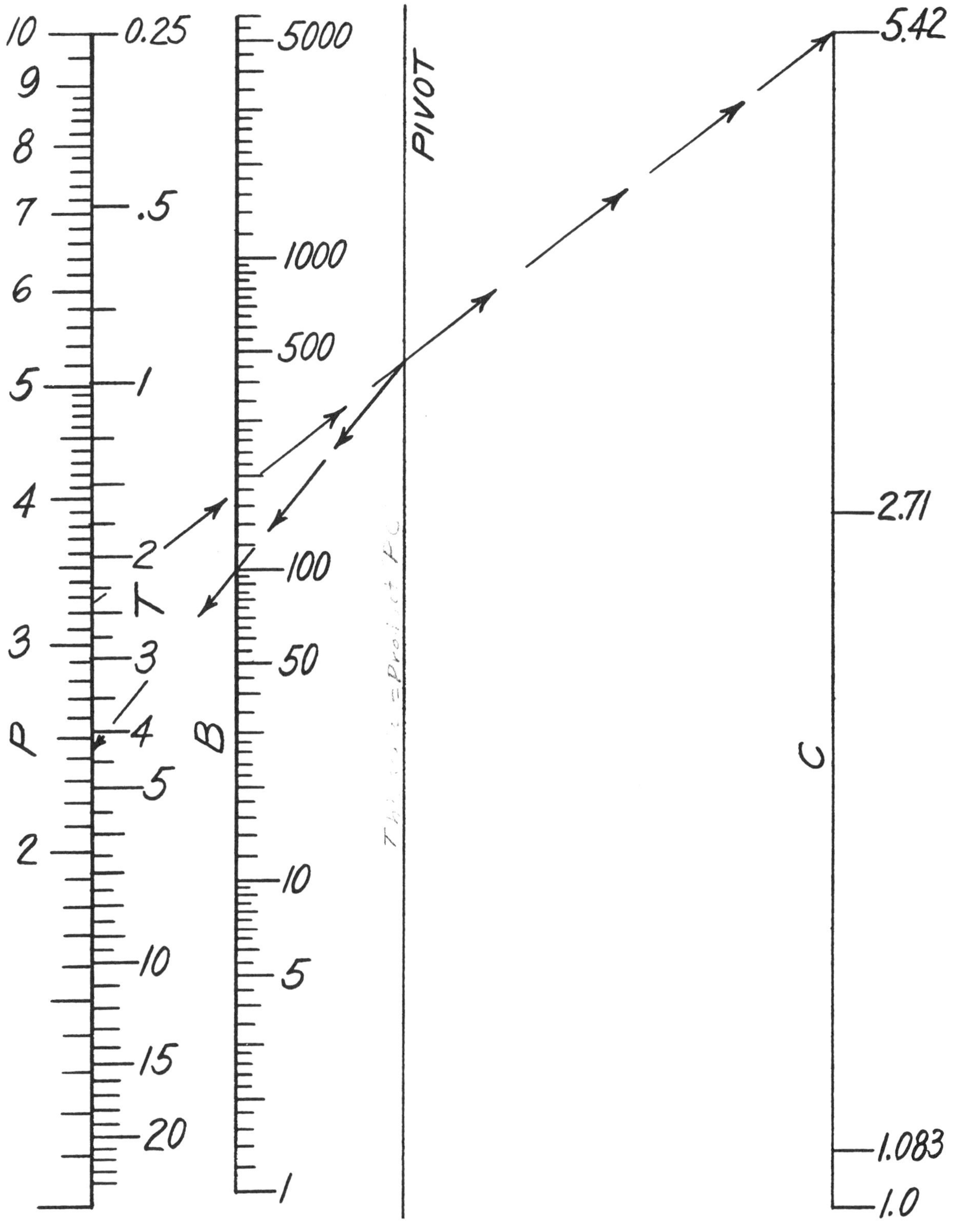

HOW TO MAKE NOMOGRAMS

by **Stephen C. Lane**

IN Parts 5 and 6, we constructed simple logarithmic charts for the product and dividend, respectively, of two variables. Now we are ready to combine the techniques learned with those charts so we can apply them to a more complex function—one that involves the product and division of several variables.

In addition to presenting helpful aids to mechanical engineers in the oil and gas industry, the present article is based on an equation submitted by a reader of this series.

The equation . . . To determine daily liquid production of a well by means of hourly tests, the following equation is useful: B = 24 (PC)/T, where B = barrels of oil or water produced per 24 hours; P = production of oil or water, in.; C = tank coefficient, bbl. per in.; T = test duration, hours.

The reader submitting this equation hopes to have field personnel use the chart to obtain quick checks on well performance. It is an excellent example of how nomograms can be used to reduce the time and effort required to solve equations of various types.

Limits . . . The reader requested that (1) the oil-production scale, P, vary from 0 to 50 in., with ¼-in. subdivisions; (2) the water-production scale to vary from 0 to 200 in., with ½-in. subdivisions; (3) the test time to vary from 0 to 24 hours, with ¼-hour subdivisions; and (4) the tank coefficients to vary from 1.000 to 5.42, including the intermediate values of 1.083 to 2.71.

Though two equations were submitted, one for oil and one for water, they are both essentially the same and can be expressed as given above. Study of the limits shows that those for oil and water production and time include maximum and minimum values, respectively, beyond the range of a chart on 8½ by 11-in. paper. So a maximum limit of 10 has been chosen for P because it permits a large scale for entering the chart and can be used for any multiple of 10 by a simple shift in the decimal point of the result.

Thus, if production is 100 in., solve for 10 in. and simply shift the decimal point one place to the right in the result. For a production rate of 200 in., solve using 2 in. and move the decimal point two places to the right in the result.

Such a shift is simple to make and permits a chart having greater accuracy throughout the ranges desired by the reader. Since test runs are seldom less than 15 minutes (0.25 hour), this value was used as the minimum point on the time scale, giving larger divisions which are easier to read and use.

Chart construction . . . Lay off the P scale, Fig. 1.

1. Make it 10 in. long, with scale divisions ranging between 1 and 10.

2. Next, lay off the C scale as shown. Its divisions range from 1.000 to 5.42.

3. Using the directions given in Part 5 for the product of two variables, locate the central pivot stem. Though the pivot stem in the finished chart will not contain scale divisions, points representing unity and 10 can be marked by means of dots during the construction. They are then used in locating the results or B scale.

4. Recalling from Part 6 that the scales on a chart for the dividend of two variables ascend in opposite directions, lay off the T scale, as shown.

5. Using the points on the pivot stem and those on the T scale, locate and subdivide the B scale as detailed in Part 6. Be sure to include the constant 24 when locating the B scale.

6. Label the scales and solve a trial example to check the chart accuracy. The example shown was submitted by the reader who requested that this chart be constructed.

Study of this chart will show it is simply a combination of the two types discussed in Parts 5 and 6. Note that the B or results scale can be located on either the left or right-hand side of the pivot stem, depending on which arrangement is preferred. It was located to the left in this construction because this position seemed to offer simpler use.

Subdivisions . . . While the exact subdivisions requested for the various scales could not be used, the results obtained are sufficiently accurate for usual field tests. For example, the longhand solution worked out by the reader for the problem illustrated gave a result of 100 bbl. per day. The chart agrees with this exactly. In another example worked out by the reader, a similar closeness of results was obtained. These two problems verify the suitability of this type of construction.

Had the chart been plotted for the ranges suggested, each subdivision would have had to be considerably smaller than those shown. This would decrease the accuracy of the results and would make the chart more difficult to use under field conditions when the light might be poor.

Other forms of charts can be used to solve this equation. In general, however, they are not as suitable as the form selected here because the scale subdivisions would not be as large as those obtained.

Next month we will construct a linear-type chart for the relation of the form a/b = c/d.

Plotting a proportional chart for a relation of a : b as c : d

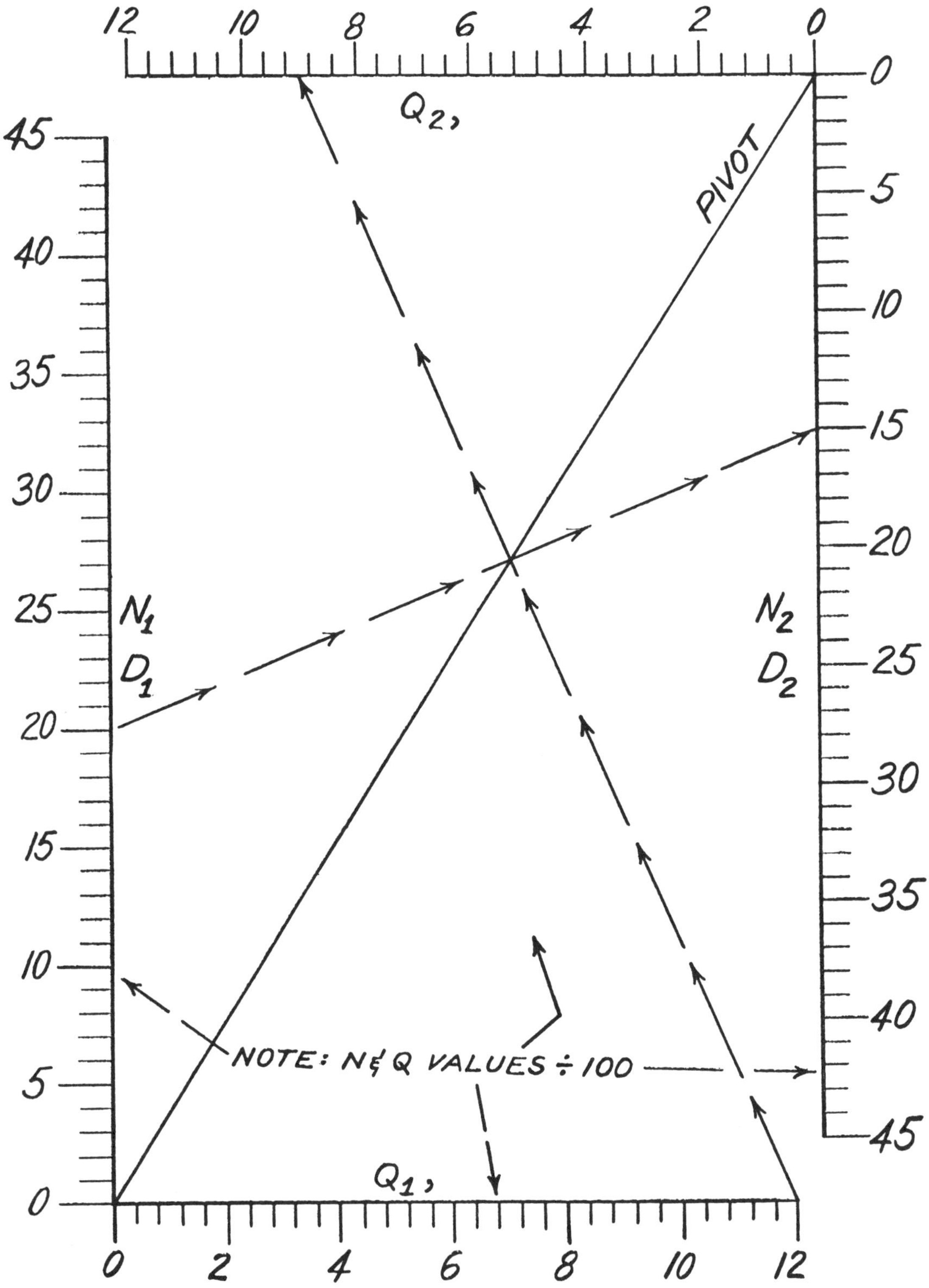

HOW TO MAKE NOMOGRAMS

by **Stephen C. Lane**

UNIFORMLY divided scales are preferred for nomograms by many engineers because they are easier to construct and can be read with greater accuracy than logarithmic scales. Nevertheless, there are many relations which could not be solved by a nomogram if logarithmic scales were not used.

This month we will plot a proportional-type chart for a relation of the form $a/b = c/d$. All its scales are uniformly divided, and it is a valuable form for many equations which have these four variables. Note that none of the variables in the relation can be to any power other than the first.

The equation . . . Centrifugal pumps deliver liquid at a rate directly proportional to speed when the impeller diameter is constant throughout the speed range being considered. They also deliver liquid at a rate directly proportional to impeller diameter when speed is held constant. Using the subscripts 1 and 2 to denote original and final speeds and diameters, these relations can be expressed as $N_1/N_2 = Q_1/Q_2$, and $Q_1/Q_2 = D_1/D_2$, where: N = pump speed, revolutions per minute; Q = pump discharge, gallons per minute; and D = pump impeller diameter, inches. This concept of pump-capacity variation with impeller speed and diameter is an important one in field application of centrifugal pumps.

Limits . . . With a.c. motor drive, most process pumps operate at 1,800 or 3,600 r.p.m. Some high-speed units run at higher speeds but these are not too common. With d.c. motor drive, pump speed may vary anywhere from 0 to 3,600 r.p.m., or higher. Knowing these facts, we can arbitrarily choose upper speed limits of 4,500 r.p.m. for N_1 and N_2.

Pump capacity can vary anywhere from zero to hundreds of thousands of gallons per minute. However, for process pumps, usual capacities are in the range of 1,000 g.p.m., or less. Of course there can be many exceptions to this, depending on the application. To cover the majority of installations, we will choose capacity limits of 0 and 1,200 g.p.m.

Should the speed or capacity for a given problem exceed the limits we have selected, a simple adjustment can be made. This will be described later.

Chart construction . . . On an 8½ by 11 in. sheet of paper, lay off a vertical scale for N_1 about 9 in. long (Fig. 1). Near the bottom of the paper, draw another scale perpendicular to N_1, as shown. Do the same in the upper righthand corner, laying off two scales. These scales may be any suitable distance apart and can be located as conveniently as possible. Connect the two corner points by means of the pivot line (Fig. 1). The pivot must always be drawn between the corner points, as shown.

1. Lay off the N_1 scale on the left-hand vertical line. Use 1 in. = 500 r.p.m., giving a total scale length of 9 in. Start at the corner point at zero and divide the scale uniformly between this point and that representing 4,500 r.p.m.

2. Subdivide the N_2 or right-hand scale in the same manner, using 1 in. = 500 r.p.m. The same scale is used for N_2 because it can vary from 0 to 4,500 r.p.m., depending on the type of motor used to drive the pump.

3. Lay off the Q_1 scale along the bottom line, using 1 in. = 200 g.p.m. This gives a total scale length of 6 in., which is enough for the desired accuracy and gives the range desired.

4. Now, using two suitable values of N_1 and N_2, draw a straight line between them across the chart from left to right. For example, N_1 = 2,000 r.p.m. = N_2, can be used. The point of intersection of this line with the pivot line = 2,000/2,000 = 1. Then we can set up the relation $N_1/N_2 = 1 = Q_1/Q_2$, or $Q_1 = Q_2$, for this point on the pivot line, but only this point.

Since, for this one point, $Q_1 = Q_2$, any line drawn from a value on Q_1 through this point on the pivot line will intersect the upper or Q_2 scale at a value equal to Q_1. Using this procedure, construct the Q_2 scale. Subdivide the main divisions in units suitable for the desired accuracy.

5. Since capacity varies directly with impeller diameter when pump speed is constant, the N_1 and N_2 scales can be used for the D_1 and D_2 scales. They could be divided as shown in Fig. 1 or another scale could be set up. If the latter is done, new Q_1 and Q_2 scales may have to be constructed, depending on where the unity point falls. If it occurs in the same place as for the original variables, new capacity scales are not necessary. But if it falls elsewhere, new capacity scales must be plotted because all values on this chart are in proportion.

6. Note that the subdivisions of the Q_1 and Q_2 scales are the same on this chart. This occurs when there is no constant in the relation on which the chart is based. When a constant occurs—for example in a relation like $N_1/N_2 = 2Q_1/Q_2$—the scale subdivisions will not be the same.

Notes . . . Where values beyond the ranges plotted occur, simply divide each by a convenient value before entering the chart. Solve as shown and multiply the result by the number divided into the original values.

The two opposite corner points of a proportional chart of this type must always start at zero for all variables. Thus, N_1, N_2, D_1, D_2, Q_1 and Q_2 all start at zero. When only higher values in a given relation are to be used, this requirement may introduce a disadvantage because less than the desired accuracy may be obtained. Therefore, under these circumstances, another chart form may prove more practical. But for the majority of relations met in engineering practice, this chart form is entirely suitable.

Where a variable to a power other than unity occurs, this form cannot be used because the scales must be divided nonuniformly. By combining a proportional chart of this type with an N-chart, Parts 3 and 4 of this series, a chart for the relation (a) (b) (c) = (x) (y) can be plotted. This is because uniformly divided scales are used throughout except for the results scale of the N-chart. Combining two proportional charts permits solving a relation of the type (a) (b) (c) = (x) (y) (z). Next month we will plot another type of proportional chart.

Plotting a double-Z proportional chart.

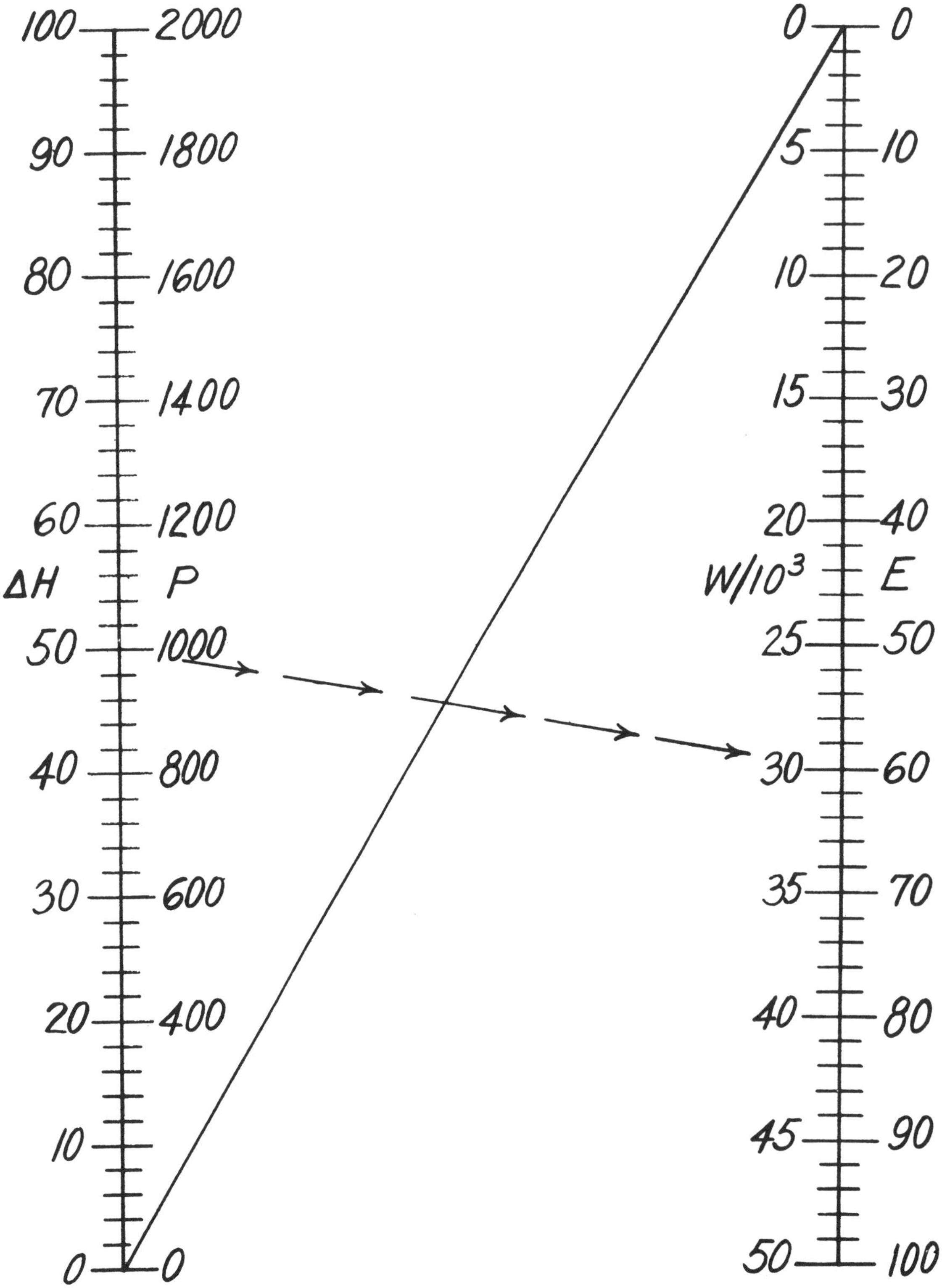

HOW TO MAKE NOMOGRAMS

by **Stephen C. Lane**

LAST month we constructed the first of two proportional charts to be presented in this series. We found that its scales are uniformly divided throughout and that it presents no real problems in construction if the relation we are plotting is of the form a/b = c/d. This month we will plot another form of proportional chart for a relation of the same form. Its principal advantages over the chart of last month are use of a single solution line and easier use with certain equations. Note that with this chart, as with the one in Part 8, that none of the variables can be to any power other than the first.

The equation . . . Compressors for raising the pressure of the so-called imperfect gases—air, butanes, butylenes, ethane, ethylene, etc.—are widely used in petrochemical processing. During preliminary estimates it is often useful to be able to make a quick computation of the approximate horsepower required to compress one of these gases from a known initial pressure to some final pressure. The general relation for any imperfect gas can be expressed as $(\Delta H/2{,}544E) = P/W$, where ΔH = enthalpy change from inlet to outlet of the compressor = outlet enthalpy minus inlet enthalpy, B.t.u. per pound; E = compressor efficiency, expressed as a decimal; P = power input to the compressor, horsepower; W = weight of gas compressed, pounds per hour. The value 2,544 is a constant, and is the number of B.t.u. per hour per horsepower.

Limits . . . The enthalpy change, ΔH, will depend on the gas being compressed, the initial and final pressure in the compressor, and the temperature of the gas entering the compressor. A change of about 100 B.t.u. per pound is usually about the greatest amount encountered in usual process work, with many applications having a much smaller range between inlet and outlet of the compressor.

Of course, there may be some applications where this range is exceeded. But the chart can still be used if a suitable adjustment is made. Therefore, a range between 0 and 100 B.t.u. per pound will be chosen for ΔH. Compressor efficiency, E, may be high or low, depending on compressor condition, construction, and load. To allow for these variations, an efficiency range of 0 to 100 per cent will be used. By including the value of 100 per cent, it is possible to use the chart to compare ideal compression with that actually achieved.

The weight of gas compressed, W, can vary from 0 to 50,000 lb. per hour or more. However, a range up to 50,000 lb. per hour will be suitable for the majority of process calculations. Horsepower input, P, can be estimated by using the maximum and minimum values of the variables being plotted. Using these, the range is found to be 0 to 2,000 hp.

Chart construction . . . Draw two vertical lines, each about 10 in. long, on a sheet of 8½ by 11-in. paper, Fig. 1. Near the bottom of the left stem mark the zero point of the ΔH scale. Near the top of the right-hand stem mark the zero point for the W and E scales. These coincide.

1. Draw a straight diagonal line connecting the two zero points, Fig. 1. This line can be made light or dark, depending on whether it is desired as a guide in the solution of problems. In general, use of the diagonal as a guide is helpful, especially to those who are not too familiar with nomogram construction.

2. Since the limits for ΔH are 0 and 100, a scale length of 10 in. will be suitable. Use subdivisions of 1 in. = 10 B.t.u. per pound and lay off the ΔH scale as shown, Fig. 1. Minor subdivisions of 2 B.t.u. per pound are satisfactory for all usual problems and give a scale which is easy to read.

3. Lay off the E scale from 0 to 100, using a length of 10 in. with major subdivisions of 1 in. = 10 per cent. (Some time can be saved by projecting horizontally across from the ΔH scale, since its subdivisions are the same as for the E scale. When doing this, be sure your T-square is always accurately aligned.)

4. Lay off the W scale on the same stem as the E scale, using the same subdivisions, if possible. In this chart it was possible to do this. Note that the two variables on the right-hand stem ascend in a direction opposite to that of the variables on the left-hand scale. Both must begin at zero if the diagonal is to be drawn as shown. If one or both variables must start at a value other than zero, use the construction method described in the inset of Part 4 of this series.

5. Earlier we found that the lower and upper limits of the unknown, P, are 0 and 2,000 hp., respectively. These are based on rounding off the constant 2,544 to 2,500 B.t.u. per hour per horsepower. This introduces an error of 44 in 2,500, or about 1.7 per cent. Considering that we will be able to use the same subdivisions for P as for H, and that the error is on the safe side (i.e., gives a higher required input than theoretically required), it appears that rounding the constant to 2,500 is justifiable. Also, the variations in compressor performance are such that an error of 1.7 per cent is not likely to introduce serious overloading of the driver at any time. By substituting values in our original equation and solving for P we find that 1 in. on the P scale = 200 hp. Using this value, and starting at the same zero point as that used for ΔH, lay off the P scale.

6. Mark the major subdivisions and letter the scales as shown, Fig. 1. Solve a typical problem by longhand and check the results with the chart. Note that only one solution line, connecting any three of the four variables, is needed. This is an outstanding advantage of this form of the proportional chart. Knowing any three of the four variables, the unknown can be found by entering at either the left or right side of the chart.

The compressor relation used in plotting this chart applies to any type of compressor handling an imperfect gas. The approximation, made by rounding off 2,544, applies to all types and has the same percentage effect. If desired, the chart could be plotted for the exact value of the constant, but the increase in accuracy would not be of great import. Next month we will plot a chart for a relation involving three reciprocal functions.

Plotting charts for reciprocal functions.

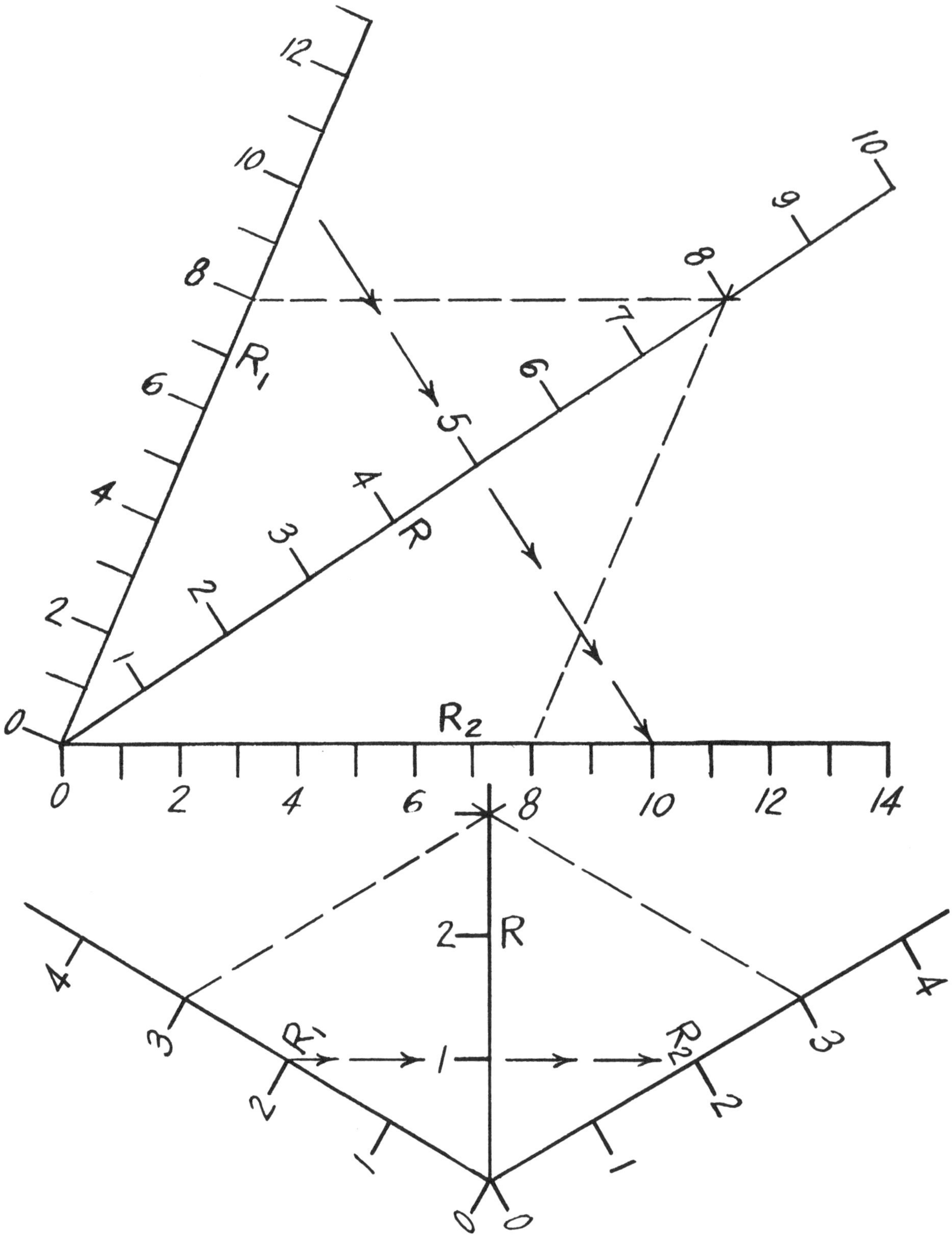

HOW TO MAKE NOMOGRAMS

by **Stephen C. Lane**

RECIPROCAL functions are expressions in which the unknown is of the form $1/a + 1/b = 1/c$. These occur in mechanical engineering during the determination of the total resistance to heat flow through a substance either made of several different materials or having air or liquid films on one or both sides. They occur in electrical engineering when the resistance of a parallel circuit is being found. And they occur in a number of other applications such as determining the focal length of lenses.

Knowing how to plot a chart involving these functions is a valuable aid because the addition of reciprocals cannot be performed on the ordinary slide rule. This means that the reciprocals must be found on the slide rule, noted on a piece of paper, and then added. When a person is not familiar with reciprocals, he is likely to be prone to errors. Charts of the type discussed this month eliminate many of these errors.

The equation . . . The resistance of two electric conductors in parallel is found by solving the relation $1/R = 1/R_1 + 1/R_2$, where: R = circuit resistance, ohms; R_1 = resistance of one branch of the electrical circuit, ohms; and R_2 = resistance of the other branch of the circuit, ohms.

To make our results as foolproof as possible, we'd like to be able to construct a chart in such a manner that the result we obtain is R, not its reciprocal. Should the chart give us $1/R$, we have an additional operation to perform with the chance of an error.

Limits . . . The limits for a chart of this type, Fig. 1, will depend on the circuits we wish to analyze with it. Since many circuits met in practice have a resistance somewhere between 0 and 10 ohms per branch, we will choose upper limits of 13 ohms for R_1 and 14 ohms for R_2. It must be realized that this selection is purely arbitrary and that higher or lower limits could be chosen, if desired.

Where this chart form is to be used in heat-transfer calculations, it is necessary to review the materials which will be encountered before the chart is constructed because so many variations exist from one material to the next. Should the resistance of one material be beyond the limits of its scale, the chart cannot be used for the combination. Also, it is not usually possible to substitute a lesser value and then multiply the result by a suitable factor because reciprocal relations are involved.

While these factors may appear to be a disadvantage, they actually are not. For, with a little planning before the chart is constructed, it is possible to secure a valuable tool in a very short time. Construction, by the method given here, is both simple and quick.

Chart Construction

1. Attach a sheet of 8½-by-11-in. paper to a drawing board or pad and, starting about 1 in. from the left-hand edge, draw a horizontal line for the R_2 scale, Fig. 1. At the left end of this line, mark a point and through it draw a sloping line for the R_1 scale. Arrange this line so it makes an angle of less than 90° with the horizontal line. Two charts are shown on Fig. 1, but at the moment we are discussing only the upper chart.

2. Subdivide the R_1 scale uniformly, using any suitable units. The unit chosen for the chart in Fig. 1 was 1 in. = 2 ohms, or ½ in. = 1 ohm. Do the same for R_2. To simplify the construction, the same subdivisions were used in Fig. 1 for both R_1 and R_2. However, this is not at all necessary. The only requirement is that both scales be subdivided uniformly. Either scale, horizontal or sloping, can be used for either variable. In other words, the sloping scale could be made the R_2 scale, provided it was suitably divided and the horizontal scale was made the R_1 scale.

3. After subdividing the horizontal and sloping scales, draw the short lines indicating each subdivision. These are perpendicular to their respective scales.

4. To locate the central sloping scale, draw a line from a selected value on the R_1 scale parallel to the R_2 scale. From the same value on the R_2 scale, draw a line parallel to the R_1 scale. The point of intersection of these two lines (shown dotted in Fig. 1) is the point corresponding to R_1 and R_2 values on the R scale.

For example, the point chosen for R_1 and R_2 in the upper chart of Fig. 1 was 8. The two dotted lines intersect at a value of 8 on the R scale. Since there is only one such value on R, a line drawn through the zero point of the R_1 and R_2 scales and the intersection of the dotted lines is the line along which the R scale must be plotted. Draw this line as shown in Fig. 1.

5. Subdivide the R scale by drawing similar parallel lines through points on the R_1 and R_2 scales. These lines will intersect at a point on the R scale, as they do at the value 8. To save time when one scale is horizontal, as the R_2 scale is here, just use a T-square and draw horizontal lines to intersect the R scale after its stem has been drawn in. Note that this scheme can be used only when one scale is horizontal and the central scale has been drawn into position.

6. Enter the proper values and solve a typical example. Thus, when $R_1 = 10$ and $R_2 = 10$, $R = 5$. Note that the correct value of R is obtained directly, not the reciprocal.

The upper chart in Fig. 1 is sometimes called an acute-angle type because the angle between its scales is less than 90°. Where space requires that a larger angle be used between the scales, an obtuse-angle chart, as shown in the lower part of Fig. 1, can be used. As the construction lines indicate, the same method and steps are followed as for the acute-angle chart.

We have now reached a point where we have constructed several charts having uniform scales. Any of these can be combined to produce a chart suitable for solving more complex equations. In constructing such nomograms, some study of the equation and its limits is necessary before the best form can be determined. During a study like this it is helpful to make a rough sketch of the chart, with its limits, to see what the final form will be. Next month we will construct an intersection chart.

Plotting simple intersection chart on graph paper.

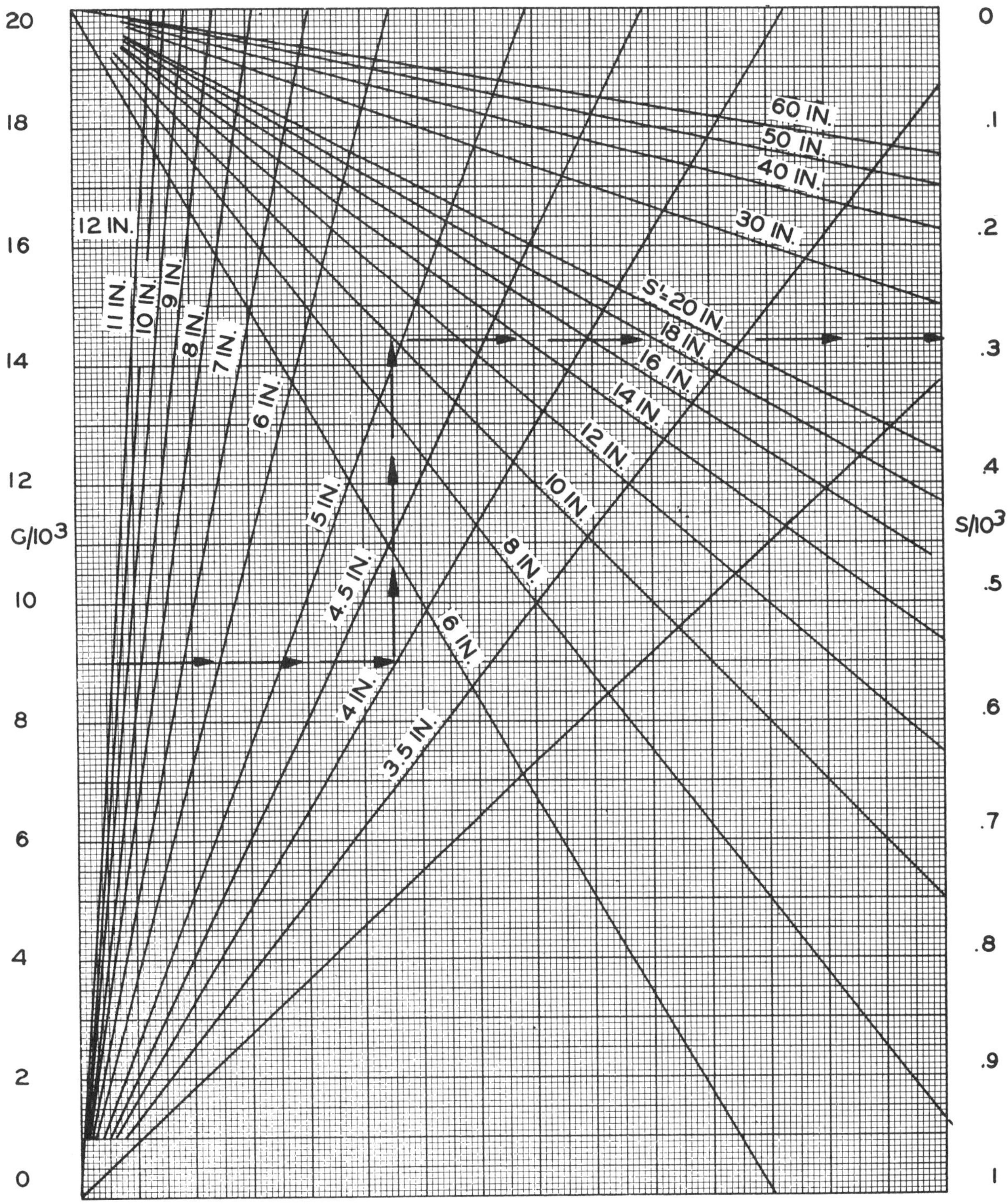

HOW TO MAKE NOMOGRAMS

by **Stephen C. Lane**

THERE are many instances where intersection charts prove simpler to construct than nomograms. And they give just as accurate results. This month we will study a typical intersection chart and learn how it is plotted for a selected relation.

Chart form . . . The intersection chart, Fig. 1, has two linear axes. Scales are plotted on these for two or more variables—depending on the equation for which the chart is constructed. The simplest way to illustrate the principles of construction is to choose a suitable equation and proceed with preparation of the chart.

Equation . . . The relation $S = 3.85G/AS'$ expresses the number of strokes made by a direct-acting steam pump in terms of capacity, water-cylinder area, and stroke length. In the relation, S = strokes per minute made by pump; G = pump capacity, gallons per hour; A = area of water cylinder, sq. in., and S′ = stroke length, in.

Limits . . . These will depend on the pump size and can be obtained from a catalog dealing with the make of direct-acting pump for which the chart is being constructed. For the chart in Fig. 1, the following limits were used: G, 0 to 20,000 gal. per hour; A, 7 to 131 sq. in., corresponding to cylinder diameters of 3 to 12 in.; S′, 6 to 60 in.

Construction . . . The first step which most engineers find extremely helpful is to make a rough sketch of the proposed chart. However, this is not always necessary, especially if you are familiar with this type of chart. But, for your first few attempts, it is advisable to make a quick sketch of the finished chart. This will help you to visualize the various curves and scales. It may also prevent time-consuming errors in construction.

1. On a sheet of scrap paper draw a rectangle with its long side vertical. On the left vertical side mark the letter G. This is a convenient value at which to enter the chart because the capacity of a pump is probably its best known characteristic. Since G varies from 0 to 20,000 gal. per hour, mark 0 in the lower left corner and 20,000 in the upper left corner. This will be the G scale and it will range between these limits.

2. Now, for a moment, ignore the S′ value in the equation and solve for S, using the minimum value of A. This works out as 3.85G/7, and for the maximum value of G, 20,000, is 3.85(20,000)/7 = 11,000. Mark the left end of the top of the rectangle 0. The right end could be marked 11,000 but it would cause the lower values of the D curves, Fig. 1, to be too far apart and the upper values to be too close together for easy reading. So, arbitrarily choose some value less than 11,000, but not less than about half of that. For this chart, a value of 7,500 was chosen because the upper scale of the graph paper used had 15 major divisions and 7500/15 = 500, a convenient number to work with.

3. We must now construct a series of curves passing through G = 0 and intersecting the temporary upper scale at 3.85G/A. Take the midpoint of the G scale where G = 10,000. Here 3.85G = 38,500. Set this up on a slide rule and divide by the area corresponding to a 3-in.-diameter water cylinder, or 7.069 sq. in. The result, 5,450, determines a point which is at the intersection of a horizontal line through 10,000 and a vertical line through 5,450 on the top horizontal scale. Draw these lines roughly on the sketch. A line through G = 0 (lower left corner) and this point determines the straight line marked D = 3. (Though three points are sometimes required to determine a straight line, two are sufficient when the relation for which it is plotted is linear, as it is in this case).

4. Secure a sheet of graph paper. The paper used for Fig. 1 has 10 by 10 divisions to the ½ in., with the fifth lines accented. This type paper is usually suitable for intersection charts. Copy on the graph paper the G and temporary upper scale. Locate the point discussed in (3) and draw the curve D = 3 in. Do the same for the other values of D shown, remembering that although the value of D is shown on the curve, the actual value used in plotting it is the area of a circle having D as its diameter. The value D is shown because pump cylinders are generally spoken of in terms of diameter instead of area.

5. We must now perform a similar step between the upper horizontal scale and the right vertical scale of the rectangle. The maximum value on the upper scale, 7,500, divided by the minimum value of S′, 6 in., is 7,500/6 = 1,256, say 1,000 since there are 20 divisions in the vertical scale of the graph paper used. Divide the right scale as shown and locate a point for each curve. Thus, when S′ = 20 and 3.85G/A = 5,000 (on upper temporary scale), S = 5,000/20 = 250. Draw a vertical line through 3.85G/A = 5,000 and S = 250. The point of intersection and the upper left corner of the chart, where 3.85G/A = 0, determine the two points for the straight line representing S′ = 20 in. Using the same method, find one point for each of the other curves and draw them as shown.

6. Finish the chart by drawing all the curves and solve a trial example. The solution shown is for the case where G = 9,000, A = 12.57 (corresponding to D = 4 in.), and S′ = 10. Enter at 9,000 and project horizontally to the right until the curve D = 4 in. is intersected. Then project vertically upward or downward until the proper S′ curve is intersected. From here project to the right to read S = 276 s.p.m.

This form of intersection chart can be used for any four-variable equation, regardless of the type of expression for any of the variables. However, when one or more of the variables is to any power other than unity, a series of points must be determined for each of its curves because they generally are not straight lines. The simplification given here applies only to those variables raised to the unity power.

Next month we will plot a more complex intersection chart, showing how it can be used for relations involving more than four variables.

Plotting intersection charts for up to 10 variables.

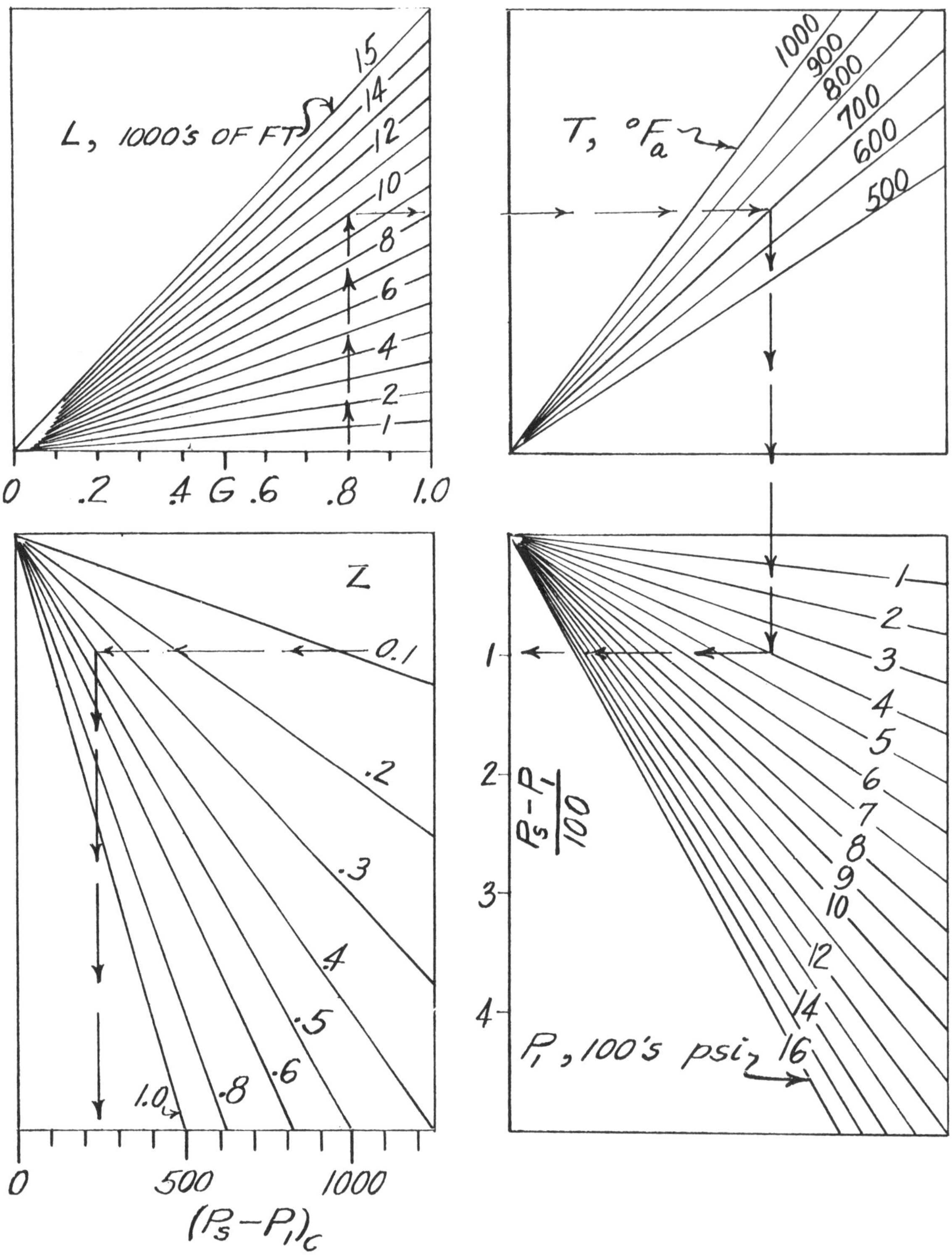

HOW TO MAKE NOMOGRAMS

by **Stephen C. Lane**

LAST month we learned how to construct a simple intersection chart. This form is often extremely useful for various types of equations. This month we will plot a similar chart but it will be of a form suitable for a large number of variables—up to 10.

The intersection chart, Fig. 1, constructed for a number of variables, consists of a series of straight lines when all elements of the equation are plotted as shown. This chart is based on a widely used equation submitted by a reader. Its construction should be of more than usual interest.

Equation . . . The relation $(P_s - P_1)_c = P_1(e^{GL/53.34T} - 1)/Z$, gives the pressure due to the weight of a static column of gas, neglecting the deviation of the gas from Boyle's law. In the relation, P_s = pressure at the bottom of the well, psia.; P_1 = pressure at the wellhead plus pressure drop due to friction, psia. (in a well where the gas column is static, the pressure drop is zero); $P_s - P_1$ = pressure due to weight of gas column, psia.; $(P_s - P_1)_c$ = pressure due to weight of a gas column, corrected for compressibility; G = specific gravity of gas (air = 1.00); L = average length of gas column = well depth, ft.; e = Napierian logarithm base = 2.71828; T = average temperature of gas column, °F. absolute = °F. + 460; 53.34 = gas constant for air; and Z = compressibility factor.

The limits submitted by the reader are: P_1, 100 to 1,500 psia.; G, 0.60 to 1.00; L, 1,000 to 15,000 ft.; T, 520° to 710° F. absolute; and Z, 0.35 to 0.995. The compressibility factor varies with P_1, T, and G. The reader submitting this equation felt that a chart would save him much time in designing gas-lift installations.

Construction . . . Make a rough sketch of the proposed chart. Studying the given relation, it is easy to see that the exponent of e must be evaluated before any other operations can be performed. The exponent is composed of three variables and a constant. The product of two of the variables is divided by the product of the constant and the third variable. Neglecting the constant for a moment, we can see that a simple chart (upper left, Fig. 1) will give us the product of GL. Then another similar chart (upper right) will give us GL/T. The constant is easily incorporated now.

1. On a sheet of 8½ by 11-in. paper, draw a rectangle with a horizontal scale of convenient length for G (upper left, Fig. 1). The values of G can be laid off from 0 to 1.0, as shown, or from 0.6 to 1.0, if desired. The method shown is easier for the beginner because all the L lines pass through G = 0. All values of L between 1,000 and 15,000 can be shown if the right-hand vertical scale of the rectangle is divided into 15 equal parts. Do this, using dividers. Then the intersection points for an L line are G = 0 and each of the points on the vertical side of the rectangle.

Thus, the right side of the rectangle represents a temporary scale varying from 0 to 15,000 in steps of 1,000. A straight line drawn through G = 0 and any one of the subdivisions on the vertical scale is the line representing L. Its value is the value intersected on the vertical scale. The line through the first subdivision is L = 1,000; through the second, L = 2,000; etc., up to and including 15,000.

2. We now have the product GL represented by the vertical scale of the rectangle. Draw a second rectangle alongside the first, as shown. Letter its lower left corner 0 and the lower right one 0.4. These are obtained by noting that the minimum value of GL, which is zero, is still zero when divided by any value of 53.34T greater than zero. (This product will always be greater than zero because the gas temperature is about that of the atmosphere, or higher.)

The value 0.4 is obtained by taking the maximum product of G and L, 15,000, and dividing by the median value of 53.34T, or about 35,000 (= 53.35 × 700). Lines representing T will pass through GL = 0, lower left corner, and a point obtained by dividing GL by some value of 53.34T. Thus, when GL = 5,334 and T = 500, the curve for T = 500° F. absolute will pass through GL = 0 and the point 0.2.

This latter point occurs on a perpendicular erected through 0.2 on the bottom scale of the second rectangle. Its vertical position is found by drawing a horizontal line through GL = 5,334 until it intersects the perpendicular through 0.2. Using this method, draw the other temperature curves, as shown.

3. The bottom scale of the right-hand rectangle now represents the value of the exponent, GL/53.34T, of the constant e, defined above. Since the values on this scale range from 0 to 0.4, the values of $e^{GL/53.34T}$ range from 1.00 to 1.492. Then, subtracting 1.00 from these, as indicated in the original relation, the range of the scale becomes 0.0 to 0.492.

Using the same methods as for the first two rectangles, draw a third in the lower right-hand corner of the paper. In it plot the product of P_1 and $e^{GL/53.34T}$, using the vertical left-hand scale for the product. As shown in Fig. 1, this gives the uncorrected value, $P_s - P_1$.

4. Draw the fourth and last rectangle in the lower left-hand corner and divide the values of $P_s - P_1$ by Z, using the bottom scale of the last rectangle as the $(P_s - P_1)_c$ scale. Its values range from 0.0 to 1,250 psia., giving a satisfactory range for the corrected pressure. However, in the event the solution should fall outside the range of the Z rectangle, the $P_s - P_1$ scale can be used and the resulting value divided by Z, using a slide rule or longhand.

5. To solve a problem, follow the trace shown. In the example illustrated, G = 0.80, L = 10,000 ft., T = 700° F. absolute, P_1 = 400 psia., and Z = 0.4. As can be seen, $(P_s - P_1)_c$ = 240 psia., which agrees well with its computed value using a slide rule.

Note that the four rectangles in this chart could be placed so their inner and upper scales overlap. However, a slight separation, say ½ in., permits the solution lines to be more readily seen.

Constructing Curved-Scale

by **Stephen C. Lane**

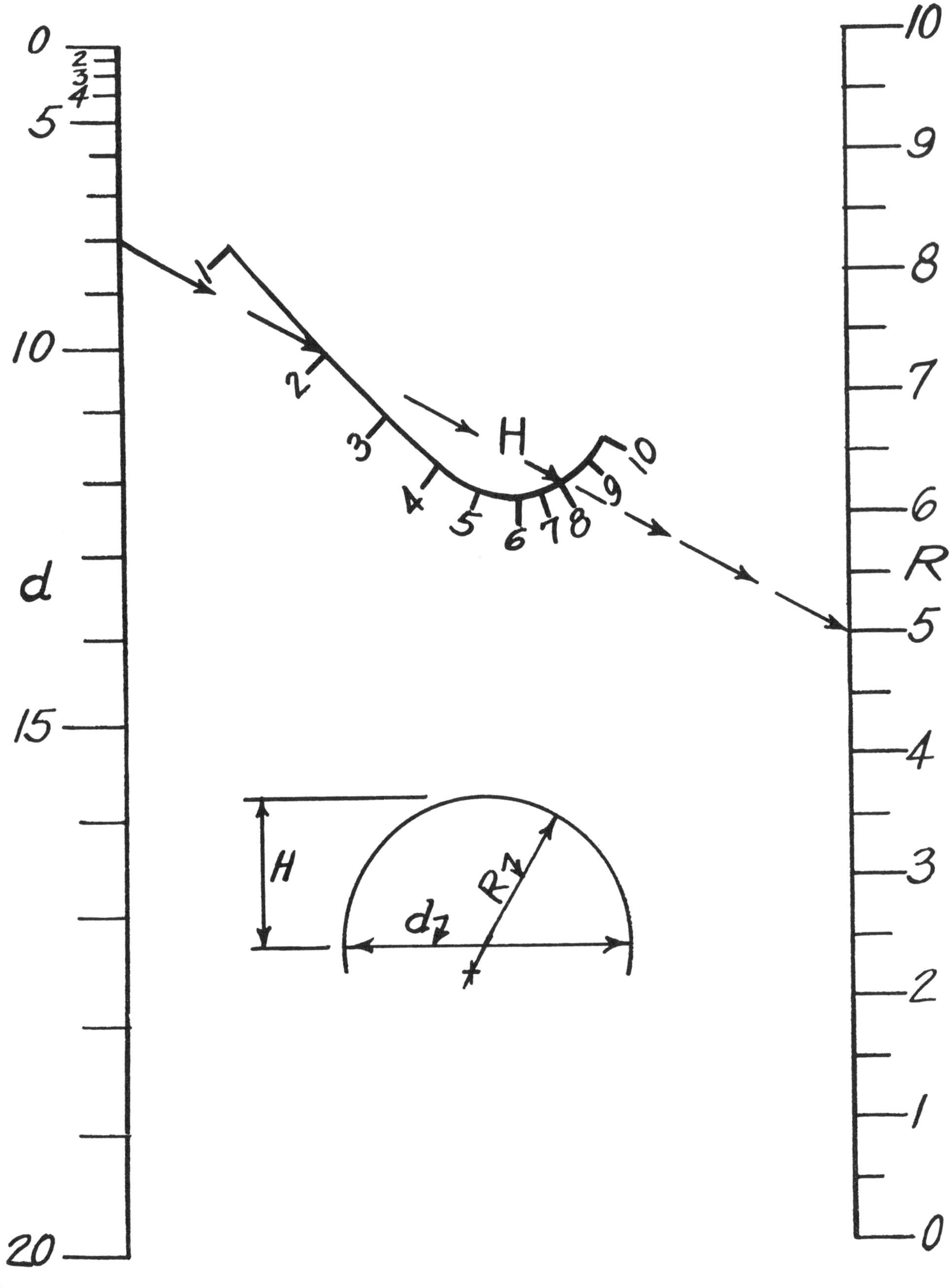

Nomograms Isn't Hard

. . . Here's how you can do it, simply and easily

NOMOGRAMS with curved scales are popular for equations of the form $u + vw = w^n$. In this relation the exponent n may have any value other than zero, and the sign between the groups of variables may be either positive or negative.

Because the general equation involves the sum and product (or the difference and dividend) of three variables, the chart is not as simple as for a relation involving only one of these operations. When linear scales are used for each of two variables, the third scale is a curved line. Its subdivisions are neither uniform nor logarithmic. The only way they can be determined is by calculation and construction.

The equation . . . The chordal height of the segment of a circle intersected by a straight line is given by:

$$8RH - 4H^2 = d^2$$

Where:

H = chordal height
R = radius of the circle through which the chord is drawn
d = chord length

Any consistent set of units may be used in this relation. Thus, if one dimension is in feet, the other two also should be in feet.

This equation offers a good opportunity to show how to construct charts of this type because it contains two variables that are power functions and also has a minus sign, instead of a plus sign, between the groups of variables.

Limits . . . Any suitable limits can be chosen for this chart because there are no ranges specified by the existing physical conditions or other factors. So a range of 0 to 20 will be chosen for d and a range of 0 to 10 for R. The range of H will be determined by substituting values for d and R in the equation.

Scales . . . A study of the equation shows that a uniformly divided linear scale can be used for R because it is to the first power, i.e., unity in the expression. The scale for d, however, must be graduated for a squared function because d is raised to the second power. This is done as follows:

First, write a scale equation for d; that is, $x = md^2$, when x = the scale length in inches for various values of d, and m = the scale modulus. Now, assuming the chart is to be plotted on 8½ by 11-in. paper and the d scale is to be 10 in. long for d = 20, the maximum value, substitute these values in the scale equation and solve for m. Or, $m = x/d^2 = 10.0/400 = 0.025$. Thus, $x = 0.025\ d^2$, a relation which gives the distance any value on the d scale is from the zero point of the scale. Hence, for d = 10, $x = 0.025\ (10)^2 = 2.5$ in. from the zero point on the d scale.

Chart Construction

1. On an 8½ by 11-in. sheet of paper draw two parallel straight lines about 7 in. apart for the d and R scales. Subdivide the R scale uniformly, using a value of 1 in. = 1 unit. (If a larger or smaller chart were being constructed, any other suitable subdivision could be chosen for R.)

2. Using the above scale equation, compute the lengths for the various subdivisions of the d scale. Lay these off as in figure. Thus, for d = 10, measure 2.5 in. down from d = 0 and mark this distance on the scale stem. Continue this process for other values of d between 0 and 20. The resulting scale will be 10 in. long, because the scale equation was based on this length.

3. The next step is to determine the location and shape of the H scale. As noted earlier, this must be done by calculation and construction. Start by assuming a value for H and for one of the other variables, say R, and solve for d. Thus, assume H = 4 and R = 2. Substitute in the equation for which the chart is being plotted, $8RH - 4H^2 = d^2$, and solve for d. Or $8\ (2)\ (4) - 4\ (4)\ (4) = d^2$, or d = 0. Therefore a line through d = 0 and R = 2 will intersect the H scale at H = 4. But as yet the H scale has not been plotted and the point at which H = 4 along the line connecting d = 0 with R = 2 is not yet known. So, using the same value for H and another value for R, again solve for d. Thus, when H = 4, and R = 4, d = 8. Draw a line from d = 8 through R = 4. The point where this line intersects the first line connecting d = 0 with R = 2 is the point on the H scale where H = 4.

Perform this operation for every value of H to be plotted on the H scale. Values of H from 1 through 10 were chosen for this chart. When all the points on the H scale have been determined, draw a smooth curve through them and mark the scale subdivisions.

4. To use the chart, draw a straight line from a value of d on the left stem through a value for R on the right-hand stem. Then read the corresponding value of H on the H scale.

Notes . . . Some solution lines will give two values of H, as shown in figure. This means that either of these values will be a solution for the original equation.

When solving the scale equation and the equation for which the chart is plotted, set up a tabulation listing all of the variables in the equations. As these equations are solved for each set of values, enter the appropriate numbers in the table. Keep the tabulation for reference as construction of the chart proceeds. Note that the d and R scales ascend in opposite directions because there is a negative sign between the groups of variables.

How to Construct

This type of chart will help solve the relation:

$$xy = z$$

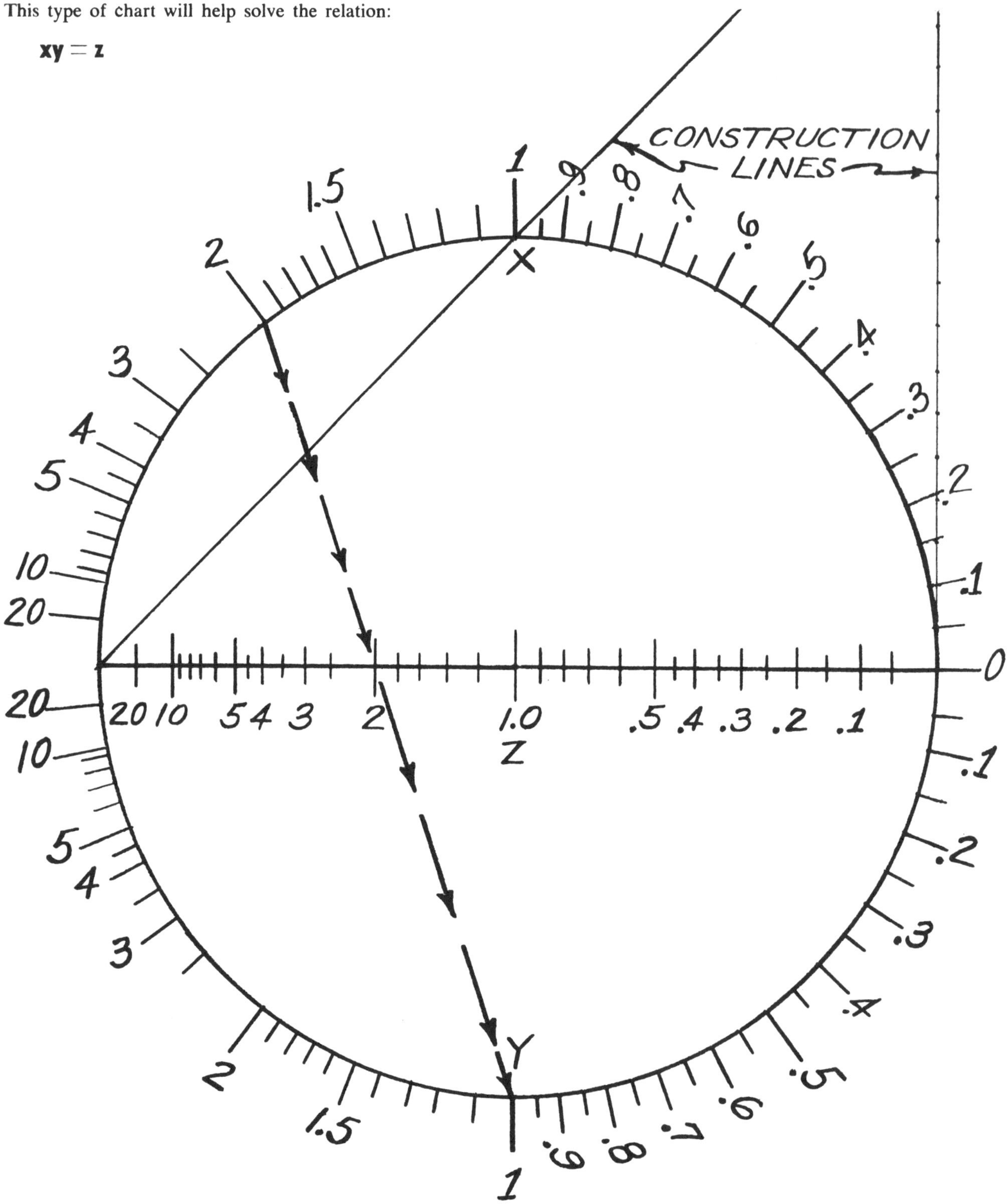

Curved-Scale Nomograms

. . . Here's an easy one to use for four-variable equations

by **Stephen C. Lane**

NET charts permit you to solve relations involving four variables. By using this type of chart, you obtain a plot that permits reading all four variables with a single solution line. This is often an advantage because you can observe the change in one or more of the variables as two of the others change. This chart is plotted by similar procedures used in constructing a chart for the sum or difference of two variables.

The equation . . . The angular displacement, D, of a uniformly accelerated rotating body starting from rest is given by:

$$D = VT + \frac{1}{2}AT^2$$

Where:

- D = angular displacement of the rotating body during the period of its acceleration for time, T, radians
- V = angular velocity at the start of the period being considered, radians per second
- T = time during which the body is uniformly accelerated, seconds
- A = angular acceleration radians per second per second

This equation is useful when making any of a number of calculations for rotating bodies being uniformly accelerated.

Limits . . . Assume that the initial angular velocity, V, varies from 0 to 10 radians per second, while the time during which acceleration takes place varies from 1 to 3.5 seconds, and the angular acceleration varies from 0 to 9 radians per second per second. These limits cover many ordinary applications and provide sufficient range to give a good illustration of chart construction.

Chart Construction

1. As noted above, the construction of this chart resembles that for the sum or difference of two variables. To convert a four-variable relation of the type given above to one involving three variables, i.e., the sum or difference of two variables, assume successive values for one of the variables and write the equation obtained. Thus, assuming T = 1, D = V + ½A; for T = 2, D = 2V + 2A; for T = 3, D = 3V + 4.5A; for T = 4, D = 4V + 8A.

2. With T = 1, D = V + ½A. Or, V − D = − ½A, when the equation is rewritten to express A in terms of V and D. Draw the V and D scales as parallel lines about 6 in. apart, as shown in Fig. 1. Subdivide each uniformly, the V scale ascending upwards and the D scale downwards, as shown.

3. Working with the equation V − D = − ½A, for T = 1, assume a value for V and one for A within the limits chosen above. Solve for D. Thus, when V = 5 and A = 10, D = 10. Next, draw a line through V = 5 and D = 10. Holding A = 10, assume V = 3. Then D = 8. Draw a line through V = 3 and D = 8. Where it intersects the line through V = 5 and D = 10, draw a vertical line. All values of A for T = 1 lie along this perpendicular, one point being that already determined for A = 10.

4. With V = 5, continue to assume values for A from 0 through 9, increasing by unity, and solve for D. Thus, for A = 0, D = 4; for A = 1, D = 5.5; for A = 2, D = 6, etc. Using these values of V and D, locate the various values of A on the vertical line for T = 1, Fig. 1. Thus, a line through V = 5 and D = 4 cuts this perpendicular at A = 0.

5. Afterward, using the equation for T = 2, perform the same steps to locate the vertical line for T = 2 shown in Fig. 1. Since the equation for T = 2 is different from that for T = 1, assume values for A from 0 through 9 and substitute in D = 2V + 2A, while using a fixed value for V. A series of points will be obtained along a vertical line.

6. Following the same procedures as in (4) and (5), plot the vertical and its points for T = 1.5, 2.5, 3.0 and 3.5.

7. For each value of A from 0 through 9 draw a smooth curve through the points representing it on the various verticals, Fig. 1. Label the verticals, the curves, and the two exterior scales with suitable symbols. The chart is now ready for use.

8. To use this chart, the values of at least two variables must be known. Or if the values of three of the variables are known, the fourth can be found immediately with a single solution line. Thus when V = 5, T = 2, and A = 1, D = 12, as shown in the example on the chart in Fig. 1.

Net charts of this type can be constructed for any four-variable equation which can be expressed as the sum or difference of two variables when one of the four variables is held constant. The general procedure is the same as described here. This type of chart permits a rapid solution of many equations and graphically shows how the various variables are related. Its chief advantages lie in its use of a single solution line and its ease in being plotted. Note that the negative sign in the right-hand member of the rewritten equation can be neglected because it results only from the form imposed upon the relation to facilitate plotting and is not an integral part of it.

How to Construct

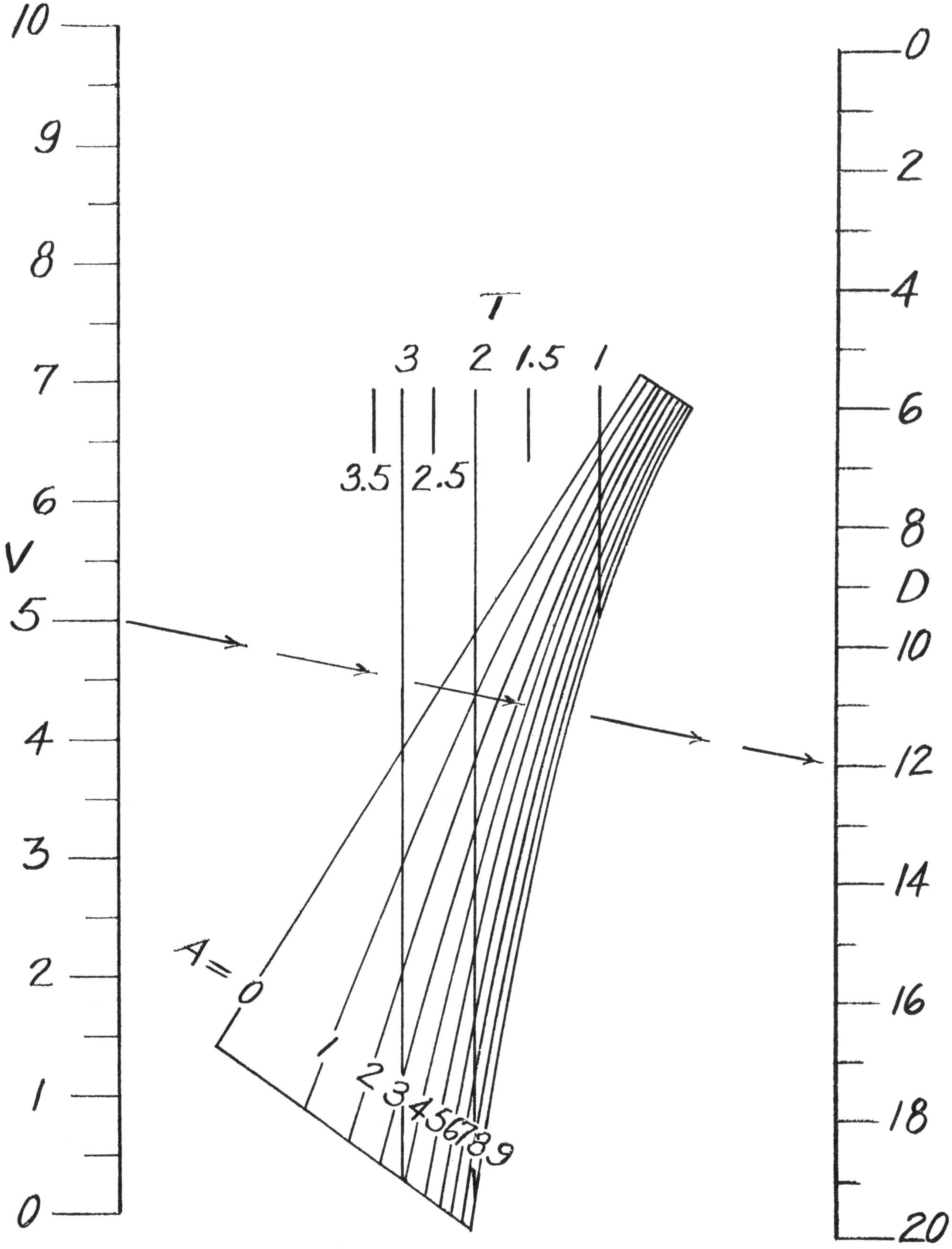

Circular Nomograms

. . . Use this chart to solve three-variable equations

by **Stephen C. Lane**

THE circular chart is often useful when plotting relations of the form $xy = z$ and variations of this basic relation, i.e., $y = z/x$, $(x - 6)(y^2 - 4) = 3z - 8$, $\log \cos P(x^3 - 8) = 7 \sin^2 z$, etc. Being relatively simple to construct, the circular nomogram or chart is easy to use. It provides a fast and accurate solution to a variety of relations that can be expressed as the product of two variables.

The equation . . . Let us plot a circular nomogram for the equation $xy = z$, where any or all of the three variables may be raised to a power other than unity. This relation could stand for any of a number of well-known equations. These include the area, z, of a square or rectangle having dimensions x and y, or the volume of a cube having area x for the base and depth y, etc.

Limits . . . Assume, for the sake of this example, that x and y have the same minimum and maximum limits (0 to 20). The units can be any consistent set producing a suitable unit for z. For instance, x and y could both be in feet, inches, or any other suitable units.

Chart Construction

1. Attach a piece of 8½ by 11-in. paper to a drawing board and locate the horizontal center at a point midway between the top and bottom edges of the paper. Draw a horizontal line through this point.

2. Choose a desirable diameter for the chart. In general, the larger the diameter, the better the chart because it is easier to read and use. A diameter of 7 in. was arbitrarily chosen for this chart to secure a circle with as large a diameter as possible. Draw the circle using the point found in (1) as its center.

3. In any circular nomogram of the type described here, the zero point for both the x and y scales is at the right-hand end of the horizontal diameter, Fig. 1. Mark this point 0. The unity points for the x and y scales of this chart lie on the intersection of the circle with a vertical line through the center of the circle, or at the 90° and 270° positions, respectively.

4. The upper semicircle forms the x scale, from zero to infinity. The lower half carries the y scale between the same limits. The horizontal diameter forms the z scale, with its unity point located at the center of the circle, Fig. 1. This is known as the product scale.

5. From the right-hand end of the horizontal centerline, erect a perpendicular tangent to the circle. Extend this upward a distance of 7 in., or a distance equal to the diameter of the circle used for the nomogram. Then, from the left-hand end of the horizontal diameter, draw a line through the value of unity on the x scale until it intersects the extreme left end of the horizontal diameter and the perpendicular tangent at the right-hand side of the chart. This intersection will be 7 in. above the 0 point on the right-hand end of the horizontal diameter. Portions of these two lines are shown lightly in Fig. 1 and are labeled "Construction Lines."

6. Subdivide the vertical tangent line into 20 equal parts. Then, from the left end of the horizontal diameter, draw straight lines through each of the subdivisions on the tangent line. Where each line intersects the upper semicircle between 0 and 1, make a mark. These 20 points indicate 20 subdivisions between 0 and unity on the x scale, Fig. 1.

7. Proceed in the same way for values of y between 0 and 1, except that the tangent is drawn vertically downward from the right-hand 0 point.

8. Subdivide the z scale by drawing lines from subdivisions on x through unity on the y scale. This will give values between 0 and unity on the z scale. Thus, a line from 0.9 on x to 1.0 on y cuts z at 0.9. This is because $z = xy$.

9. To locate points on x and y between 1 and 20, choose a value on z and divide it by a suitable value on x and y to produce the desired value. Thus, with $z = 1$ and x or $y = 0.2$, x or $y = 5.0$. Draw a line through x or $y = 0.2$ and $z = 1$. Extend it to intersect the x or y scale and mark this point. Continue this process until the desired number of points between 1 and 20 are secured.

10. Subdivide z from unity to 20 by the same method as given in (8), i.e., by drawing lines from values on x through unity on y. The point where this line cuts the z scale corresponds to the value on x from which the line is drawn.

Note several characteristics of the circular nomogram. Values of unity on the x, y, and z scales always lie where shown on Fig. 1, if the relation is of the form $xy = z$. Where powers or other functions of the variables are involved in this relation, the position of the unity point will usually be different. In general, it is better to have the horizontal diameter as the product scale because the chart is then easier to use.

Here's a Handy Chart for Gas Column Pressure

by **Stephen C. Lane**

THIS chart permits you to easily determine the pressure resulting from a vertical column of gas. It is based on the relation:

$$P_v = P_w (e)^{GL/53.3\,TZ}$$

Where:

P_v = pressure exerted by gas column, psia.
P_w = static pressure at the wellhead, psia.
G = gas gravity
L = well depth, ft.
T = absolute temperature of the gas (°F. + 460)
Z = compressibility factor for the gas
e = 2.72

Example: What is the pressure due to the gas in a well 9,000 ft. deep if the gas gravity is 0.7, the compressibility factor is 0.8, gas temperature is 140° F., and the wellhead pressure is 2,000 psi.?

Solution: Project from G = 0.7 on the bottom of the chart up to L = 9,000 ft. on the depth curve. From here project horizontally to T = 600° R = 140° F. + 460°. From this point project vertically to Z = 0.8 and then horizontally to P_w = 2,000 psia. At the top of the chart read P_v = 2,550 psia. All other problems are solved in a similar manner on this chart.

THIS CHART IS A PLOT

of the equation . . .

$$P_v = P_w (e)^{GL/53.3\,TZ}$$

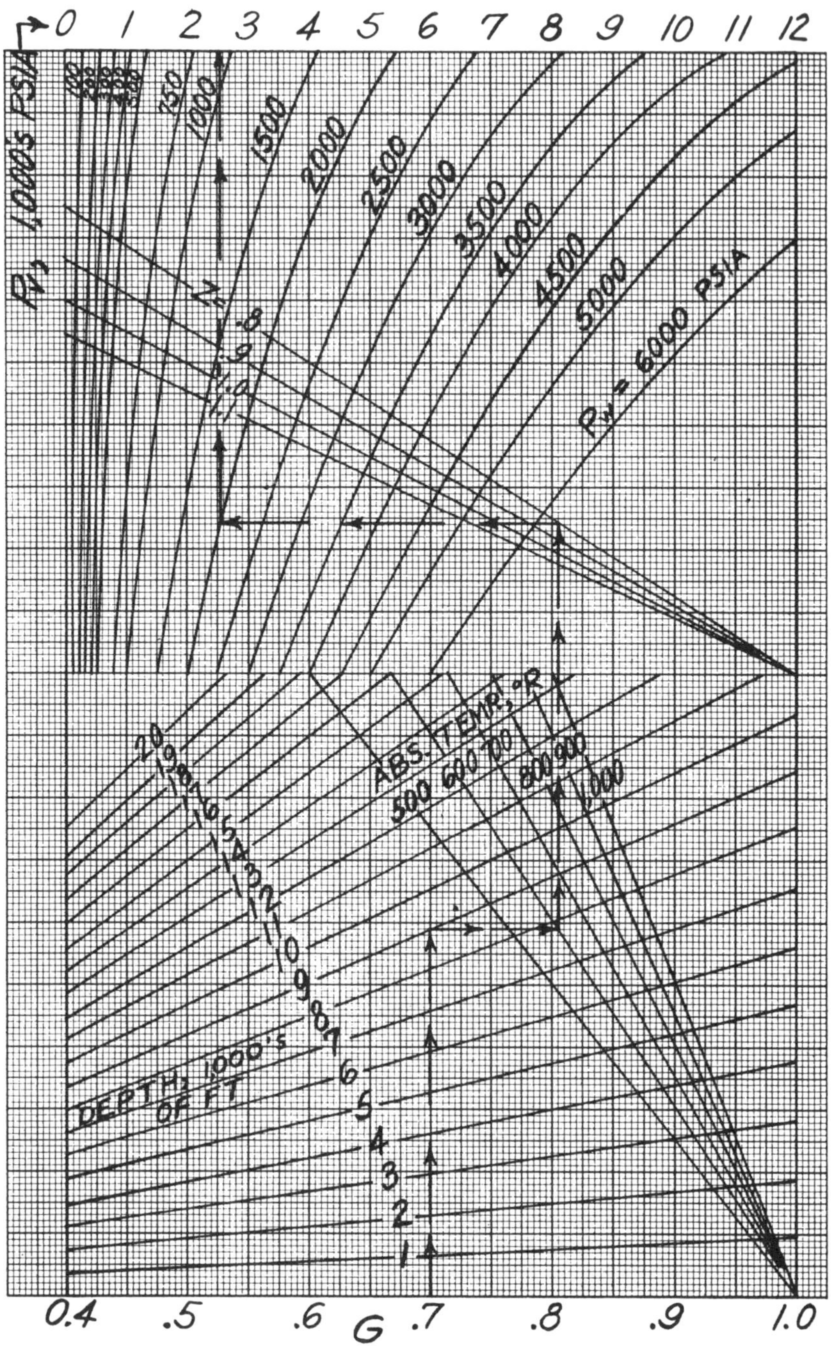

Nomogram corrects gas volume for water vapor

Bill Sisson
Superintendent of Operations
Nipak, Inc.
Pryor, Okla.

COMPRESSOR specifications normally show the volume of air or gas received at the suction as dry volume. However, in many installations, the air or gas is saturated with water vapor when it enters the compressor. With this nomogram, you can find the saturated volume directly when the dry volume and actual temperature and pressure are known.

Example 1. A volume of gas measured dry at 60° F. and 30 in. Hg is 15,000 cfm. What is the corrected wet volume, if the actual temperature is 115° F. and the pressure is 14 psia?

Align 14 psia on Scale B with 115° F. on Scale E (Step 1). Mark where the line crosses Scale D. From this point on Scale D align with 15,000 cfm on Scale C. Extend the line to Scale A and read the corrected wet volume as 19,500 cfm (Step 2).

Example 2. The saturated (wet) volume of a gas at 120° F. and 13.4 psia is 8,430 cfm. What is the dry volume at 60° F. and 30 in. Hg?

Align 13.4 psia on Scale B with 120° F. on Scale E and mark where the line crosses Scale D (Step 3). Next, align the known saturated (wet) volume of 8,430 cfm on Scale A with the point marked on Scale D in Step 3 and read the dry volume as 6,000 cfm at 60° F. and 30 in. Hg where the line crosses Scale C (Step 4).

END

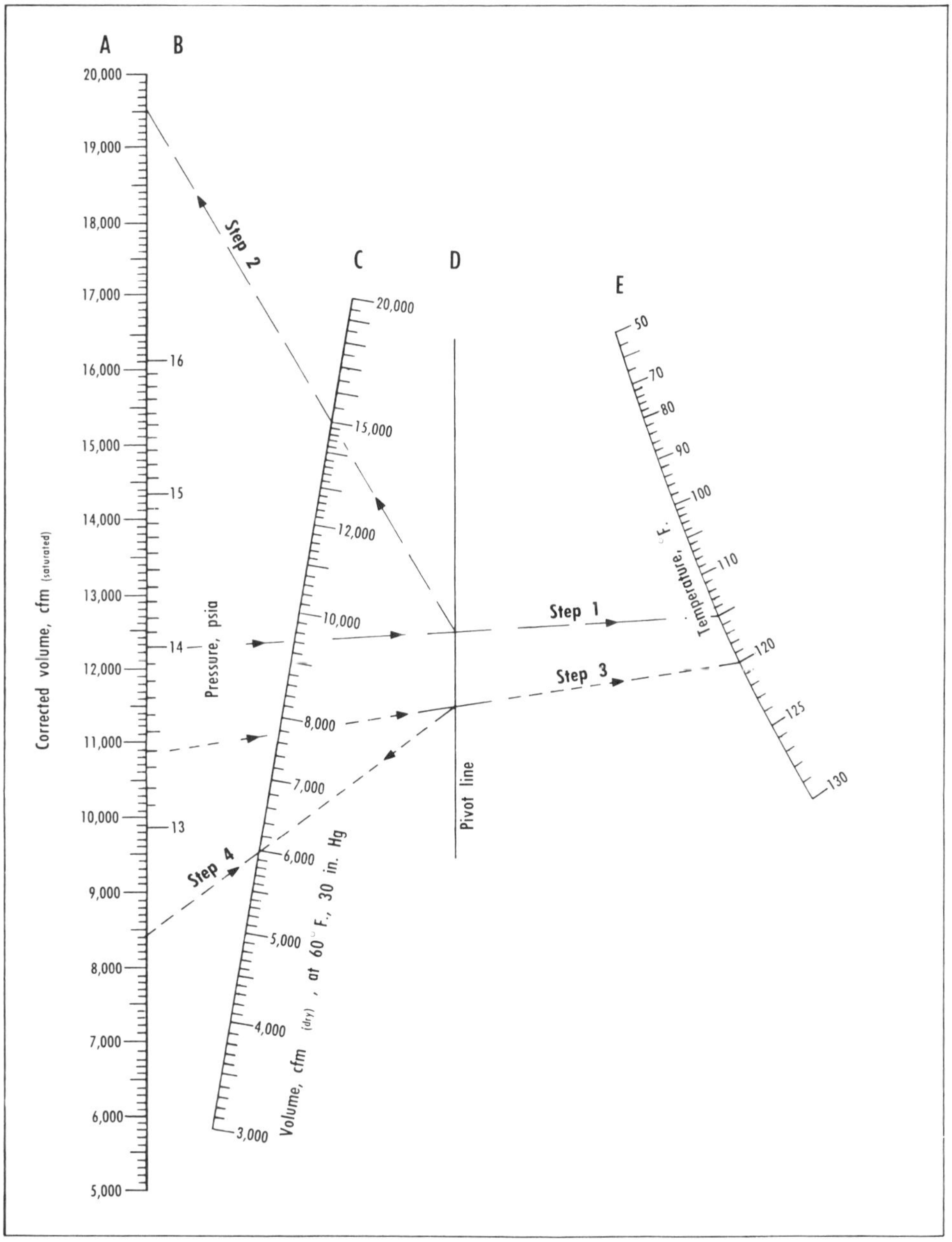

Nomogram gives bottom-hole gas pressure

Bill Sisson
Superintendent of Operations
Nipak, Inc.
Pryor, Okla.

THE nomogram shown here permits you to rapidly calculate the pressure resulting from a vertical column of gas, such as the bottom-hole pressure of a gas well. It is based on the relation:

$$P_v = P_w e^{GL/53.3TZ}$$

Where:

P_v = pressure exerted by gas column, psia
P_w = static pressure at wellhead, psia
G = gas specific gravity
L = well depth, ft
T = absolute temperature of gas (°F. + 460)
Z = compressibility factor for the gas
e = 2.72

Example. What is the bottom-hole pressure due to the gas in a well 8,000 ft deep, if the gas specific gravity is 0.75, the compressibility factor is 0.8, gas temperature is 140° F., and the wellhead pressure is 3,000 psia?

Solution. Step 1, align 0.75 on the specific gravity (G) scale with 8,000 ft on the well-depth (L) scale, extend the line to Pivot Line 1, and mark the intersection. Step 2, from the marked point on Pivot Line 1, align with 140° F. on the temperature (T) scale, and mark the intersection with Pivot Line 2. Step 3, align the point on Pivot Line 2 with 0.8 on the compressibility factor (Z) scale, extend the line to Pivot Line 3, and mark the intersection. Step 4, align from the marked point on Pivot Line 3 with 3,000 psia on the wellhead pressure (P_w) scale, extend the line to the bottom-hole pressure (P_v) scale, and read 3,770 psia bottom-hole pressure.

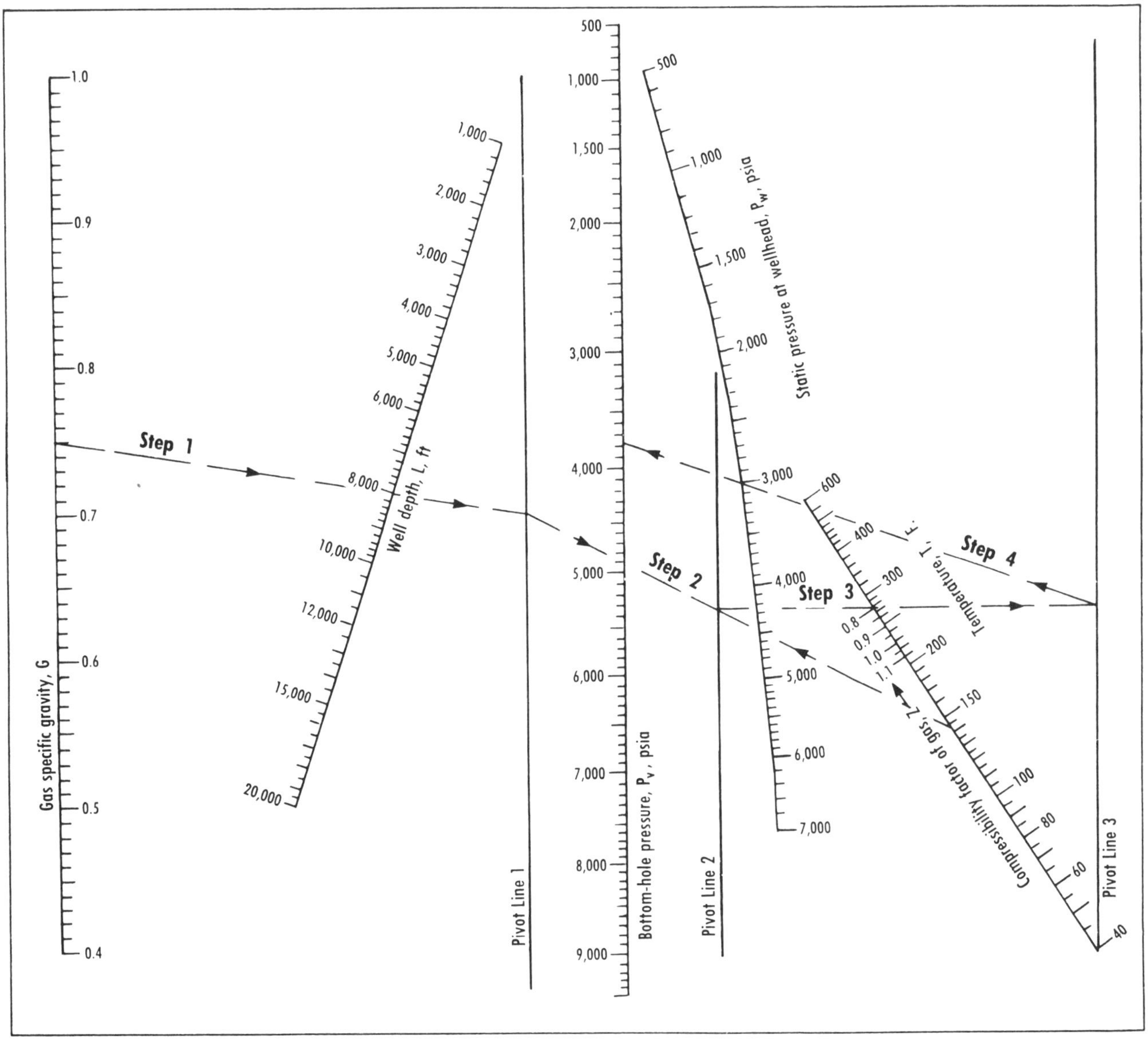

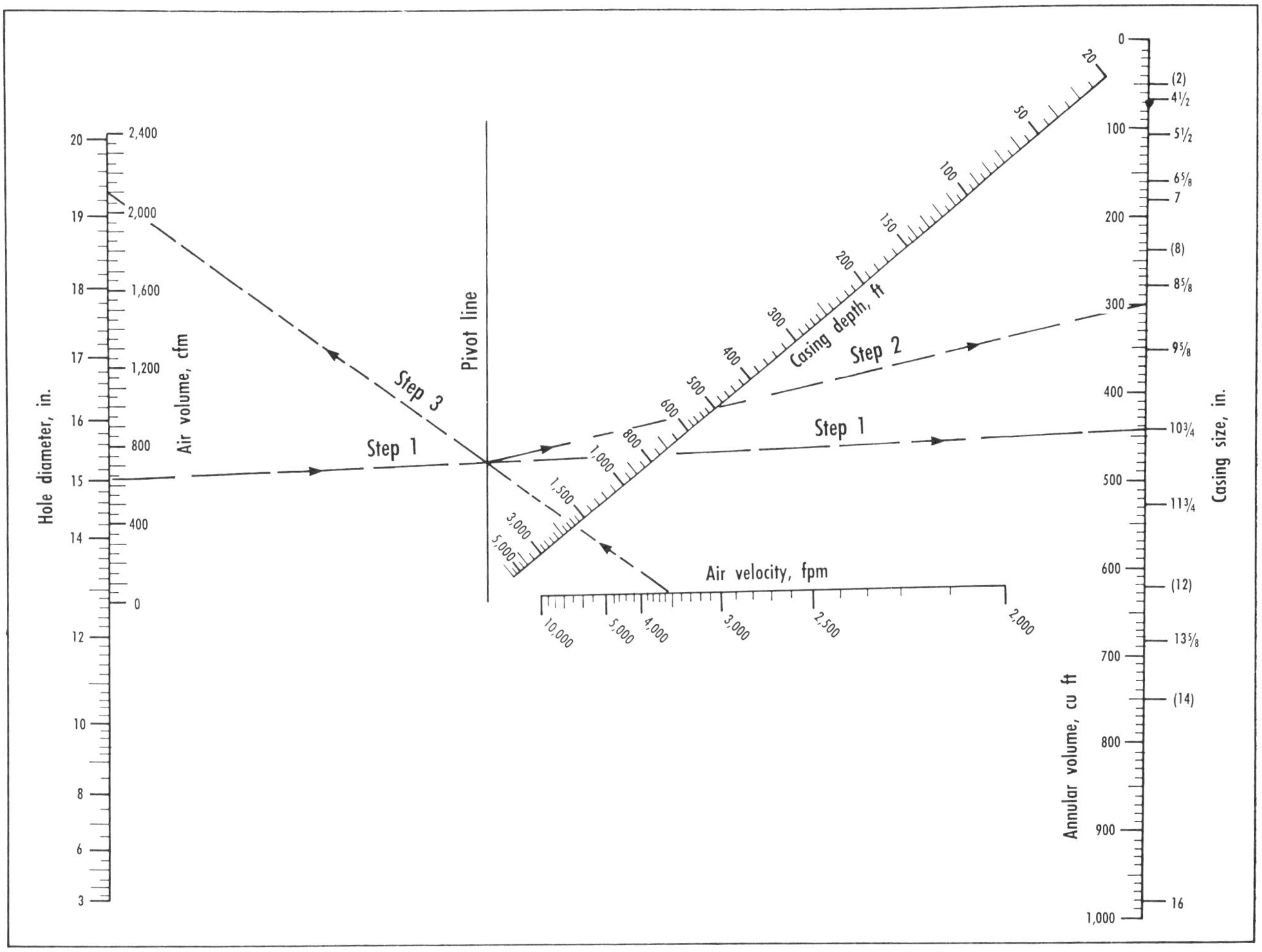

Nomogram tells down-hole volume, velocity between casing and open hole

Bill Sisson
Superintendent of Operations
Nipak, Inc.
Pryor, Okla.

TO PROVIDE efficient air-drilling performance, the volume of air produced by the compressor must be enough to provide the necessary annular velocity for proper hole cleaning. For percussion tools, that volume should be in accord with manufacturer's recommendation. The drill pipe must be sized to provide adequate return-air velocity.

Return velocity of hole cleaning varies according to drilling conditions—weight of cuttings, hole depth, amount of moisture encountered, etc. Many authorities agree that 3,000 fpm represents the minimum. When heavy cuttings result from drilling, excess hole depth, or when moisture is encountered, higher velocities are required. Annular volume and annular velocity can be determined by the following formulas:

$$\text{Annular volume, cu ft} = [0.7854\ (D^2 - d^2)\ L]/144$$

$$\text{Annular velocity, fpm} = (\text{cfm} \times 144)/0.7854\ (D^2 - d^2)$$

Where:

cfm = air volume

D = diameter of hole, in.

d = outside diameter of casing or pipe, in.

L = depth of casing, ft

A convenient way to determine annular volumes and velocities is by use of the nomograph shown here and based on the above formulas. Note that values in parentheses on casing-size scale represent nominal pipe sizes.

Example. If open-hole diameter is 15 in., casing size is 10¾ in., and casing depth is 500 ft, what is the annular volume? If it is desired to have a return hole-cleaning air velocity of 3,500 fpm, what is the volume required?

Solution. Line 15 in. on hole-diameter scale with 10¾ in. on casing-size scale and mark where line crosses pivot line (Step 1). From this point, line with 500 ft on casing-depth scale, extend line to annular-volume scale, and read 300 cu ft (Step 2). Next, line 3,500 fpm on air-velocity scale with marked point on pivot line, extend to air-volume scale, and read 2,100 cfm (Step 3). END

Here's a Handy Pipe-Weight Estimator

. . . You can use it to estimate the weight of plain-end steel pipe

by **Frank E. Hangs**
Pipe Line Technologists, Inc.
Houston, Tex.

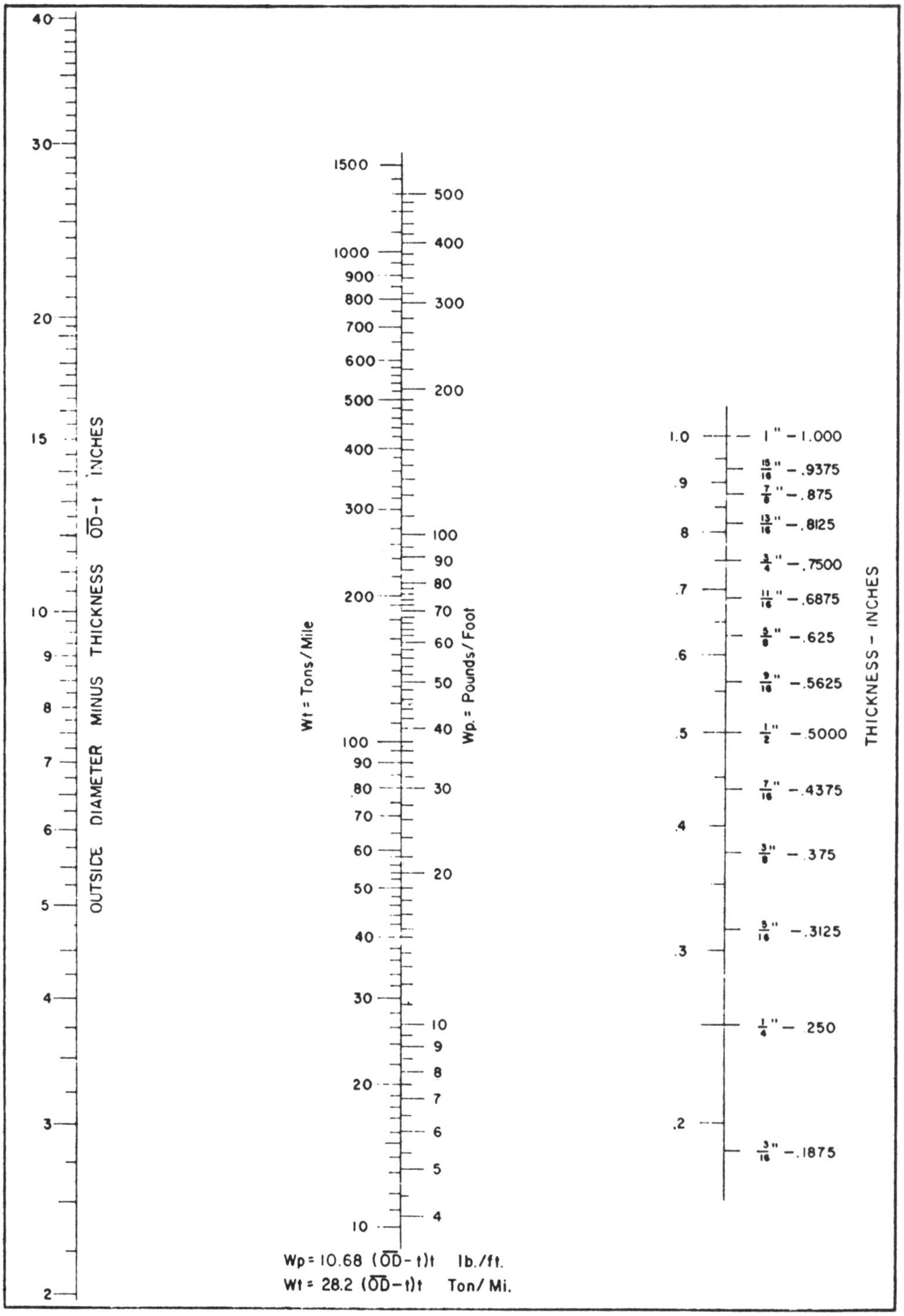

This nomogram...

Determines thermal expansion of pipe

Bill Sisson
Superintendent of Operations
Nipak Inc.
Pryor, Okla.

DETERMINATION of the expansion of any given segment of a pipeline of a number of different piping materials is simplified by the use of this nomogram. It is not implied that all materials are suitable for the temperatures shown.

Example 1. If 210 ft of carbon-steel pipeline is carrying steam at 50 psig (298° F.) with 22° superheat and the minimum anticipated temperature is 0° F., what is the expansion?

Solution. The total temperature span is: 298 + 22 + 0 = 320° F. On the nomogram, connect the carbon-steel reference point on M scale (Step 1) with 320° F. on T scale, and extend to Pivot line. From point found on Pivot line (Step 2), align with 210 ft on L scale, extend to E scale, and read expansion at 5.3 in.

Example 2. The dotted lines (Steps 3 and 4) show an example for stainless steel (18-8) pipe where the total temperature span is 600° F. and the length of line is 95 ft. END

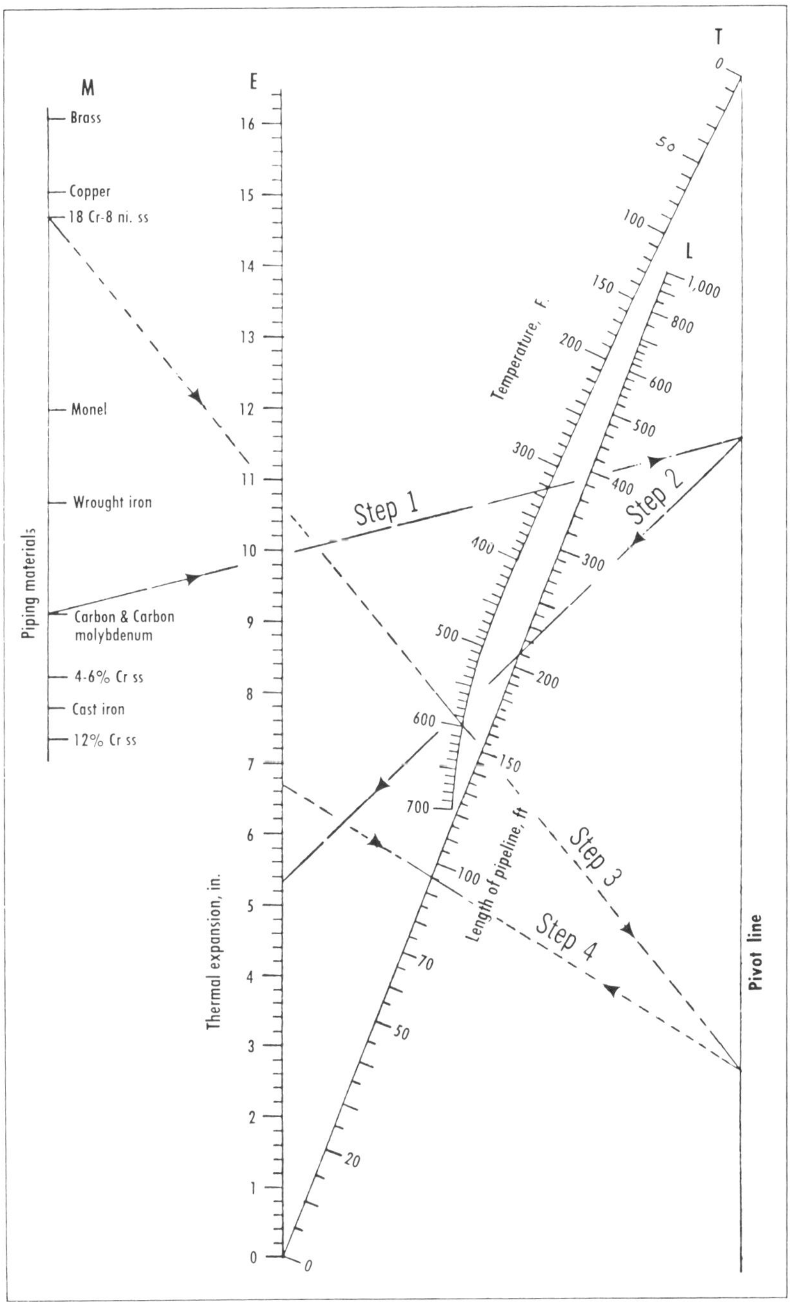

Nomogram determines thermal expansion, length of pipe loops

Bill Sisson
Superintendent of Operations
Nipak Inc.
Pryor, Okla.

RECOGNIZING that pipe expands when heated and contracts when cooled, the first requirement in the design of a piping system is to determine the amount of thermal expansion in each piping link, and that each link has adequate flexibility.

If investigation shows that insufficient natural flexibility exists, then additional flexibility must be provided. One of the commonly used methods is the pipe loop. For conditions not requiring exact analysis, it is permissible to use expansion loops designed by means of this nomogram based on the equation:

$$L = 7.76\,(De)^{1/2}$$

Where L = length of pipe loop, ft.; D = outside diameter of pipe, in., and e = expansion, in. (with fiber stress of 10,000 psi for Schedule 40 pipe).

In this formula, the bends are designed with a radius of 6 times the pipe diameter, resulting in a stress intensification factor of one. Should long-radius elbows be used, the stress intensification factor will increase to about 2.5, the height (H) of the loop and width (W) must be increased considerably.

Example. A preliminary check shows that the natural flexibility of a 180-ft section of 8-in. carbon steel pipe that carries 150 psig saturated steam (366° F.) is insufficient. Determine the thermal expansion and the minimum length of pipe loop when the pipe is installed at 70° F.

Solution. Step 1, connect 70° F. on t_i scale on nomogram with 150 psig (366° F.) on t_o scale, extend to Pivot line, and mark intersection. Step 2, connect this point with 180 ft on L_1 scale, extend to E scale, and read expansion as 4.3 in. Step 3, connect 4.3 in. on E scale with 8 in. on D scale, extend to L_2 scale, and read 47 ft as the minimum required length. Assuming a 2-to-1 ratio of width to height, L = 2.5H, or H = 47/2.5 = 18.8 ft, and W = 9.4 ft. END

Pipe loop expansion, length

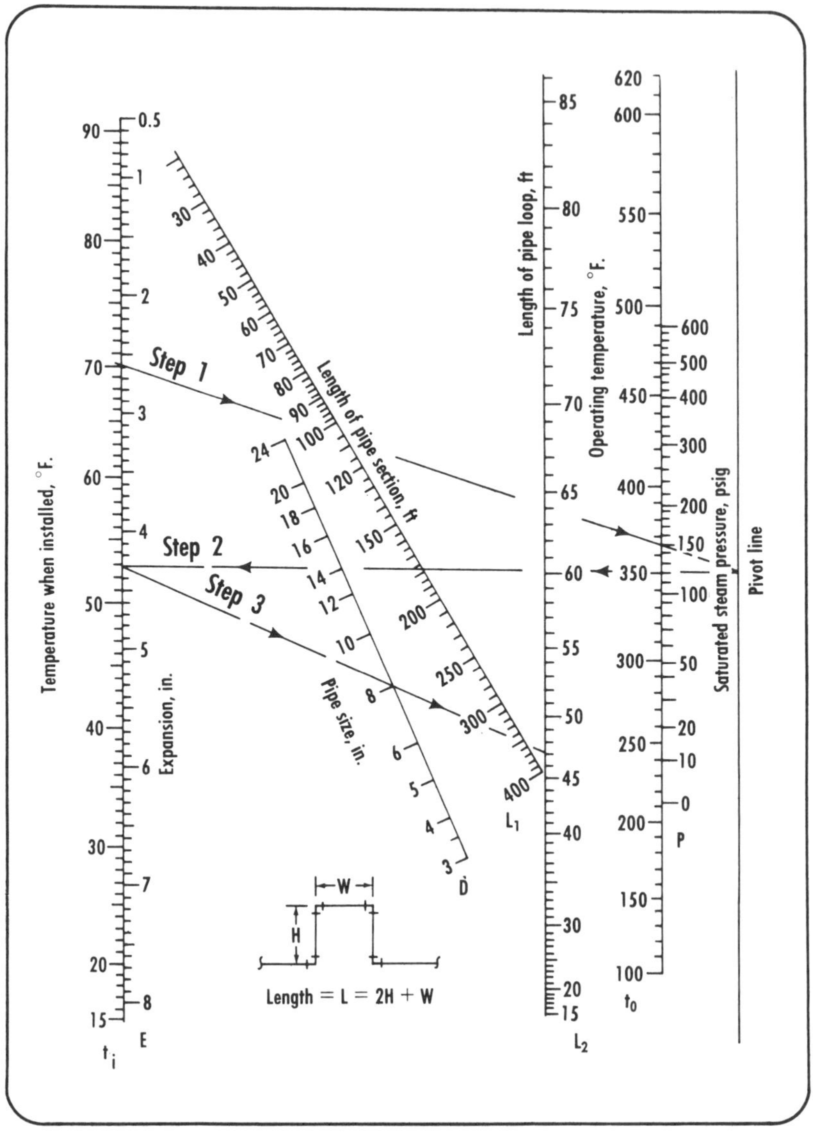

Estimate insulation thickness for various surfaces and pipe

BILL SISSON
Superintendent of Operations
Nipak, Inc.
Pryor, Okla.

A COMMON method for calculating heat losses from exposed surfaces to air is to use the combined heat-transfer coefficients for convection and radiation ($h_c + h_r$).

This nomogram, based on the combined film coefficients given in charts by Nelson[1] will estimate insulation thickness for various surfaces in still air.

Example 1. If thermal conductivity of the insulating material is 0.05 BTU/hr-ft-°F., what is the approximate thickness for the vertical walls of a furnace when fluid surface and air temperatures are 650° F., 100° F., and 50° F., respectively?

Solution. $\Delta t_o = 100-50 = 50°$ F., and $\Delta t_i = 650-100 = 550°$ F. On the nomogram, Step 1, connect 50° F. on Δt_o scale with 550° F. on t_i scale, and mark at intersection with Pivot Line No. 1. Step 2, connect this marked point with 0.05 on K scale, and mark at intersection with Pivot Line No. 2. Step 3, connect the marked point on Pivot Line No. 2 with the reference point on S scale for a vertical flat surface and read insulation thickness as 3.7 in. where line crosses X_f scale for flat surfaces.

Example 2. Estimate insulation thickness for 10-in. horizontal pipe with same conditions as Example 1.

Solution. Steps 1 and 2 are the same as in Example 1. Step 4, connect the marked point on Pivot Line No. 2 with the reference point on S scale for horizontal cylinder and mark at Pivot Line No. 1. Step 5, connect this marked point with 10-in. pipe size on D scale and read insulation thickness = 3.6 in. where line crosses X_p scale for pipe insulation.

Reference

1. Nelson, W. L., Petroleum Refinery Engineering, McGraw-Hill Book Co. Inc., New York, 1958, pp 553-554.

Insulation thickness

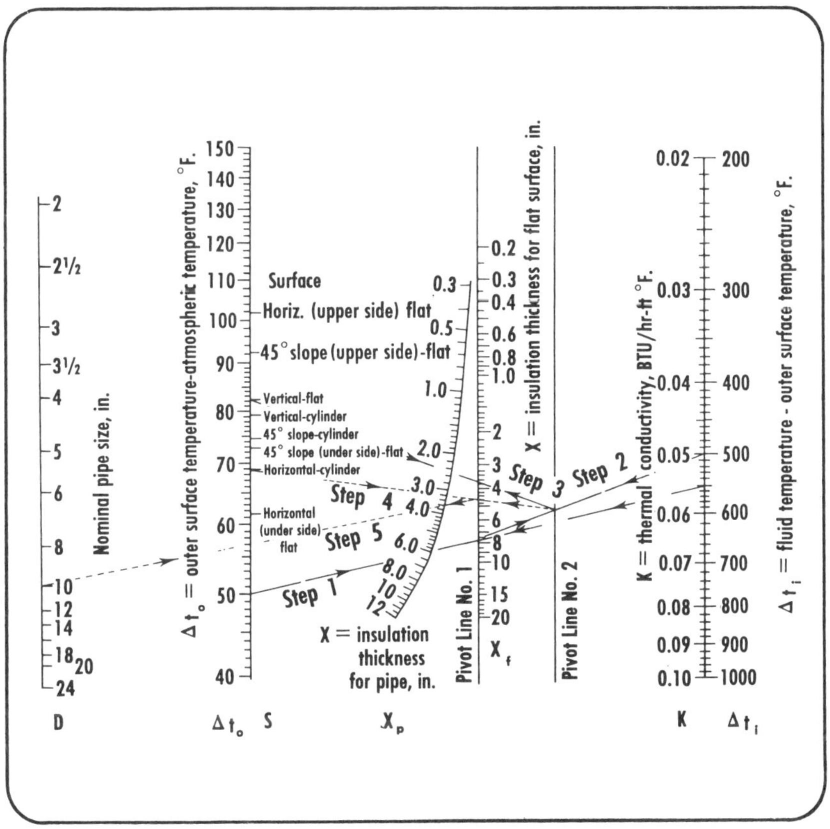

Determine heat losses through insulated pipe

Bill Sisson
Superintendent of Operations
Nipak Inc.
Pryor, Okla.

THE calculation of heat losses through insulated pipe surfaces involves the summation of the several resistances to heat flow. Heat flows through the fluid inside the pipe by convection, and by conduction through the pipe wall and insulation. The heat then passes into the surrounding air by a combination of convection and radiation. The surface resistance is usually much less than that of the insulation and need only be estimated. It is represented by the formula:

$$R_s = 1/(h_c + h_r)A$$

Where: R_s = surface resistance, BTU/hr/sq ft/°F.; h_c and h_r = convection and radiation coefficients, BTU/hr/sq ft/°F., and A = outside surface area, sq ft.

Values of the combined coefficient $h_c + h_r$ for horizontal insulated pipes can be estimated from the equation:

$$h_c + h_r = 564 / (d^{0.19})(273 - \Delta t)$$

Where: d = outside diameter of insulation, in., and Δt = insulation surface temperature minus air temperature, °F. The equation is based on that presented by Heilman (Transactions ASME, Vol. 44, p. 308, 1922).

The nomogram presented will determine both the combined coefficient and the surface resistance with a single straight line.

Example: A nominal 3-in. pipe (3.5-in. OD) is insulated with two layers of insulation, each 1½ in. thick. The outside surface temperature is estimated to be 150° F. and room temperature is 70° F. What is the combined coefficient, and what is the surface resistance?

Solution. The outside diameter of the insulation is 3.5 + 6 = 9.5 in., and Δt = 150° − 70° = 80° F. Thus connect 80° on Δt scale with 9.5 on d scale and extend to R_s scale. Read the combined coefficient as 1.91 BTU/hr/sq ft/°F. where the line crosses the $h_c + h_r$ scale, and read the surface resistance as 0.21 BTU/hr/sq ft/°F. where the line intersects the R_s scale. END

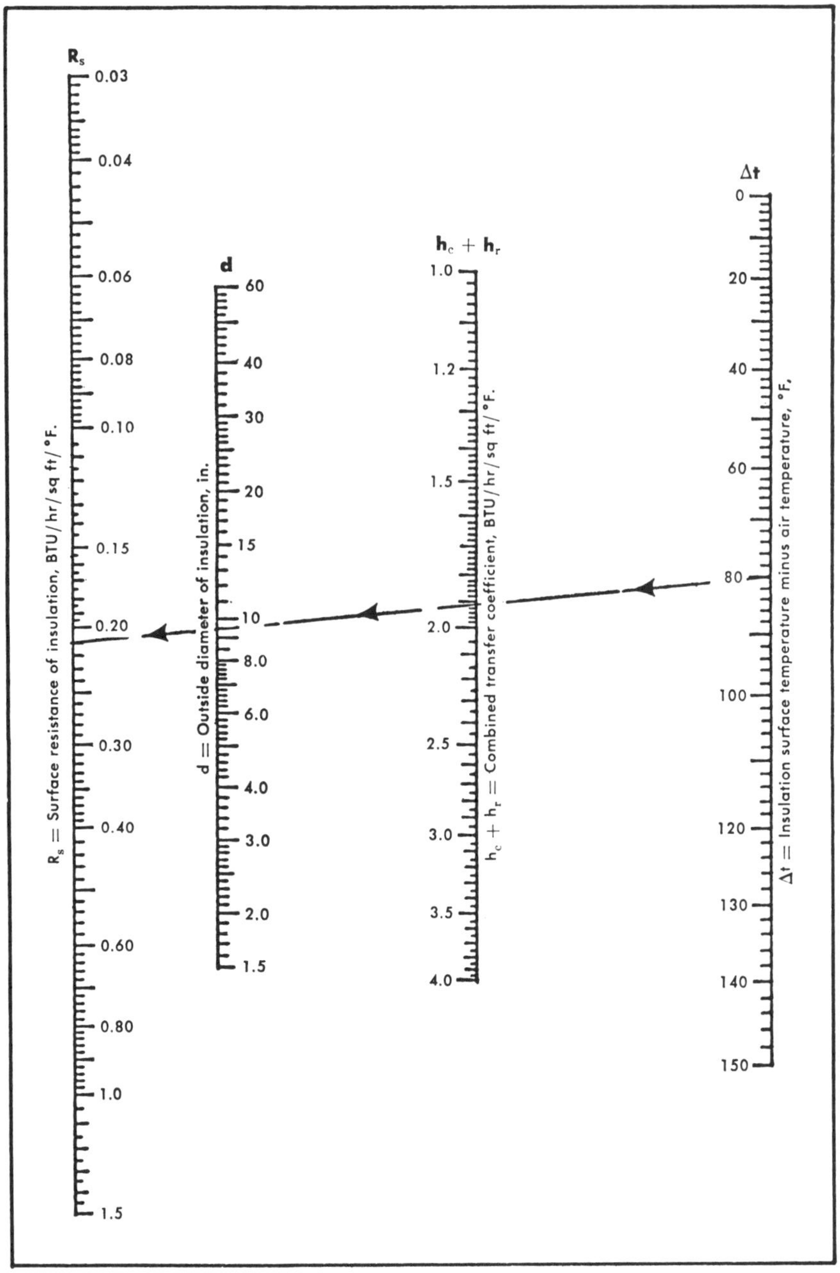

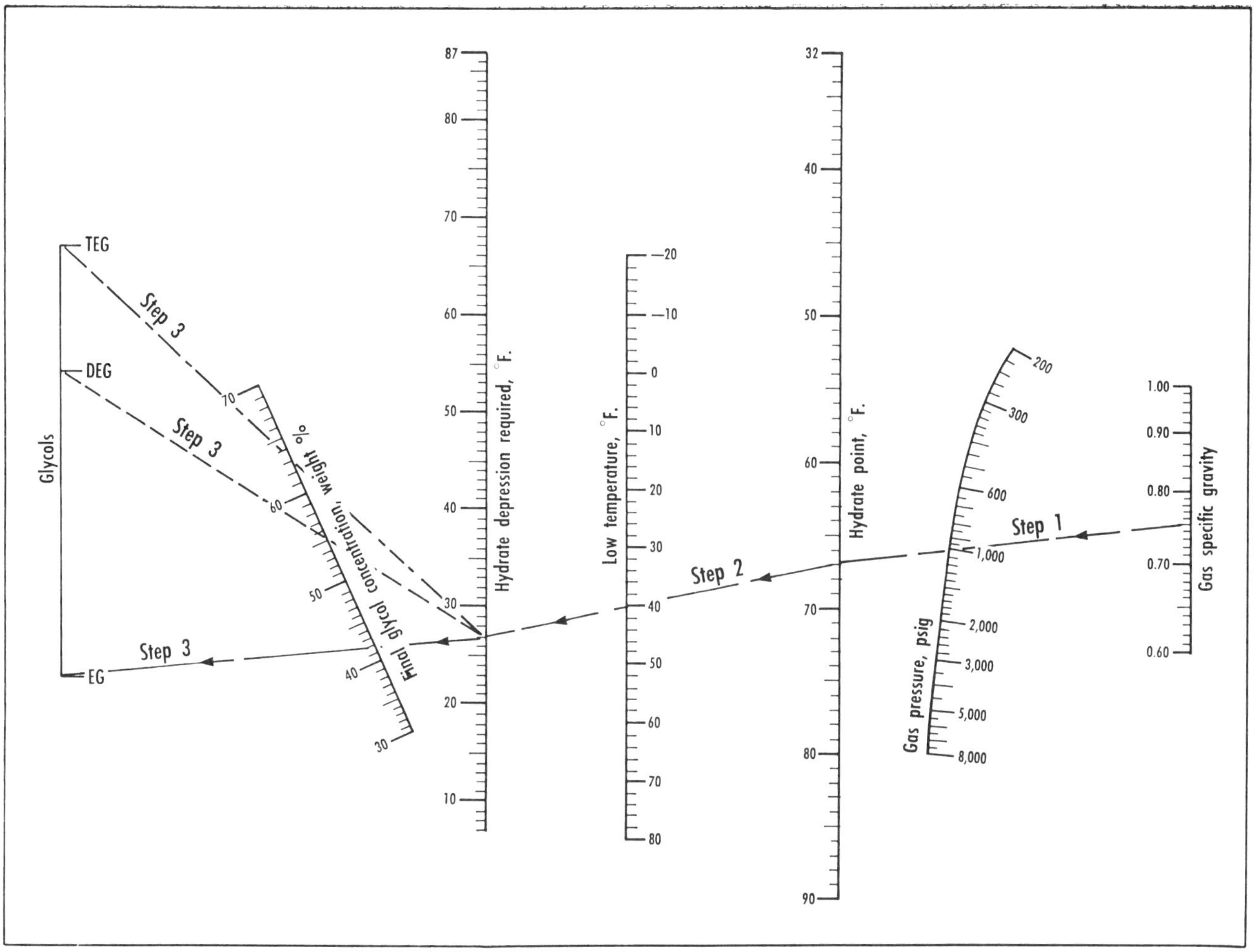

Nomogram tells when hydrates form in gas lines

Bill Sisson
Superintendent of Operations
Nipak, Inc.
Pryor, Okla.

FORMATION of gas hydrates in natural-gas lines is a common problem. Such hydrates are composed principally of hydrocarbons and water. Their formation results in decreased flow, frozen control valves, plugged orifices, and other difficulties.

A common method of preventing hydrate formation is to introduce a glycol inhibitor into the system. The choice of which glycol to use depends largely upon the hydrocarbon-stream composition.

Regardless of the glycol inhibitor used, the nomogram shown here determines hydrate point, hydrate depression required, and glycol-withdrawn concentrations when pressure and temperatures are known. It is based on the equation:

$$d = 2{,}335W/(100M - MW)$$

Where:

d = °F. by which freezing point of gas hydrate is lowered
M = molecular weight of glycol
W = weight percent of final concentration of glycol

The information for construction of conditions for hydrate-formation-of-natural-gas portion of nomogram was obtained from "Gas Conditioning Fact Book," Dow Chemical Co., Midland, Mich.

Example. For the following conditions, specific gravity of gas 0.75, gas pressure 1,000 psig, and ground temperature 40°F., determine hydrate point, hydrate depression required, and weight percent of final glycol concentration for ethylene glycol (EG), diethylene glycol (DEG), and triethylene glycol (TEG).

Solution. Connect 0.75 on specific-gravity scale with 1,000 psig on gas-pressure scale, extend to hydrate-point scale, and read 67°F. (Step 1). From this point, line with 40°F. on low-temperature scale, extend to hydrate-depression-required scale, and read 27°F. (Step 2). Now, connect from 27°F. with EG, DEG, and TEG points on glycol scale, and read final concentration as 41.8 wt % for EG, 55.1 % for DEG, and 63.5 % for TEG where appropriate line crosses final-concentration scale (Step 3). END

Effect of air velocity on heat transfer from bare pipe

Bill Sisson
Superintendent of Operations
Nipak Inc.
Pryor, Okla.

AS MANY and conflicting claims of efficiency have been made for various commercial heat insulations, it is obvious that the efficiencies and savings cannot be compared unless the same heat losses from the bare surfaces are used as a basis.

Heat losses from uninsulated pipe surfaces may be of considerable magnitude if the temperature of the surrounding air differs appreciably from that of the surface. Also, the rate of heat loss is greatly increased if the surrounding air is being moved by external force (i.e. wind). The effect of air velocity on the rate of heat transfer from bare steel pipe can be found, approximately, with this nomogram. Note that construction of nomogram is based on data from Piping Handbook, 5th ed., Sabin Crocker, McGraw-Hill Book Co., p. 6-4.

Example. Find the hourly heat loss from 50 ft of 6-in. bare pipe when the temperature difference is 300° F. and the wind velocity is 800 fpm (about 9 mph).

Solution. Step 1, on the nomogram, connect 800 fpm on V scale with 300° F. on Δt scale, and extend to pivot line. Step 2, from point found on pivot line, connect with 6 in. on D scale, extend to H scale, and read heat loss as about 3,300 BTU/hr/linear foot of pipe. Then 3,300×50=165,000 BTU/hr.

Pipe air velocity effect

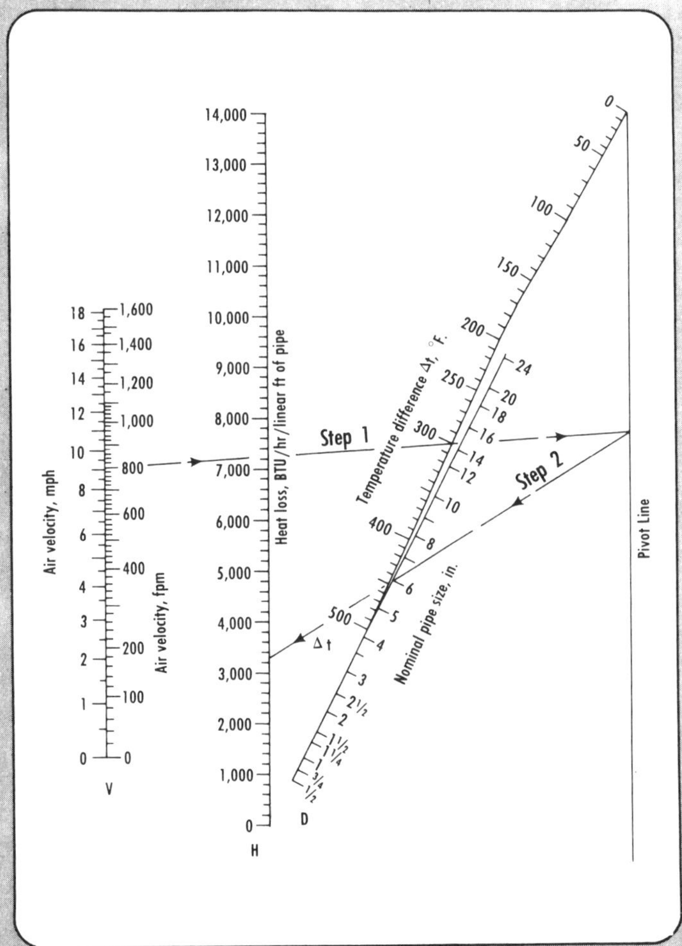

Find gasoline-line friction loss

Bill Sisson
Superintendent of Operations
Nipak Inc.
Pryor, Okla.

GASOLINE has a relatively low viscosity. The decreasing effect which viscosity has on fluid motion in the turbulent range makes it possible to solve gasoline-flow problems with reasonable accuracy by empirical formulas. A modification of the Williams-Hazen hydraulic formula has been employed extensively for this. Modified to include the specific gravity of gasoline and expressing the friction loss Δp as psi/1,000 ft of line and flow b in barrels (42 gal)/hr, the Williams-Hazen formula may be written as:

$$\Delta p = 2{,}324\, b^{1.85}\, s/c^{1.85}\, d^{4.87}$$

Where c = coefficient representing roughness of interior surface of the pipe; s = specific gravity; and d = inside diameter of pipe, in. This form of the equation contains the pertinent factors except viscosity which is assumed to be constant.

Numerical values of the Williams-Hazen coefficient c usually obtained in clean gasoline lines range from 125 to 150. It is said that average values for 4, 6, and 8-in. nominal-diameter lines are 140, 135, and 130 respectively. This is contrary to most published data which show an increase in c with pipe diameter. The nomogram based on the formula determines friction loss in gasoline lines as shown by the following example.

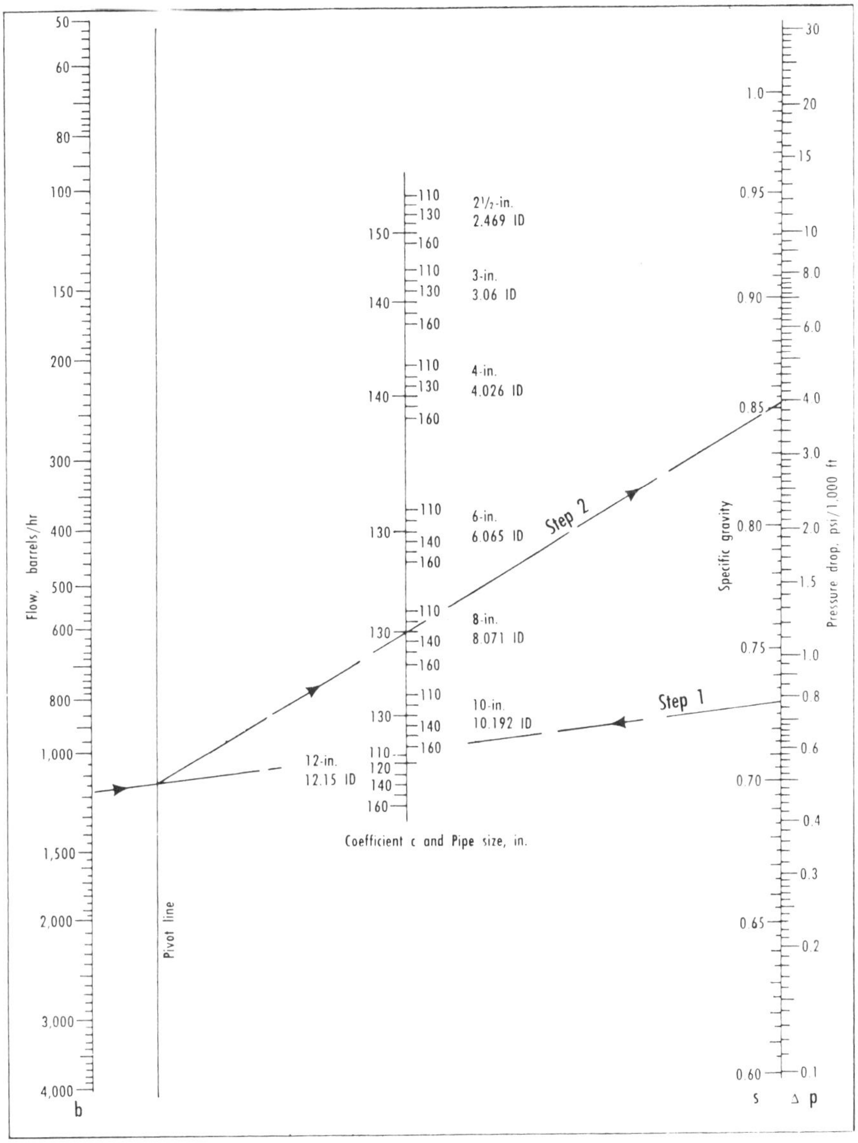

Example. If 1,175 barrels/hr of gasoline having a specific gravity of 0.73 flows through an 8-in. (8.07 ID) line and the c value is 130, what is the pressure drop/1,000 ft of line?

Solution. Connect 1,175 on b scale with 0.73 on s scale (Step 1) and mark at intersect with Pivot line. From this point, line with 8-in. (c = 130) point on Pipe-size scale (Step 2), extend to Δp scale, and read 4 psi/1,000 ft. END

Nomogram determines flow plus pressure drop for low-pressure gas

Bill Sisson
Superintendent of Operations
Nipak Inc.
Pryor, Okla.

PIPING for gas appliances should be big enough and installed properly to supply the quantity of gas necessary to meet maximum demand without undue loss of pressure between the supply point, meter, and appliance. The gas-piping size required depends upon the maximum gas consumption to be provided, length of pipe, number of fittings, allowable pressure loss from outlet of meter to appliance, and the gas specific gravity. To obtain the cu ft/hr of gas required by the appliance, divide the BTU input rate at which the appliance will be adjusted by the average BTU heating value/cu ft of gas.

The following formula is frequently used for estimating pipe sizes for low-pressure gas (pressure less than 1 psig):

$$Q=3{,}550\ (hd^{5}/sL\ [1+(3.6/d)+0.03d]^{1/2}$$

Where Q=quantity of gas, cu ft/hr; h=pressure drop, in. water; s=specific gravity of gas at flowing conditions; d=pipe ID, in.; and L=pipe length, ft.

Pipe capacities or pressure drop can be found with the nomogram presented here as shown by the following examples:

Example 1. A 2-in. pipe must carry 4,200 cfh of a gas with a specific gravity of 0.50. What is the pressure drop, if the pipe length is 75 ft?

Solution. Connect 4,200 cfh on Q scale (Step 1) with 2 in. on d scale and extend to Pivot line 2. From this point, line with 0.50 on S scale (Step 2) and extend to Pivot line 1. Connect this point with 75 ft. on L scale (Step 3), extend to H scale, and read pressure drop as 4 in. of water per 75 ft. of pipe length.

Example 2. If the pressure drop for 300 ft of 5-in. pipe is 0.80 in. of water and the specific gravity of the gas is 0.65, what is the gas flow rate?

Solution. Connect 300 ft on L scale (Step 4) with 0.80 on H scale, and mark at intersection with Pivot line 1. Connect this point with 0.65 on S scale (Step 5) and mark where line crosses Pivot line 2. Line this point with 5 in. on d scale (Step 6), extend to Q scale, and read flow rate as 9,150 cfh. END

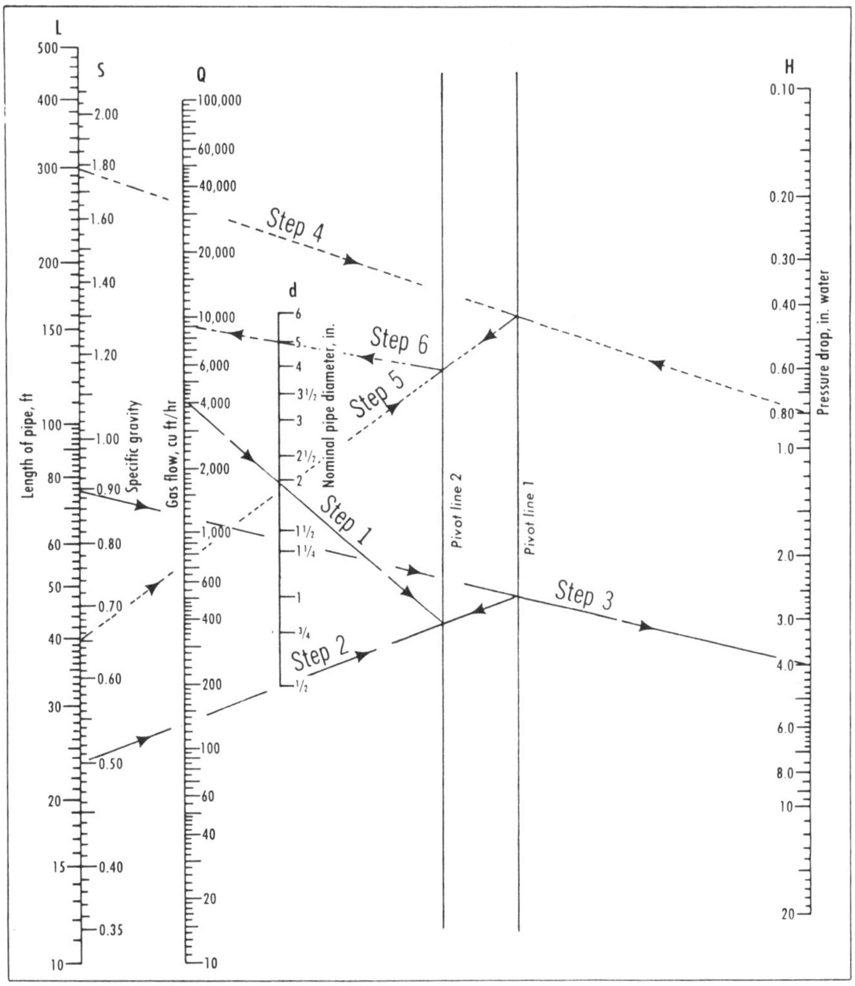

Use this nomogram...

To find friction loss for water in steel pipe

Bill Sisson
Superintendent of Operations
Nipak Inc.
Pryor, Okla.

MANY charts and nomograms have been published for finding friction loss for water in steel pipe. However, most of them are based on a specified water temperature, usually 140° F. This nomogram shows the relation of the flow of water in thousands of pounds per hour to the friction loss in feet per 100 ft of Schedule 40 pipe in sizes from ½ to 12 in., for water at temperatures from 40° to 460° F.

Example. If 40,000 lb/hour of water at 400° F. is flowing through a 3-in., Schedule 40 pipe, what is the friction loss per 100 ft of pipe?

Solution. Step 1, line 40 on Water Flow, 1,000 lb/hr, scale with 3 in. on Pipe-Size scale, extend to Pivot Line and mark. Step 2, connect marked point on Pivot Line with 400° F. on Water Temperature scale, and read friction loss as 1.775 ft/100 ft of pipe where line intersects Friction-Loss scale. END

Water friction loss

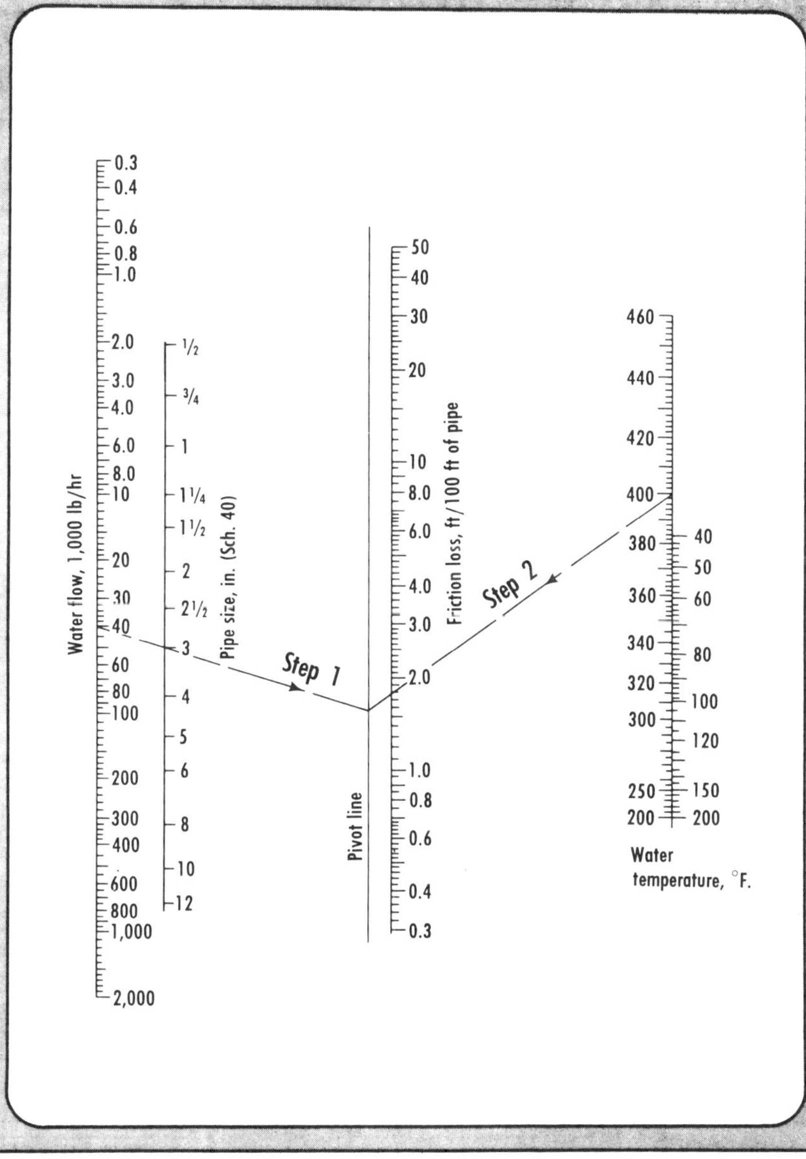

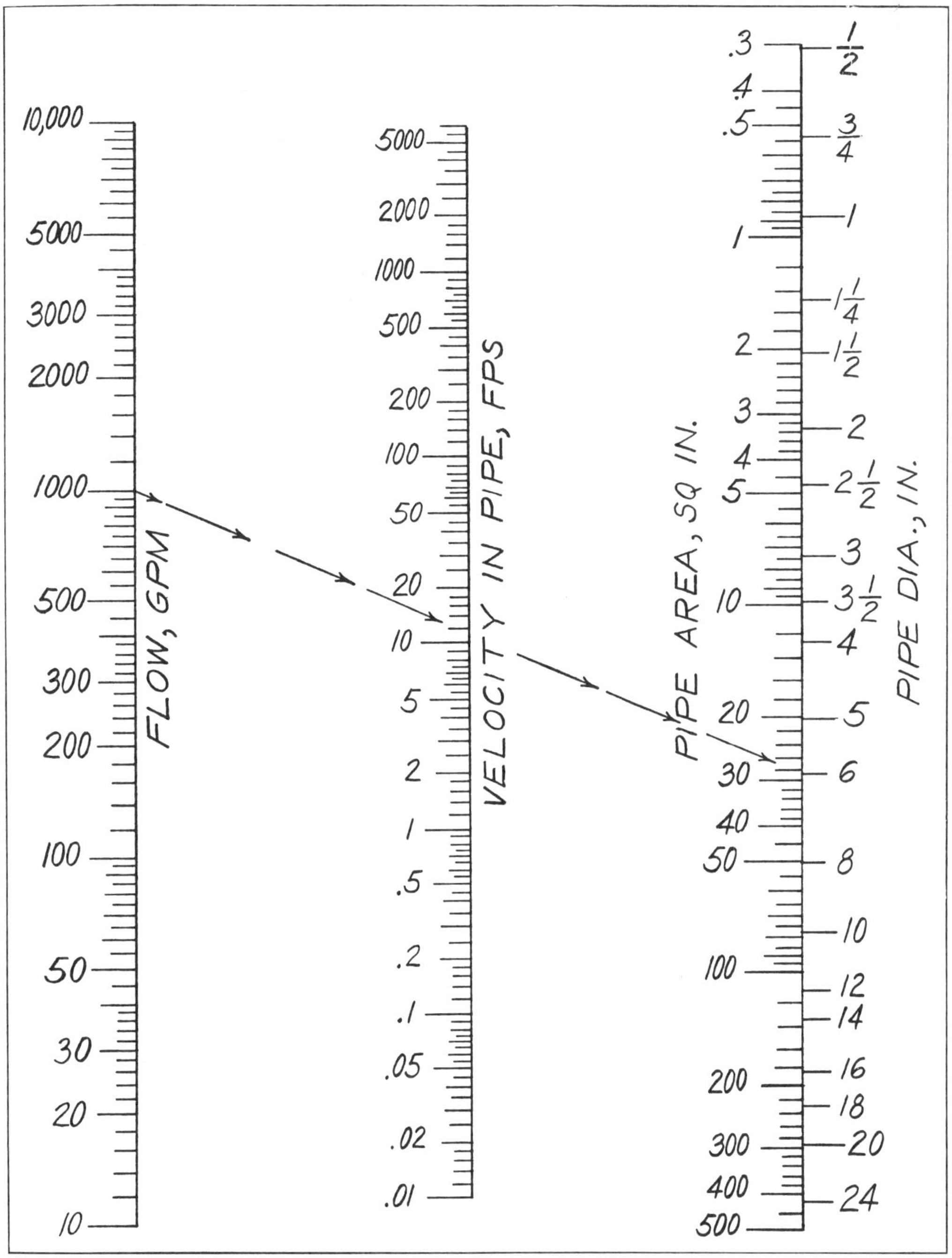

Compute line-flow velocity

HERE IS A CHART that enables you to quickly compute the velocity of any liquid in pipes of various sizes. This chart is extremely useful for a variety of velocity calculations. It is an important time-saver because velocity calculations require the computation of pipe area—a tedious job where a number of velocities are to be computed.

Example: What is the velocity in a 6-in. schedule-40 pipe conveying 1,000 g.p.m.?

Solution: Enter the chart on the left at 1,000 g.p.m. and draw a straight line through 6-in. diameter on the right. At the intersection with the central scale read the flow velocity as 11 ft. per second.

Note that the diameter scale is set up for schedule-40 pipe. This is the most common weight pipe met in practice. For other schedules, enter the chart right-hand scale, with the transverse area of the pipe instead of the pipe diameter. This is often termed the flow area and does not include the area of the metal in the pipe. You can find the flow area in any table of pipe properties.

This chart is based on $V = 0.321$ g.p.m./A, where V = flow velocity of liquid, feet per second; g.p.m. = flow rate, gallons per minute; A = pipe flow area, square inches.

Nomogram tells horizontal pipe fluid discharge

Bill Sisson
Superintendent of Operations
Nipak, Inc.
Pryor, Okla.

THE NOMOGRAM shown here provides a quick and surprisingly accurate method of determining the number of gallons per minute being discharged from an open-end horizontal pipe. Measure any convenient horizontal distance out from the open end of the pipe. This distance, in inches, will be called H. Now measure down (vertically) from the end of H to a point just touching the flowing liquid. This distance, in inches, will be called V. The following formula gives the liquid discharged:

$$\text{gpm} = 2.56\ Hd^2/V^{1/2}$$

Where:

H = distance defined above, in.
V = distance defined above, in.
d = pipe diameter, in.

See the diagram on nomogram for method of making measurements.

Example. Assume distance H measures 10 in. and distance V measures 15 in. What is the discharge from the open end of a 4-in. horizontal pipe?

Solution. Align 10 in. on H-and-V scale with 4 in. on Pipe-size scale, extend to pivot line, and mark (Step 1). Now line the point found on pivot line with 15 in. on H-and-V scale, and read 106 gpm discharge rate where line crosses Discharge-from-pipe scale (Step 2). END

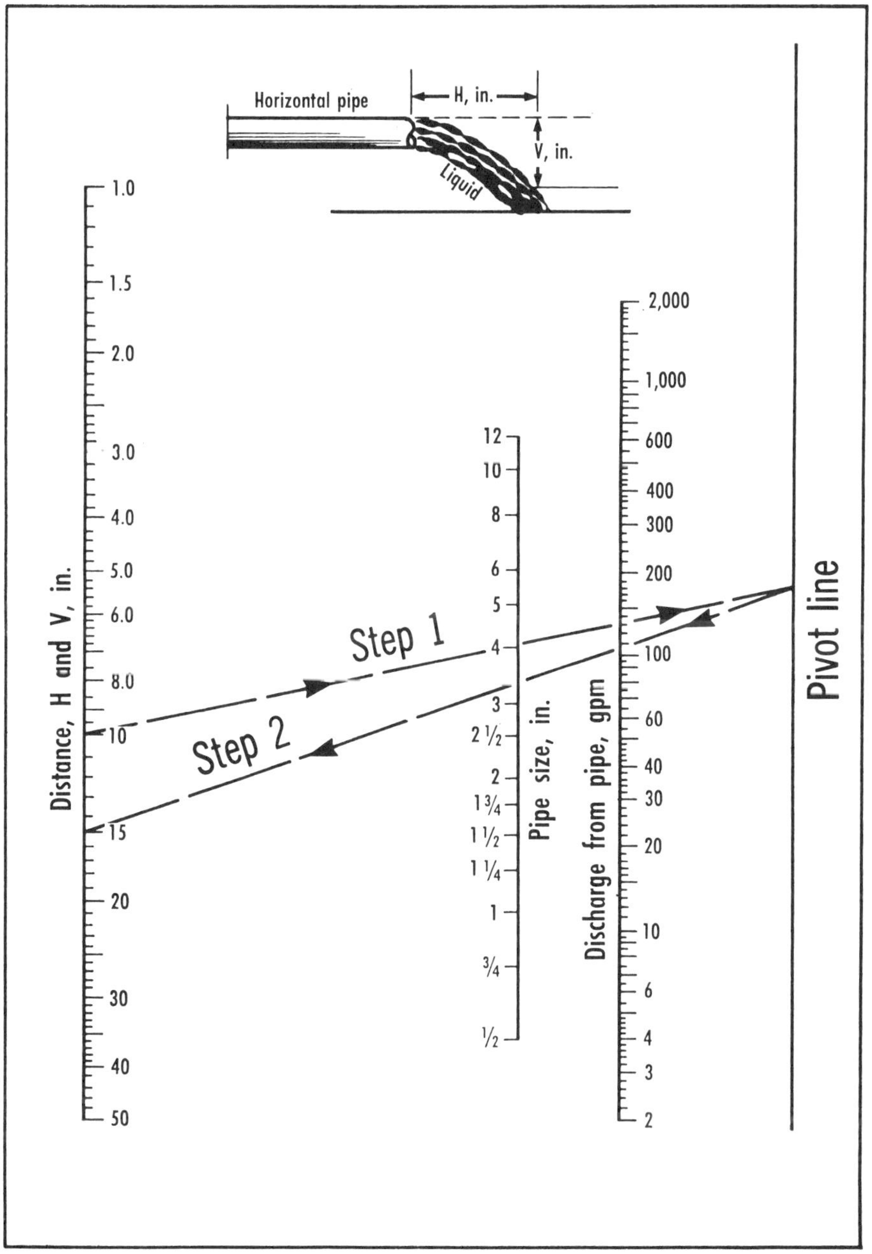

Nomogram determines compressor horsepower

Bill Sisson
Superintendent of Operations
Nipak Inc.
Pryor, Okla.

THIS nomogram gives the horsepower required to compress a given gas flow at various compression ratios or the capacity of a compressor of known brake-horsepower for various kinds of compression curves. Theoretical adiabatic horsepower required by a single-stage compressor for an ideal gas is calculated from this formula:

$$hp = \frac{144 P_1 V_1}{33,000} \left(\frac{k-1}{k}\right) \left[\left(\frac{P_2}{P_1}\right)^{(k-1)/k} - 1\right]$$

Where:

P_1 = suction pressure, psia.
P_2 = discharge pressure, psia.
k = ratio of specific heats.
V_1 = actual cfm of gas at intake conditions.

Some typical reciprocating compressor efficiencies are given in Table 1.

Example. Assume 2,000 cfm (at intake conditions) of a gas with a k value of 1.26 enters a compressor at 50 psia, compression ratio is 2.9, and compressor overall efficiency is 82%. What is the brake horsepower required?

Solution. Connect 1.26 on k scale with 2.9 on R scale (Step 1), and mark where line crosses Index line. Line marked point on Index line with 50 psia on P_1 scale (Step 2), and extend to Pivot line. From this point, connect with 2,000 cfm on V_1 scale (Step 3), and project to Index line. Align this point with 82% on E scale (Step 4), extend to H scale, and read 634 bhp required.

If you know brake horsepower, k value, suction pressure, and overall efficiency, you can find capacity by Steps 1 and 2 as above. For Step 3, connect known brake horsepower with efficiency and mark at Index line. For Step 4, connect point found by Steps 1 and 2 on Pivot line with point found by Step 3 on Index line, and read capacity where line crosses V_1 scale. END

Table 1 — Reciprocating compressor efficiencies

Compression ratio	Type of drive: Engine	Type of drive: Motor
1.1	50-60	47.50-57.00
1.2	60-70	57.00-66.50
1.3	65-80	61.75-76.00
1.5	70-85	66.50-80.75
2.0	75-88	71.25-83.60
2.5	80-89	76.00-84.55
3.0	82-90	77.90-85.50
4.0	85-95	80.75-85.50

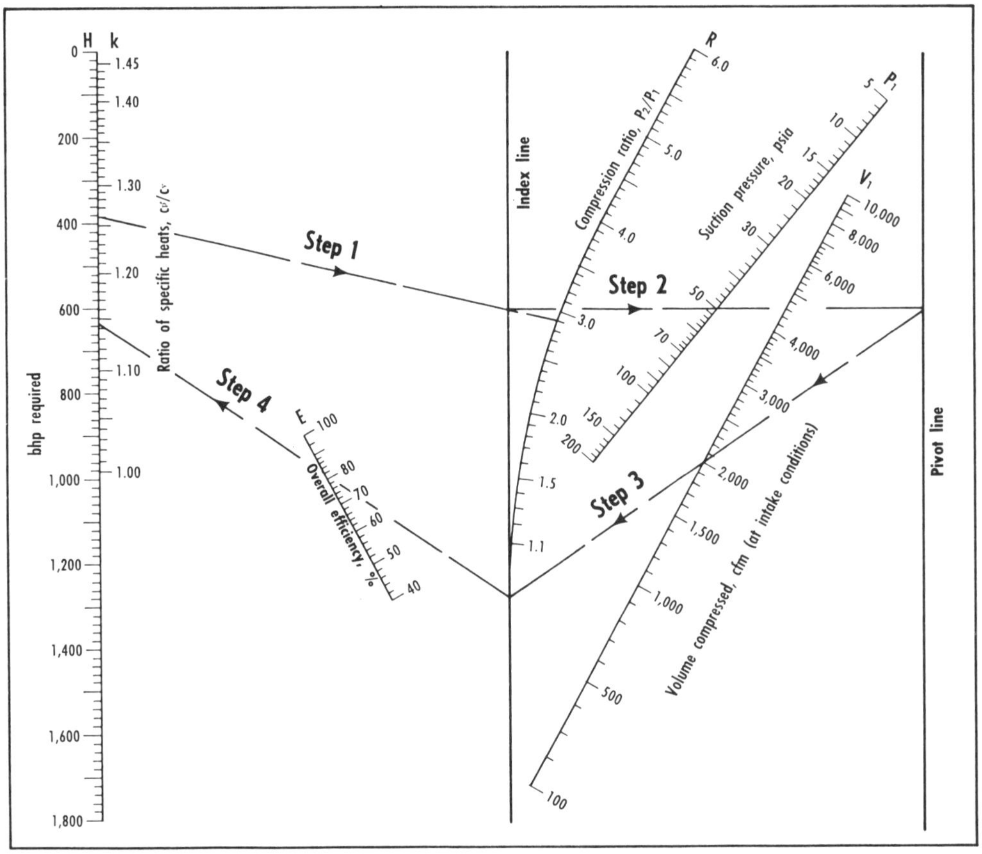

Nomogram calculates compressor head

Bill Sisson
Superintendent of Operations
Nipak Inc.
Pryor, Okla.

POLYTROPIC head is an expression used to denote the foot-pound of work required per pound of gas, and polytropic efficiency is the ratio of the polytropic compression energy transferred to the gas to actual energy transferred to the gas. Centrifugal compressor calculations are usually based on these relations. Polytropic head is found with the formula:

$$H_p=[(Z_1+Z_2)/2]\,(1{,}544/MW)T_1\,[(P_2/P_1)^{(n-1)/n}-1]/(n-1)/n$$

$$(n-1)/n=[(k-1)/k]1/\eta_p$$

Where H_p=polytropic head, ft-lb/lb; MW=molecular weight of gas, lb; T_1=inlet temperature, (° F.+460); Z =compressibility factor; k=adiabatic exponent, c_p/c_v; c_p=specific heat at constant pressure; c_v=specific heat at constant volume; (n-1)/n=polytropic ratio; P_1=suction pressure, psia; P_2=discharge pressure, psia; P_2/P_1= compression ratio, and η_p=polytropic efficiency, % (as decimal).

This nomogram will determine polytropic head and the polytropic ratio (n-1)/n as shown by the following example. Note that the term, $(Z_1+Z_2/Z$, is not included in the nomogram. However it is only necessary to multiply the head found with the nomogram by $(Z_1+Z_2)/Z$ if required.

Example. Given a gas-dry air in which k=1.395, molecular weight= 28.95; inlet pressure=14.7 psia; discharge pressure=44.7 psia; inlet temperature=100° F., and polytropic efficiency=75.6%, determine the polytropic ratio and polytropic head.

Solution. First, calculate the compression ratio, P_2/P_1=44.7/14.7=3.04. Then on the nomogram, Step 1, connect 1.395 on k scale with 75.6% on η_p scale, and read 0.378 where line crosses (n-1)/n scale. Step 2, align 0.378 with 3.04 on P_2/P_1 scale, extend to Pivot Line 1, and mark intersection. Step 3, connect marked point with 100° F. on t_1 scale, and mark where line crosses Pivot Line 2. Step 4, connect this marked point with 28.95 on MW scale, extend to H_p scale, and read polytropic head as 41,200 ft-lb/lb.

END

Compressor head

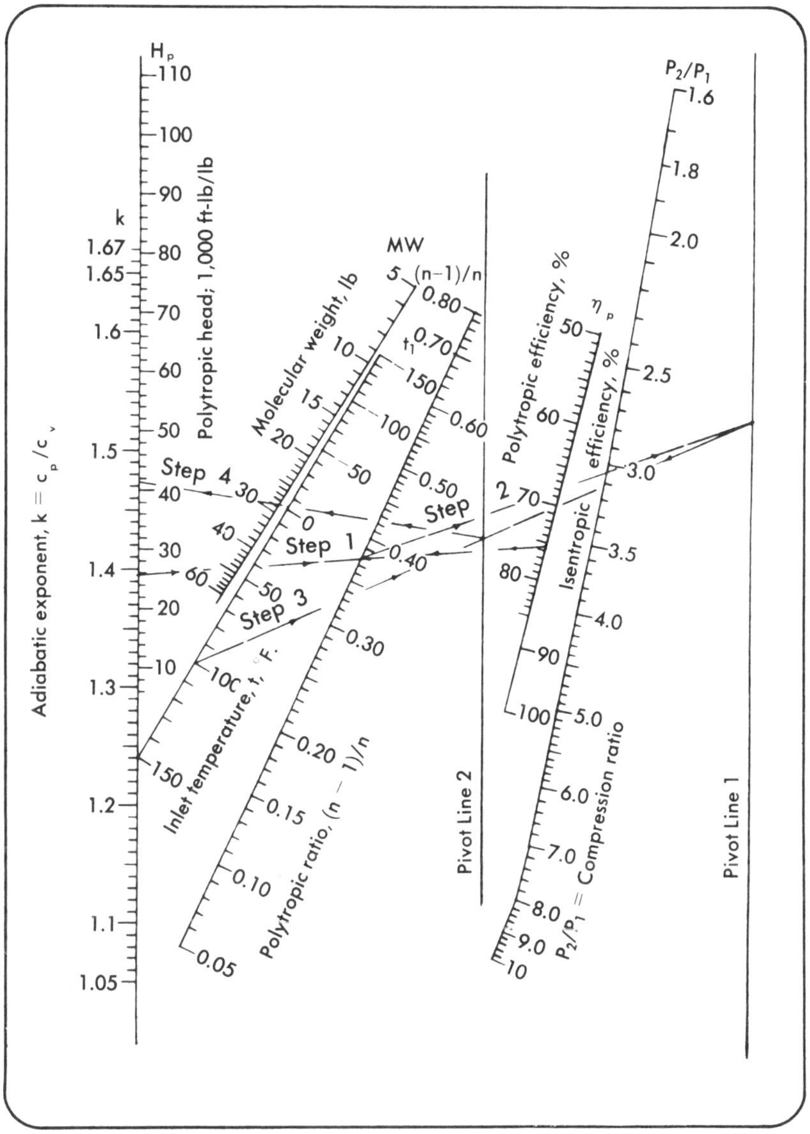

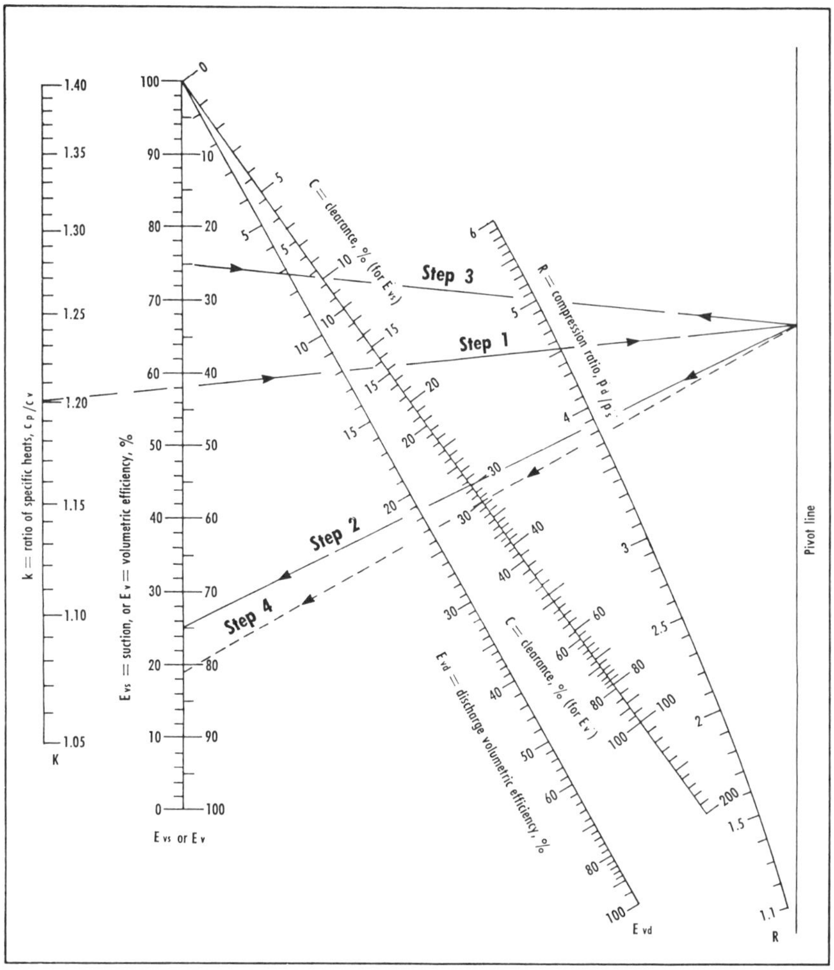

Nomogram gives volumetric efficiency of reciprocating compressors

Bill Sisson
Superintendent of Operations
Nipak Inc.
Pryor, Okla.

VOLUMETRIC efficiency (E_v) measures the volumetric efficiency of a compressor cylinder. Were there no clearance volume to reexpand and delay opening of the suction valve, the cylinder could deliver its entire piston-displacement as capacity. The effect of clearance volume on E_v depends on compression ratio and characteristics of the gas. In this case, the k value of the gas. E_v may be found from the equation:

$$\% \; E_v = 100 - R - C \, (R^{1/k} - 1)$$

Where: R = compression ratio, P_d/P_s; P_d = discharge pressure, psia; P_s = suction pressure, psia; C = clearance, %; K = ratio of specific heats, c_p/c_v; c_p = specific heat at constant pressure; and c_v = specific heat at constant volume.

To determine pipe sizes and size of a surge drum for a compressor, it is necessary to know suction and discharge volumetric efficiencies of the machine. Suction and discharge volumetric efficiencies can be found from these equations:

$$\% \; E_{vs} = [\, 1 - C \, (R^{1/k} - 1) \,] \, (100)$$

$$\% \; E_{vd} = [\, (1 + C) \, R^{-1/k} - C \,] \, (100)$$

$$\text{or } E_{vd} = \frac{E_{vs}}{R^{1/k}}$$

Volumetric efficiencies can be obtained by employing the formulas. But for a faster approach the nomograph presented here, based on the preceding formulas, has been developed.

Example. Ratio of specific heats (k) is 1.2, compression ratio (R) is 4.5, and clearance (C) is 30%. What is the volumetric efficiency and suction and discharge volumetric efficiencies?

Solution. Connect 1.2 on K scale with 4.5 on R scale, extend to Pivot Line and mark (Step 1). From marked point, line with 30% on C scale (For E_{vs}), extend to E_{vs} or E_v scale, and read suction volumetric efficiency as 25% (Step 2). From the last point found (25), move to 25 on the opposite scale, connect with marked point on Pivot Line; and read discharge volumetric efficiency as 7.15 where line crosses E_{vd} scale (Step 3). Connect marked point on Pivot Line with 30% on C scale (For E_v), project to E_v scale, and read volumetric efficiency as 19% (Step 4). END

Find compressor polytropic ratio, discharge temperature

Bill Sisson
Superintendent of Operations
Nipak Inc.
Pryor, Okla.

THE performance of centrifugal compressors can be based on either adiabatic or polytropic relations. Recently polytropic relations have generally replaced adiabatic relations for this purpose. The basic theoretical power formulas are the same as those for the adiabatic cycle except that the exponent n replaces k. Although the exponent n is seldom required, the quantity (n-1)/n is frequently needed. This can be obtained from the following equation, although it is necessary that the polytropic efficiency (η_p) be known (or approximated). The k value of any gas or gas mixture is either calculable or known:

$$(n-1)/n = [(k-1)/k]\ 1/\eta_p$$

When (n-1)/n has been determined, the following formula can be used to estimate the discharge temperatures:

$$T_2 = T_1\ (p_2/p_1)^{(n-1)/n}$$

Where: T_1 = inlet temperature, °R = °F. + 460; P_1 = inlet pressure, psia; P_2 = discharge pressure, psia; and T_2 = discharge temperature, °R.

The nomogram will quickly determine the values of (n-1)/n and T_1 as shown by the following example:

Example. If the K value of air is 1.395, inlet temperature is 110° F., compression ratio (P_2/P_1) = 3.04, and polytropic efficiency is known to be 75.6%, what is the polytropic ratio (n-1)/n and the discharge temperature?

Solution. Connect 1.395 (Step 1) on K scale with 75.6% on η_p scale and read polytropic ratio as 0.376 where line crosses (n-1)/n scale. Next (Step 2) line 0.376 on (n−1)/n scale with 3.04 on R scale, extend to Pivot Line and mark. From marked point (Step 3), line with 110° F. on T_1 scale, extend to T_2 scale, and read discharge temperature as 406° F. END

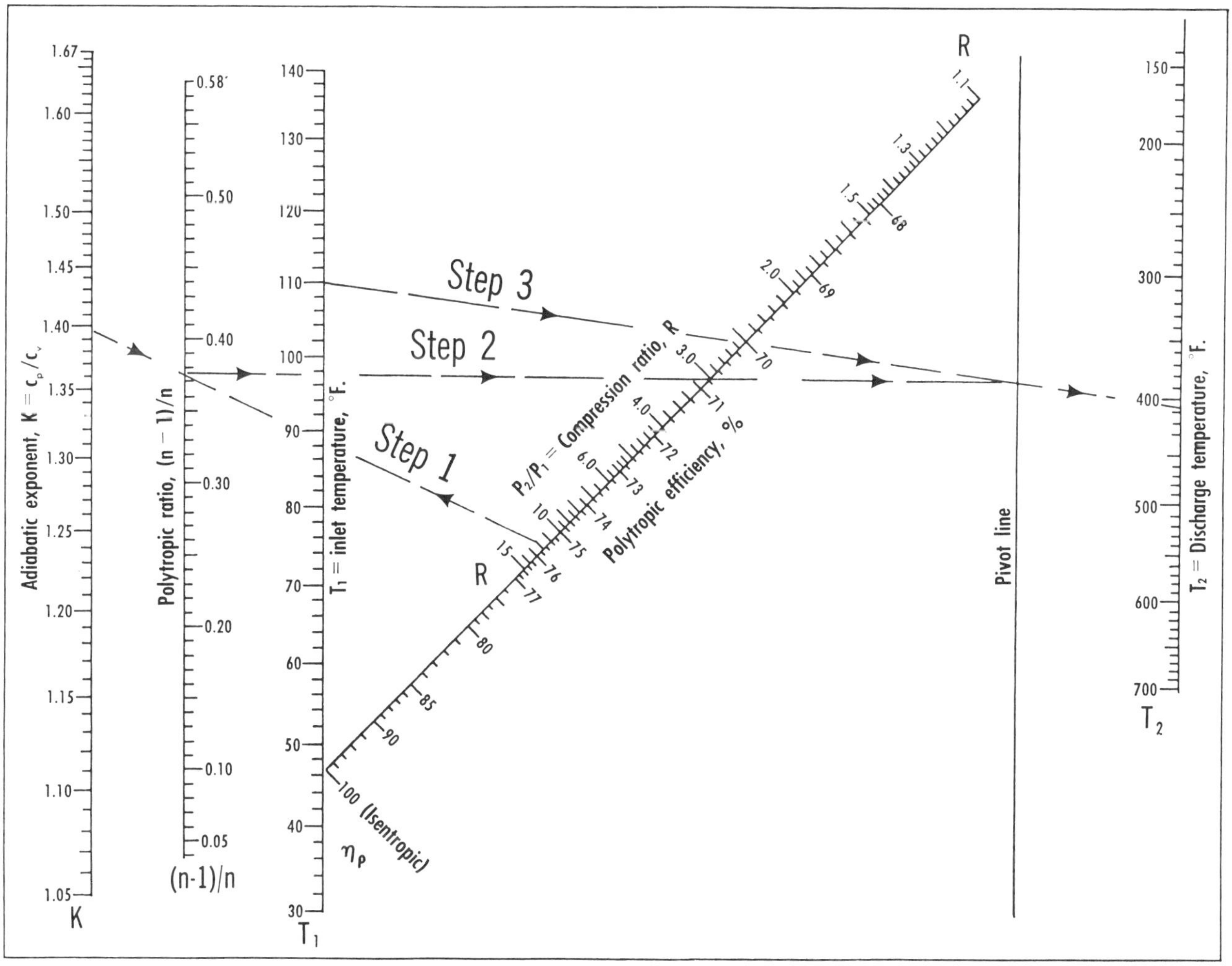

Nomogram gives gas compression temperature

Bill Sisson
Superintendent of Operations
Nipak, Inc.
Pryor, Okla.

THE NOMOGRAM presented here permits you to rapidly determine the compression temperatures of gases according to the equation,

$$T_2 = T_1 \, (P_2/P_1)^{(n-1)/n}$$

Where:

T_2 = absolute discharge temperature, (° F.+460)
T_1 = suction temperature, (° F.+460)
P_2 = discharge pressure, psia
P_1 = suction pressure, psia
n = ratio of specific heat at constant pressure to specific heat at constant volume, (c_p/c_v)
P_2/P_1 = compression ratio, R

The ratio of specific heats, or n-value, of gases can be obtained from the known, calculated, or estimated specific gravity or percentage molecular weight of the fractions constituting the mixture of gases under consideration.

Example. What is the discharge temperature of a gas, if the compression ratio is 5, suction temperature is 80° F., and the n-value of the gas is 1.3?

Solution. On Scale N align 1.3 with 5 on Scale R, extend the line to the pivot line, and mark the intersection with the pivot line. (Step 1). From the marked point on the pivot line align with 80° F. on Scale T_1 and read the discharge temperature as 323° F. where the line intersects Scale T_2. (Step 2). END

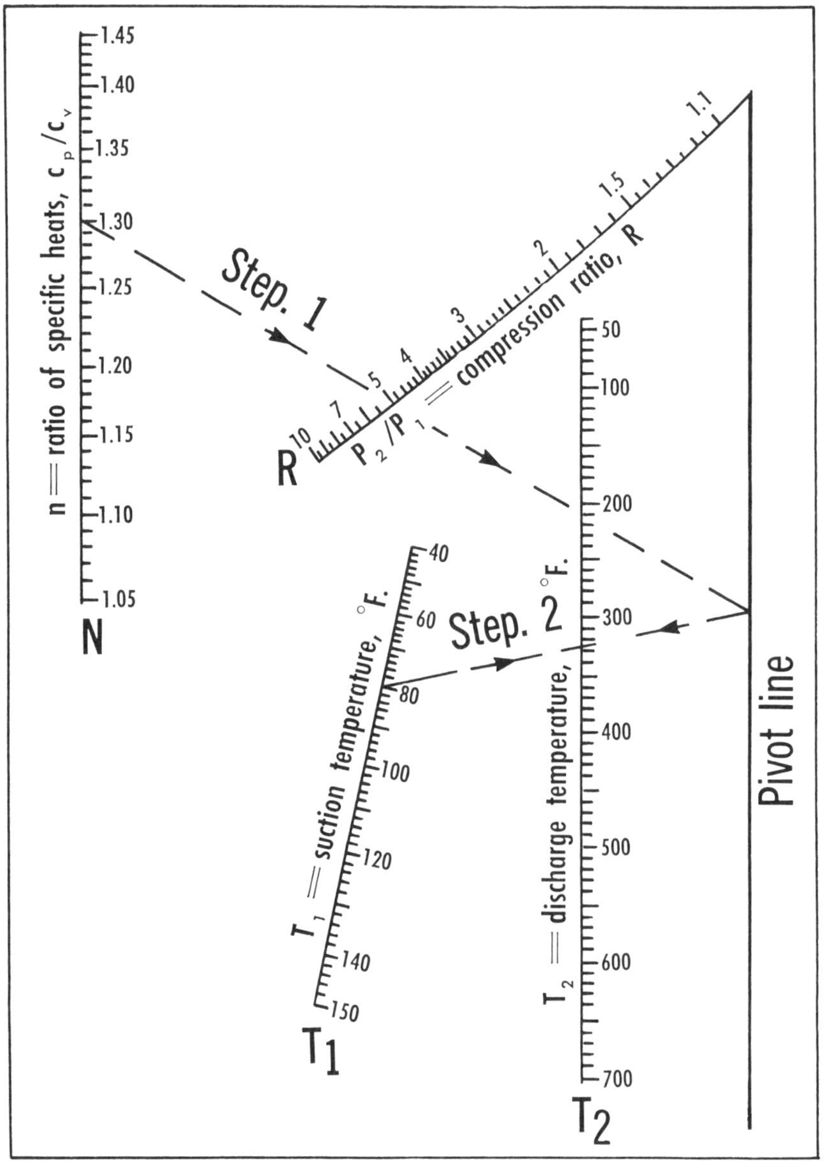

Air-compressor discharge temperature with altitude

Bill Sisson
Superintendent of Operations
Nipak Inc.
Pryor, Okla.

THE altitude at which an air compressor is operated must always be considered because barometric pressure decreases with altitude, and this results in a reduction in the density of atmospheric air.

This condition of lowered barometric pressure affects the performance of reciprocating compressors that take air from the atmosphere. This fact is well understood and allowed for with process compressors, but it is not always considered with portable air compressors that are moved from one site to another without thought to design conditions.

For a given discharge pressure, one effect of higher elevation and accompanying lower barometric pressure is to increase the compression ratio imposed on the machine. This increased compression ratio results in increased heating of the air which may be objectionable.

This nomogram will determine the theoretical discharge temperature of air compressors at various suction temperatures and altitudes. It is based on the formula:

$$T_2 = T_1 (P_2/P_1)^{0.285}$$

Where:

T_2 = discharge temperature, °R (°F. + 460)

T_1 = suction temperature, °R

P_1 = suction pressure, psia

P_2 = discharge pressure, psia

Example. What is the compression ratio and discharge temperature for an air compressor at sea level, if the suction temperature is 70° F. and the discharge pressure is 40 psig? What is the compression ratio and discharge temperature, if the compressor is operating at 6,000 ft altitude?

Solution. Step 1, connect 0 altitude on A scale with 40 psig on P_2 scale, and read compression ratio as 3.72 where line intersects R scale. Step 2, from 3.72 on R scale, connect with 70° F. on T_1 scale, and read discharge temperature as 314 °F. where line crosses T_2 scale.

Compressor discharge temperature

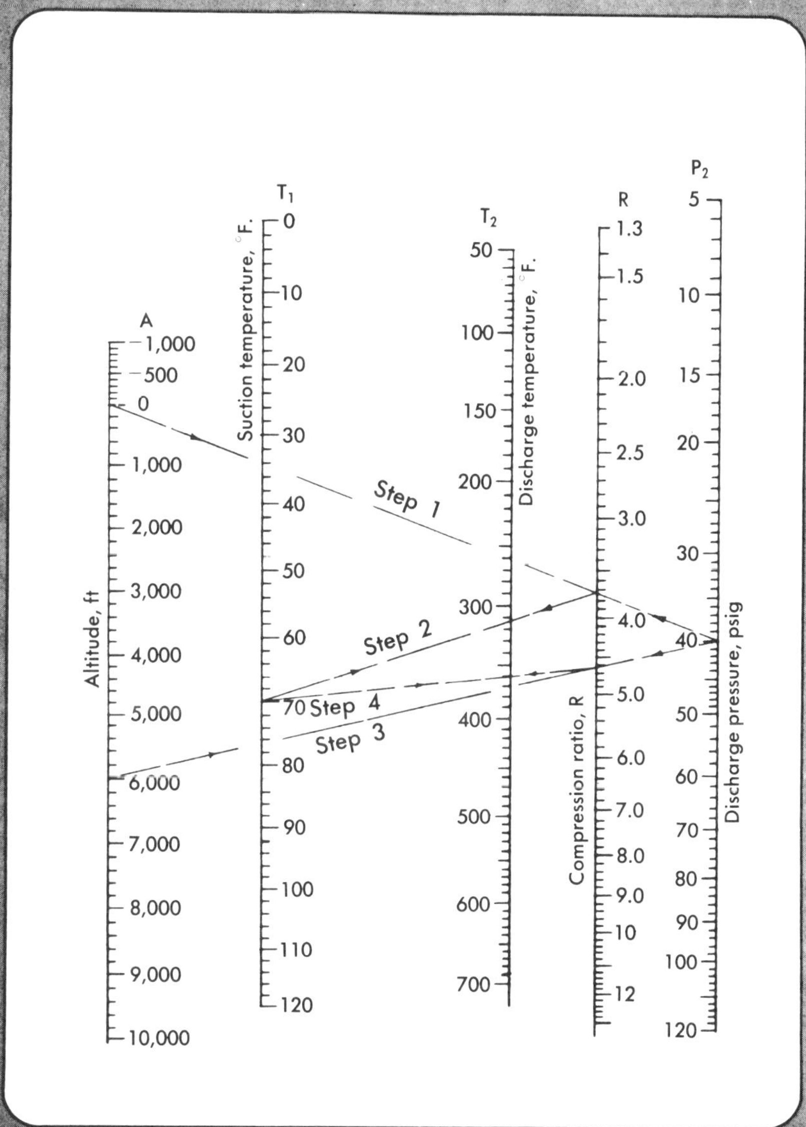

For 6,000 ft altitude, Step 3, connect 6,000 ft on A scale with 40 psig on P_2 scale, and read compression ratio as 4.64 where line crosses R scale. Step 4, from this point, connect with 70° F. on T_1 scale, and read discharge temperature as 361 °F. where line crosses T_2 scale. END

Nomogram gives venting time for natural-gas pipelines

BILL SISSON
Superintendent of Operations
Nipak, Inc.
Pryor, Okla.

WHEN it is necessary to take a natural-gas line or segment of a line out of service for repair, it is important to know the time required for venting in order to estimate total downtime.

The following Walworth Co. formula is frequently used for this purpose:

$$T_m = 0.0588\ F\ (P)^{1/3}\ (D)^2\ (G)^{1/2}\ L/d^2$$

Where:

T_m = blowdown time, min
P = initial line pressure, psig
G = specific gravity of gas
D = inside diameter of line, in.
L = length of line or section of line, miles
d = inside diameter of blowdown stack, in.
F = choke factor that has the following values: ideal nozzle = 1.0; through-port gate valve = 1.6; regular gate valve = 1.8; regular lubricated plug valve = 2.0; and venturi lubricated plug valve = 3.2

This nomogram eliminates the calculations required by the formula as shown by the following example.

Example. Determine the time required to vent a 30-in. gas-pipeline segment 1 mile long through a 10-in. blowdown stack using a regular lubricated plug valve, when the specific gravity of gas is 0.60 and initial line pressure is 1,200 psig.

Solution. The choke factor for a regular lubricated plug valve is 2.0. On the nomogram, Step 1, connect 2.0 on F scale with 0.60 on G scale, and mark at intersection with X_1. Step 2, from marked point on X_1, connect with 1,200 psig on P scale, and mark at X_2. Step 3, line point found on X_2 with 30-in. on D scale, and extend line to X_1. Step 4, line this point on X_1 with 10-in. on d scale, extend line to T_m scale, and read about 8.7 minutes. If line was, say 8 miles long, $T_m = 8.7 \times 8 = 69.6$ min.

END

Venting time

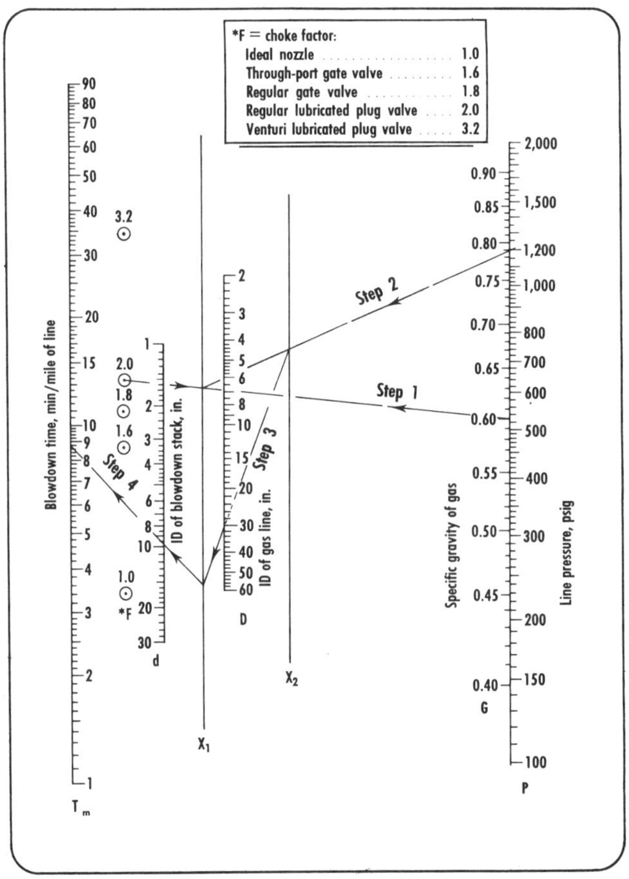

Nomogram determines pipe blind thickness

BILL SISSON
Superintendent of Operations
Nipak Inc.
Pryor, Okla.

BLINDS are used as assurance against leakage, to protect personnel engaged in maintenance and inspection, and to avoid loss of valuable pipe contents. The thickness of blinds may be determined by this nomogram based on the formula:

$$t = d\,(3P/16S)^{1/2}$$

Where d = inside diameter of gasket or, for ring-type or tongue-and-groove and similar retaining gaskets, the center line of gasket, in.; P = internal design pressure, psi; S = applicable allowable stress, psi; and t = thickness of blind, in.

Example. Determine blind thickness for a 4-in. ring joint flange at 400 psi, when allowable stress is 20,000 psi.

Solution. For a ring-joint flange of 4-in. diameter, the distance from center to center of ring periphery would be about 5⅞ in. Step 1, connect 400 psi on P scale with 20,000 psi on S scale, and mark at intersection with Pivot line. Step 2, connect point found on Pivot line with 5.875 in. on d scale, and read 0.36 in. thickness required where line crosses t scale. Use ⅜-in. thickness. END

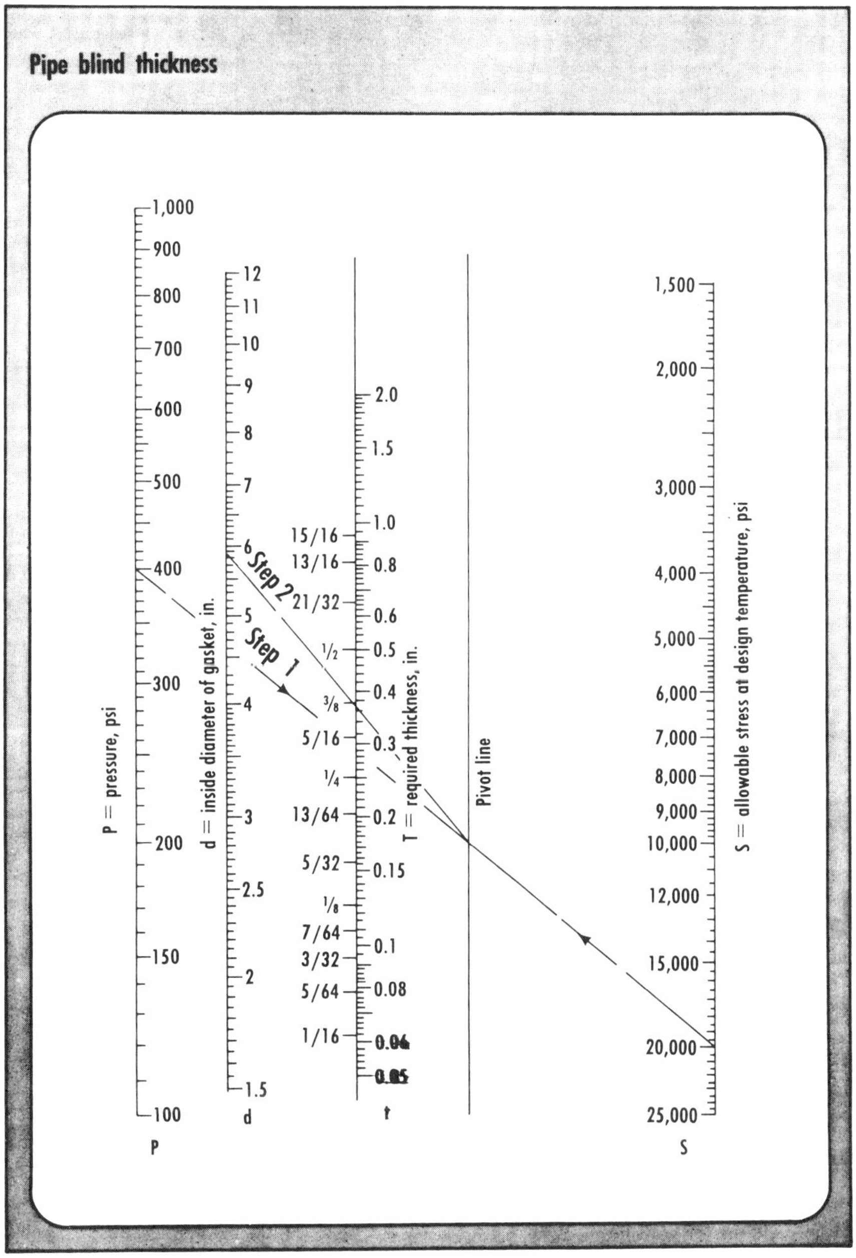

L. Templeton
Bechtel & Co.
Montreal, Canada

Use this chart to estimate pressure drop

When you change pipe size

WITH this chart you can read directly fifth-power diameter ratios. These are used in estimating a new pressure drop for a given flow rate when you change pipe size. This method of estimating pressure drop uses the Darcy equation for turbulent flow. It is rearranged to relate pressure drop with pipe diameter for a given flow rate as follows:

$$\Delta P_s \cong \Delta P_L \, (d_L/d_s)^5 \cong \Delta P_L \, D$$

Where:

ΔP_s = pressure drop in smaller pipe
ΔP_L = pressure drop in larger pipe
d_L = pipe diameter of larger pipe
d_s = pipe diameter of smaller pipe
D = fifth-power diameter ratio, $(d_L/d_s)^5$

The following examples are presented to show how to read the D factor from the chart which is based on Schedule 40 pipe sizes:

Example A. When reducing pipe size from 4-in. to 3-in., what will be the new pressure drop for a given flow rate? Enter the chart at the abscissa for 4-in. pipe and move vertically to the intersection with the 3-in. curve and read D = 4.0. Multiply this by the pressure drop for 4-in. pipe for a given flow rate to estimate the pressure drop for 3-in. pipe.

Example B. When expanding a line size from 4-in. to 6-in., what will be the pressure drop for a given flow rate? Enter the chart at the abscissa for 6-in. pipe and move vertically to the intersection with the 4-in. curve and read D = 7.5. Divide the pressure drop for 4-in. pipe by 7.5 to determine the pressure drop for 6-in. pipe. **End**

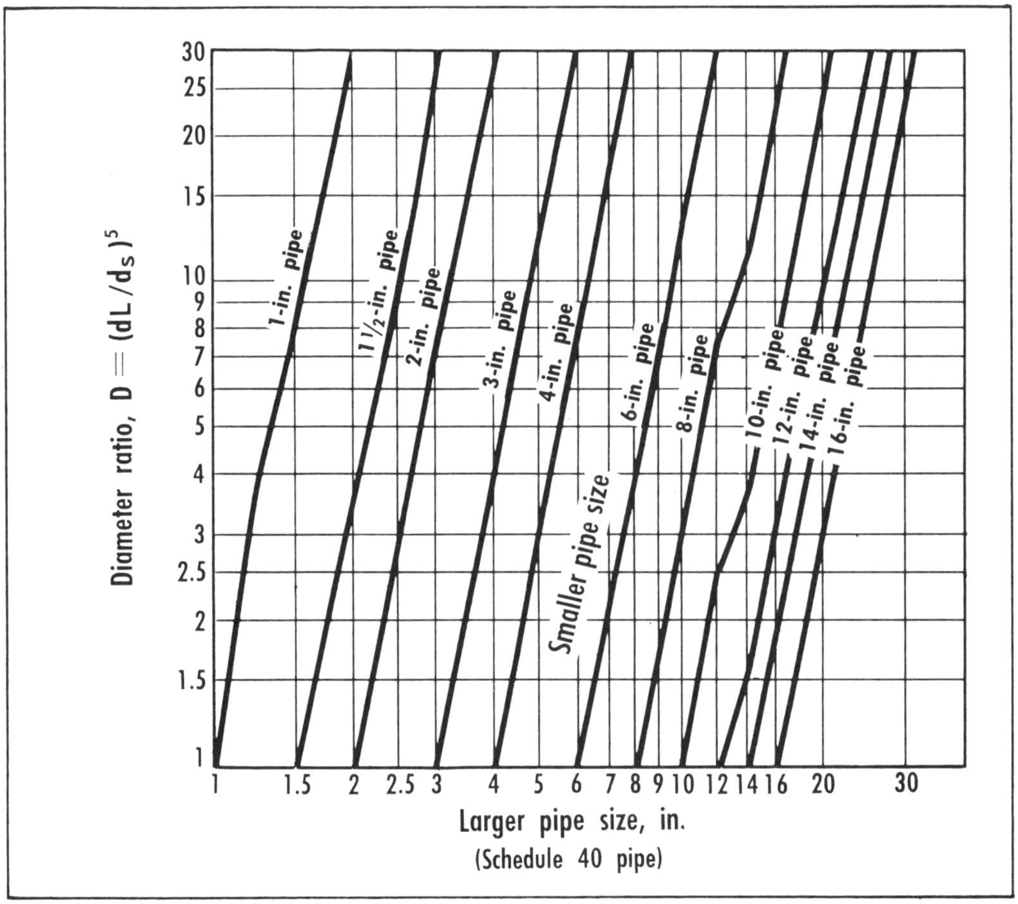

F. Caplan
Kaiser Engineers

Solving Barlow's wall-thickness formula

BARLOW'S FORMULA is frequently used to determine the wall thickness of a pipe or other cylindrical element, so as to limit the tensile stress to an acceptable value. If the tensile stress is taken to be the ultimate tensile strength of the material, this nomogram also can be used to determine the bursting pressure.

Barlow's formula is as follows:

$t = P D / 2 S$

Where:

t = wall thickness, in.
P = pressure, psi
D = outside diameter, in.
S = tensile stress, psi

Note that the D or outside diameter scale values can be divided by 10 (or 100) if, simultaneously, the P or internal pressure scale values are multiplied by 10 (or 100).

Example. What wall thickness is required for an 80-in. OD tank if the design pressure is 150 psig and the hoop stress shall be limited to 20,000 psi? Using a straightedge, align S = 20,000 with P = 150 psi and mark the intersection with the pivot line (Step 1).

Next align the mark on the pivot line with D = 80 in. and read wall thickness, t = 0.30 in. (Step 2).

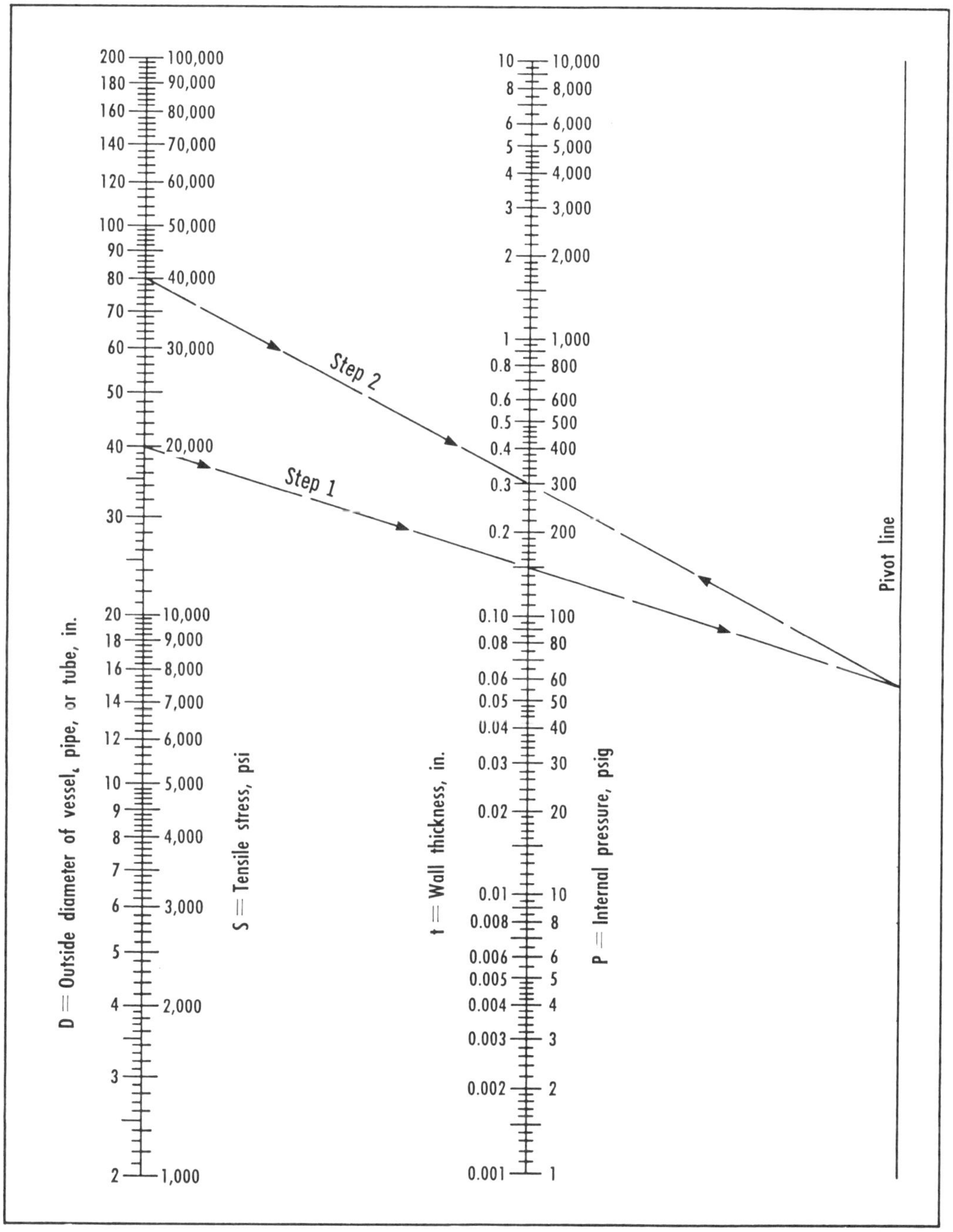

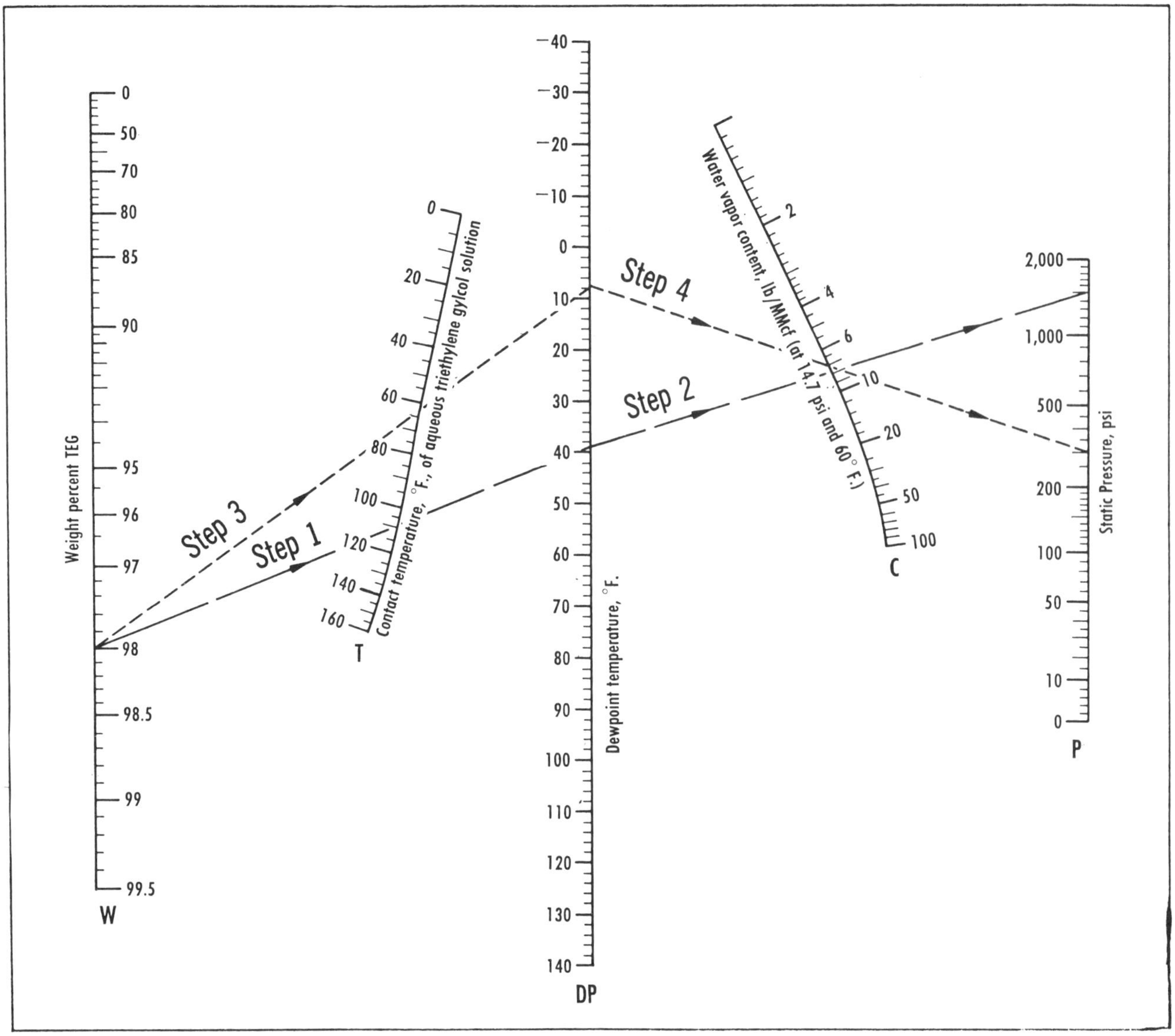

Nomogram determines dewpoint, water content for gas dehydration with TEG

Bill Sisson
Superintendent of Operations
Nipak, Inc.
Pryor, Okla.

THE GAS-STREAM dewpoint depression obtainable with a triethylene glycol dehydration plant depends on temperature at which contact with gas occurs, and the concentration of glycol in circulated solution. With pressures up to 1,500 psig, contact temperatures from 40° to 100° F. are applicable to glycol dehydration, but temperatures in excess of 100° F. require additional design features such as vacuum distillation and additional condensation facilities.

Operators of gas-dehydration plants employing triethylene glycon (TEG) as the liquid absorbent will find the nomogram shown here a valuable aid to determine dewpoint and water content of gas at various temperatures and pressures.

Example 1. If TEG is 98 wt %, contact temperature is 110° F., and static pressure is 1,500 psi, what is the dew point and water content (at 14.7 psi and 60° F.) of gas leaving an absorber?

Solution. Line 98% on W scale with 110° F. on T scale extend line to DP scale, and read dewpoint as 39° F. (Step 1). From this point on DP Scale, line with 1,500 psi on P scale, and read water-vapor content as 8.0 lb/MMcf of gas where line crosses C scale (Step 2).

Example 2. For the following conditions, TEG, 98%, contact temperature 64° F. and static pressure 300 psi, find dewpoint and water content.

Solution. Following same procedure as in Example 1, dotted lines (Step 3 and 4) show dewpoint as 8° F. and water content as 7 lb/MMcf. END

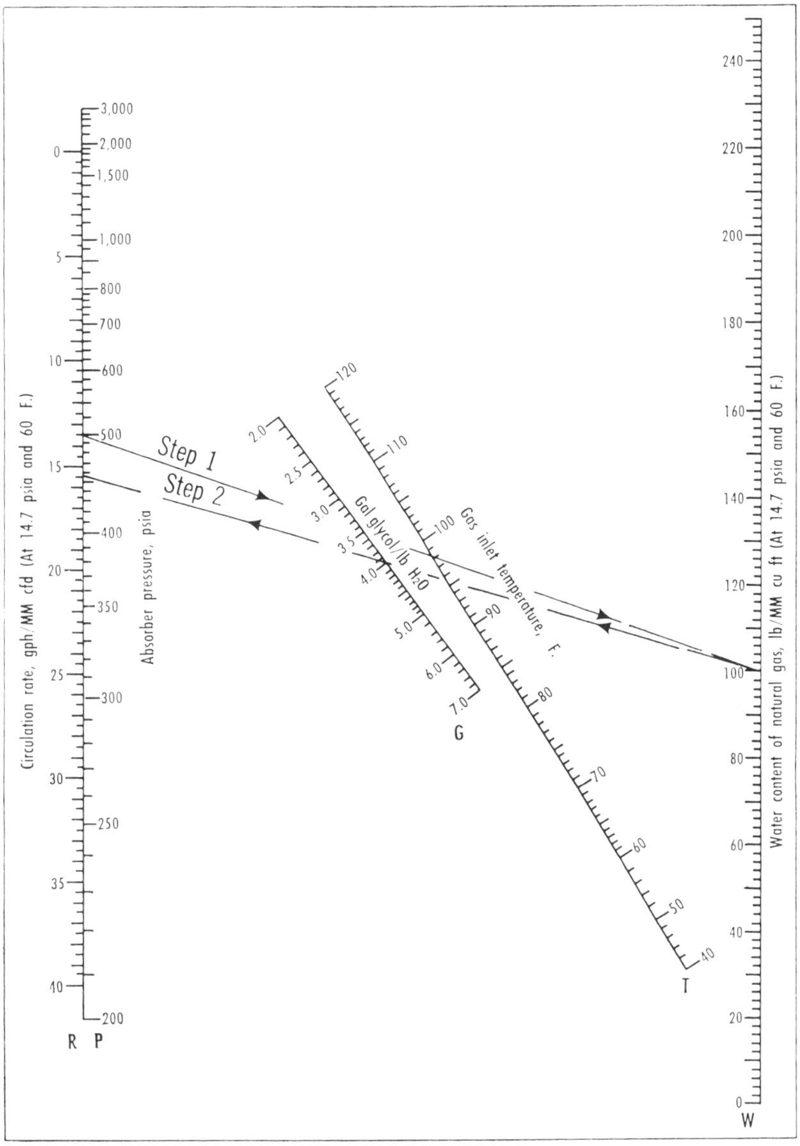

Nomogram gives glycol rate for gas dehydration

Bill Sisson
Superintendent of Operations
Nipak Inc.
Pryor, Okla.

WHEN the number of contact trays and glycol concentration are fixed, the dew-point depression of a saturated gas is a function of the glycol circulation rate. The dehydration obtained depends upon the aqueous partial pressure of the solution on the top tray and the degree of equilibrium attained.

Equilibrium can be approached in two ways: (1) increasing circulation, and (2) providing sufficient contact. Assuming infinite contact, a circulation rate of 2.5 gal of glycol per lb of water to be removed will establish equilibrium without resulting in an excessive lean-rich concentration difference. Operating experience has shown that a glycol circulation rate of 3 to 5 gal of glycol per lb of water is adequate. Circulation rates beyond these values gives diminishing returns and can seldom be justified.

It has been found that 4 trays in the contactor are adequate. However, for maximum dehydration, additional trays will be of some benefit. It is seldom possible to justify the use of more than 7 trays.

As a general rule, a circulation rate of 4 gal of glycol per lb of water is adequate with a four or five-tray absorber. Seven gal is about the maximum rate.

This nomogram provides a rapid method for determining water content of natural gas and triethylene-glycol circulation rate at various pressures and temperatures.

Example. Determine water content of natural gas at 500 psia and 98° F., and TEG circulation rate required per MMcfd (at 14.7 psia and 60° F.), based on 4 gal of glycol per lb of water.

Solution. Connect 500 psia on P scale with 98° F. on T scale (Step 1), extend to W scale, and read water content as 99.5 lb/MM cu ft. Line this point with 4 on G scale (Step 2), extend to R scale, and read 15.4 gph/MMcfd of natural gas as required circulation rate. END

Use this nomogram to find...

Density of MEA-water solutions

Bill Sisson
Superintendent of Operations
Nipak Inc.
Pryor, Okla.

MONOETHANOLAMINE (MEA) is used for treating natural gas, for purifying hydrogen, and for concentrating carbon dioxide from flue gas to dry ice.

Aqueous solutions of MEA react with carbon dioxide and hydrogen sulfide at atmospheric temperature to form compounds that, when exposed to elevated temperatures, release the acid gas and regenerate the amine for reuse.

The density of aqueous MEA solutions at various temperatures is often required in making calculations relating to MEA systems. Determination of this density is made easy with this nomogram.

Example. What is the density of 50% MEA solution at 90° F.?

Solution. Connect 90° F. on temperature scale with 50% on MEA, percent-by-weight scale, and read density as 1.0152 grams/cubic centimeter or about 8.47 lb/gal. where line intersects density scale. Note, also, that this will be the density of an 89% MEA solution at 90° F. END

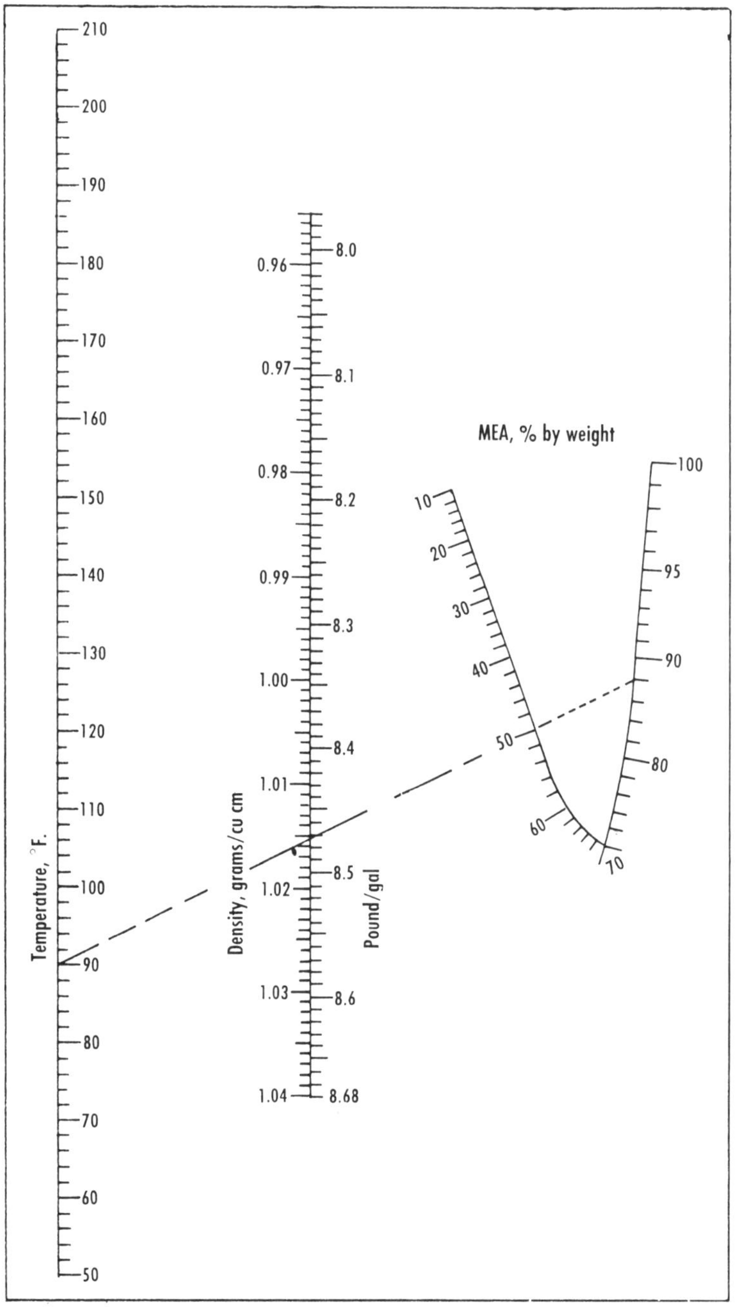

How to convert actual gas volume to standard volume

F. Caplan
Kaiser Engineers

THIS nomogram will find the volume or flow rate of a gas at standard conditions or at actual conditions. It solves the equation:

$$V_s = V_a\,(P/14.7)\,[520/(T+460)]$$

Where:

V_s = standard volume, cfm or cu ft, at 14.7 psia and 60° F.
V_a = volume, cfm or cu ft, at actual pressure, psia, and temperature, °F.
P = actual pressure, psia.
T = actual temperature, °F.

Example. What is the flow rate in standard cfm of a gas flowing at the rate of 20 cfm at 100 psig (114.7 psia) and 540° F.?

To determine this, use a straight edge and, Step (1), connect actual pressure, 114.7 psia, on the pressure scale with flow rate at actual conditions, 20 cfm, on the volume scale. Mark the intersection of the straight edge with the pivot line. Then, step (2), connect the mark on the pivot line with the actual gas temperature, 540° F. on the temperature scale. Read 81 scfm (14.7 and 60° F.) on the volume scale.

By reversing this procedure, you can determine the flow rate at actual conditions. You can also extend the flow-rate or volume scale range by simultaneously multiplying the actual and standard volumes or flow rates by such values as 0.1, 2, 10, 100, etc.

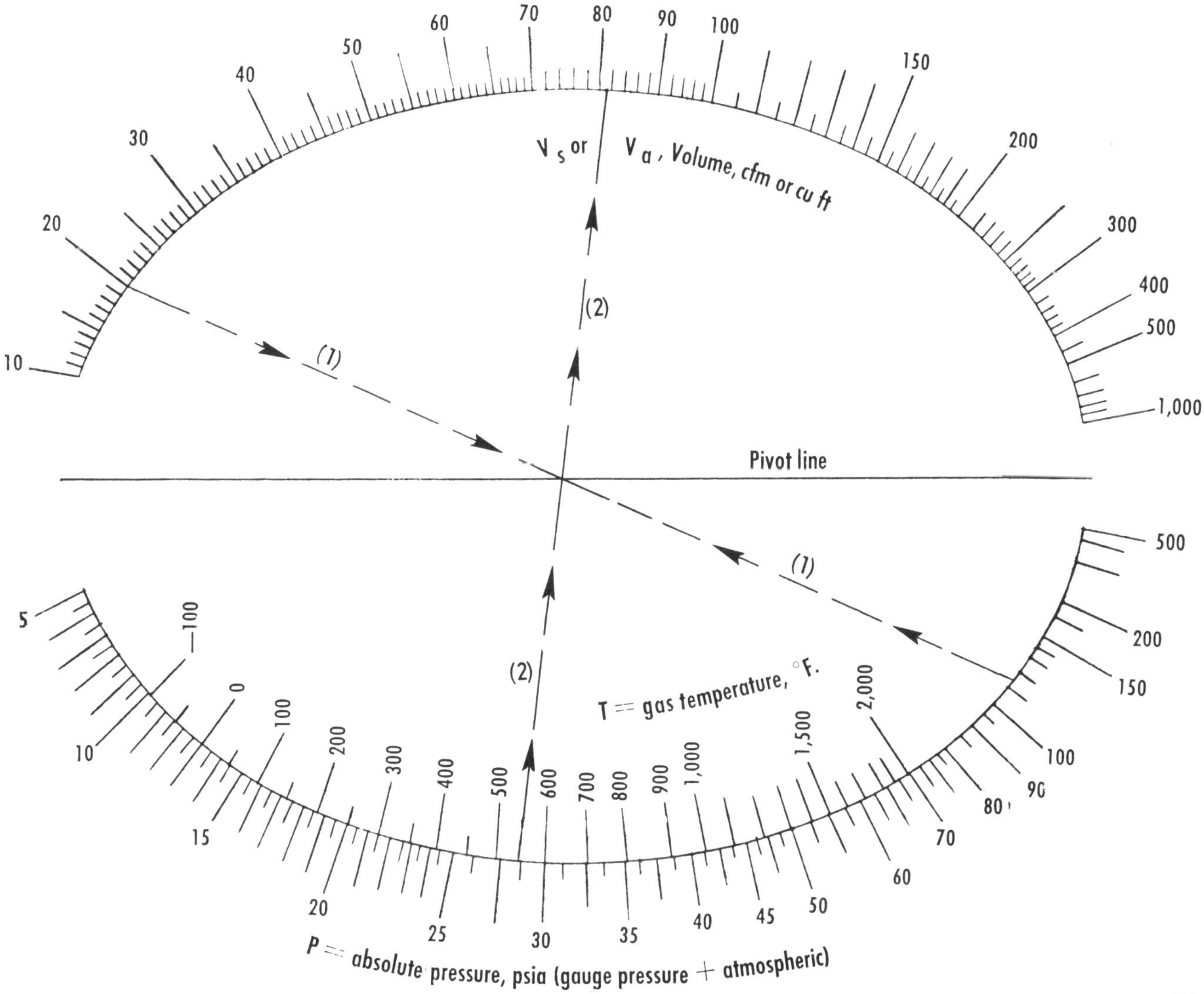

Use this nomogram to find

Partial volume of spheres

F. Caplan
Kaiser Engineers

THIS nomogram is designed to give the volume of fluid in a partially filled sphere. It is based on the equation:

$$V = 7.48\ \pi\ h^2\ [(D/) - (h/3)]$$

Where:

V = volume of liquid, gal
h = liquid height in sphere, ft
D = sphere diameter, ft
$\pi = 3.1416$

Example. How much liquid is in a sphere with a diameter of 50 ft and containing liquid to a height of 25 ft above bottom? To find the answer, align 50 ft on the sphere-diameter scale with 25 ft on the liquid-height scale and read 245,000 gal on the capacity scale.

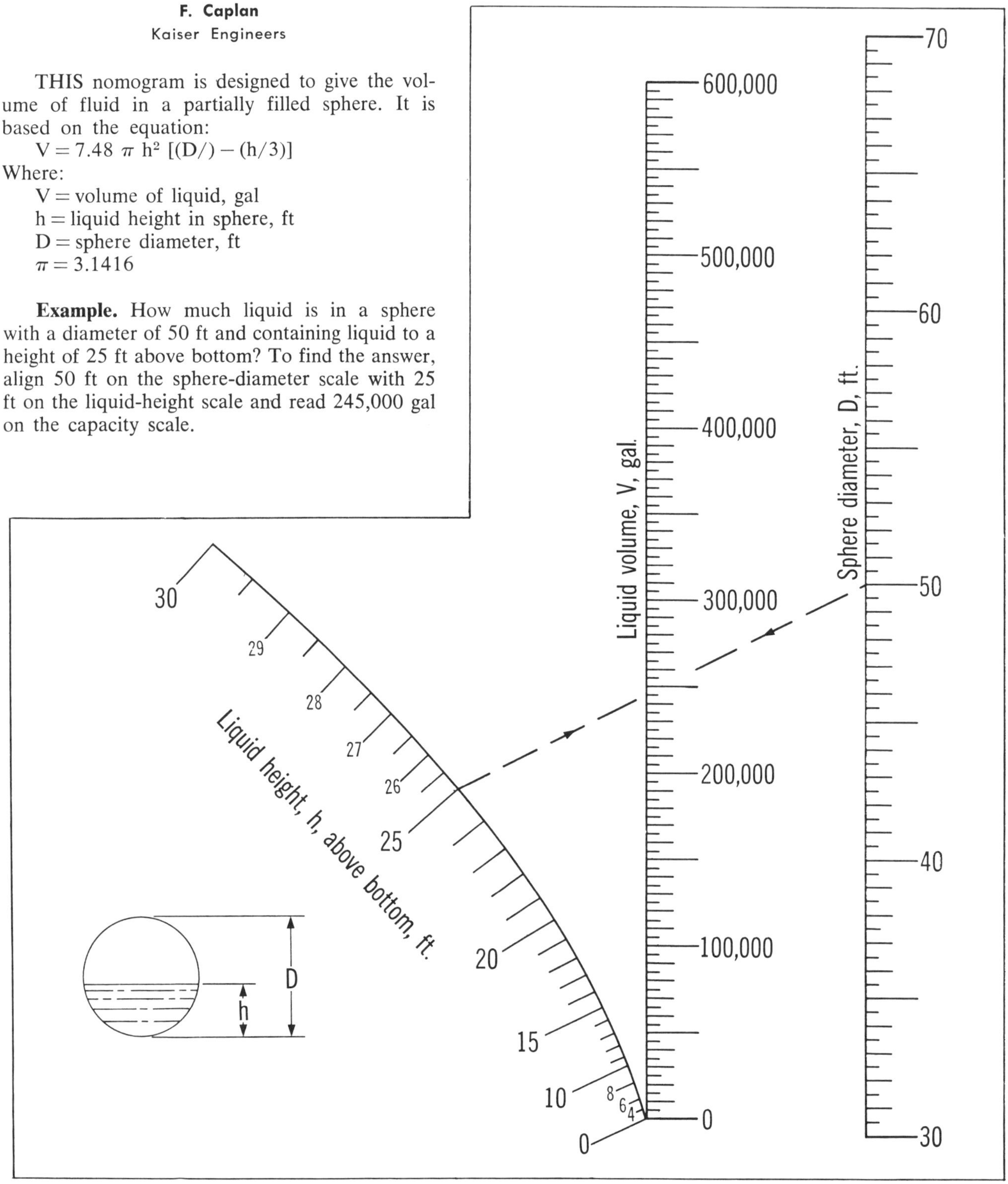

This Nomogram Speeds Combustion Calculations

by **Stephen C. Lane**

THIS chart permits you to easily evaluate a widely used equation for carbon consumption: $C = 1.5A\,(CO_2+CO)/N$, where C = carbon burned, lb. per hour; A = combustion air used, s.c.f.m.; CO_2 = per cent of carbon dioxide in flue gas; CO = per cent of carbon monoxide in flue gas; and N = per cent of nitrogen in flue gas.

Example: How much carbon will be burned when 18,000 s.c.f.m. of air are supplied, and the flue-gas analysis shows (CO_2+CO) = 15 per cent and N = 85 per cent?

Solution: Enter the chart on the left at A = 18,000 s.c.f.m. and draw a straight line through (CO_2+CO) = 15 per cent on the right. Then, from the intersection with the pivot stem, draw a straight line through N = 85 per cent on the left. Read the result on the C scale as 4,800 lb. per hour of carbon burned.

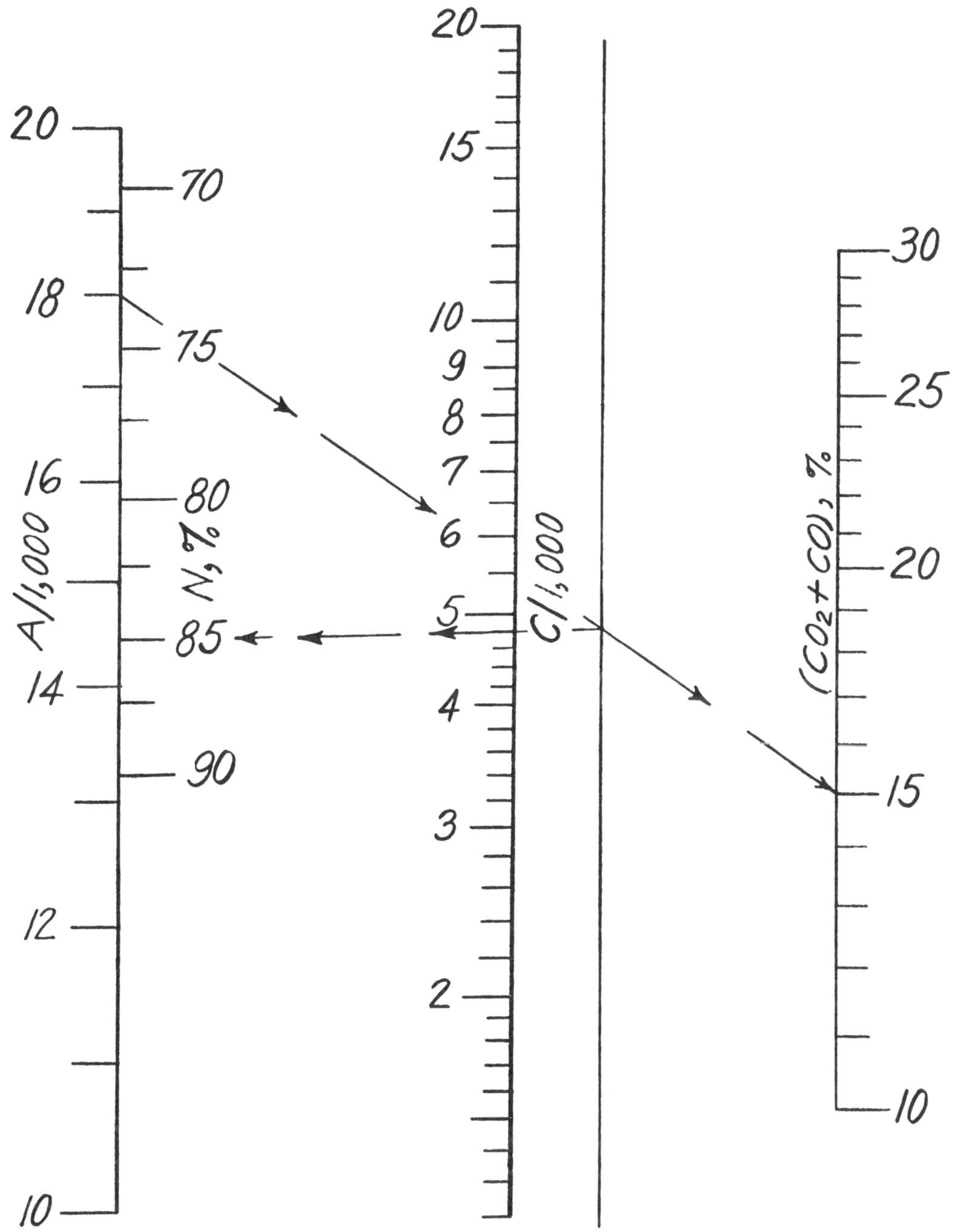

Speed Up Your Heat-Transfer Calculations!

THIS chart permits you to rapidly determine the "Logarithmic Mean Temperature Difference" in heaters and heat exchangers:

$$L.M.T.D. = (\Delta T_1 - \Delta T_2)/(\text{Log}_e\ \Delta T_1/\Delta T_2)$$

where ΔT_1 = inlet temperature difference and ΔT_2 = outlet temperature difference between the heating medium and the fluid being heated.

Example: What is the L.M.T.D. for a heater with an inlet-gas temperature of 90° F., an outlet-gas temperature of 135° F., and a water-bath temperature of 180° F.? First find ΔT_1 and ΔT_2: $\Delta T_1 = 180 - 90 = 90°$ F., and $\Delta T_2 = 180 - 135 = 45°$ F.

Solution: Enter the chart on the ΔT_1 scale at left at ΔT_1 = 90° F. With a straight edge, connect this point with the ΔT_2 scale at ΔT_2 = 45° F. On the center scale, read the L.M.T.D. as 65° F.

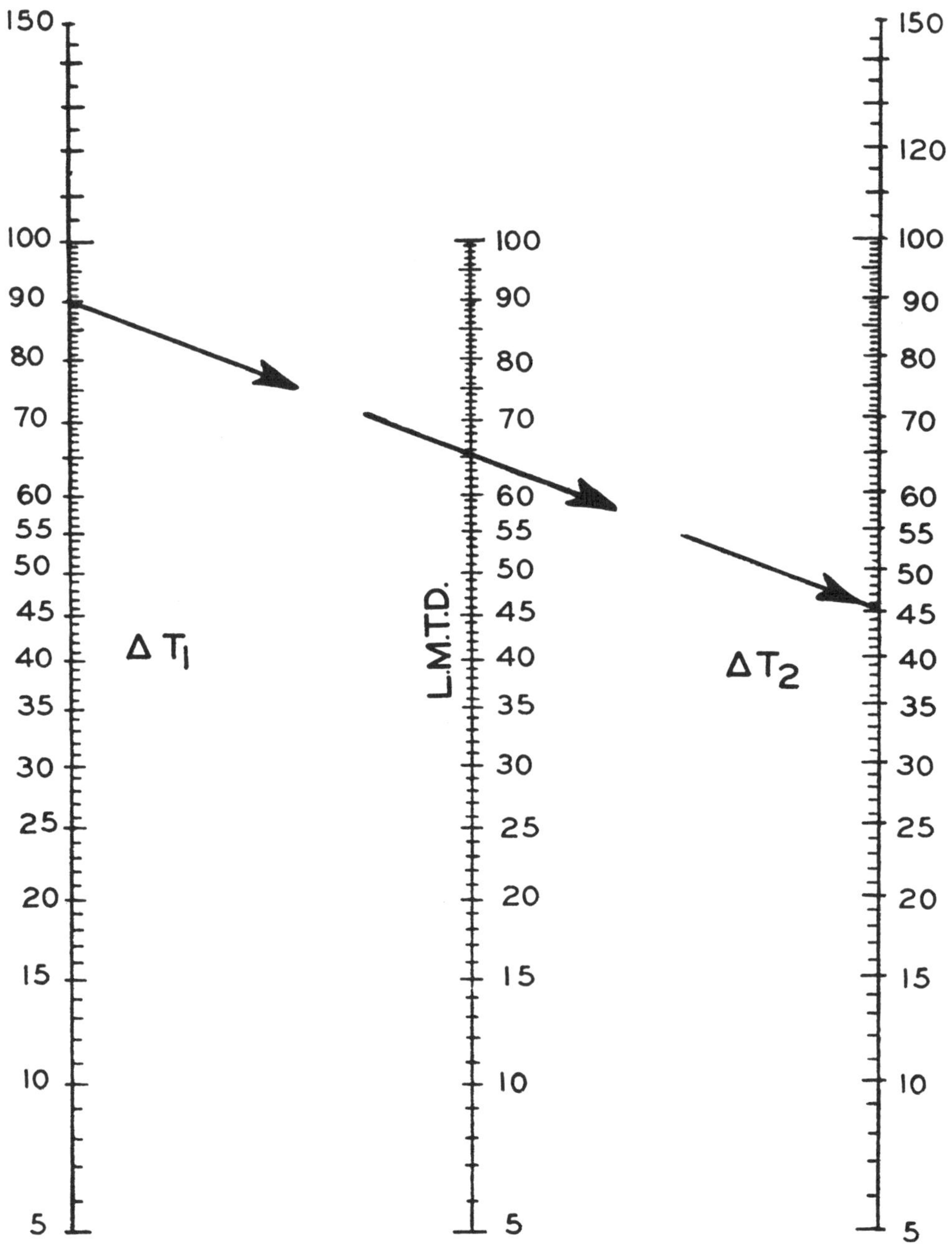

F. Caplan
Kaiser Engineers

Nomogram finds heat exchanger LMTD

This nomogram solves the equation for logarithmic mean temperature difference for heat exchangers:

$$\text{LMTD}: \frac{\Delta t_L - \Delta t_s}{\ln_e(\Delta t_L/\Delta t_s)}$$

Where:

Δt_L = Larger temperature difference
Δt_s = Smaller temperature difference
$\ln_e$ = Natural logarithm

Example. Find the logarithmic mean temperature difference (LMTD) when a hot fluid is cooled from 160° to 95° F. by a counter-current cold fluid heated from 80° to 120° F.

The larger temperature difference $\Delta t_L = 160 - 120 =$ 40° F. The smaller temperature difference $\Delta t_s = 95 - 80$ = 15° F.

To find the LMTD, align Δt_L = 40° F. with Δt_s = 15° F. and read LMTD = 25.5° F.

Note that all three scale values can be simultaneously multiplied by 0.1, 2, 10, 100, etc. to extend the range of the nomogram. END

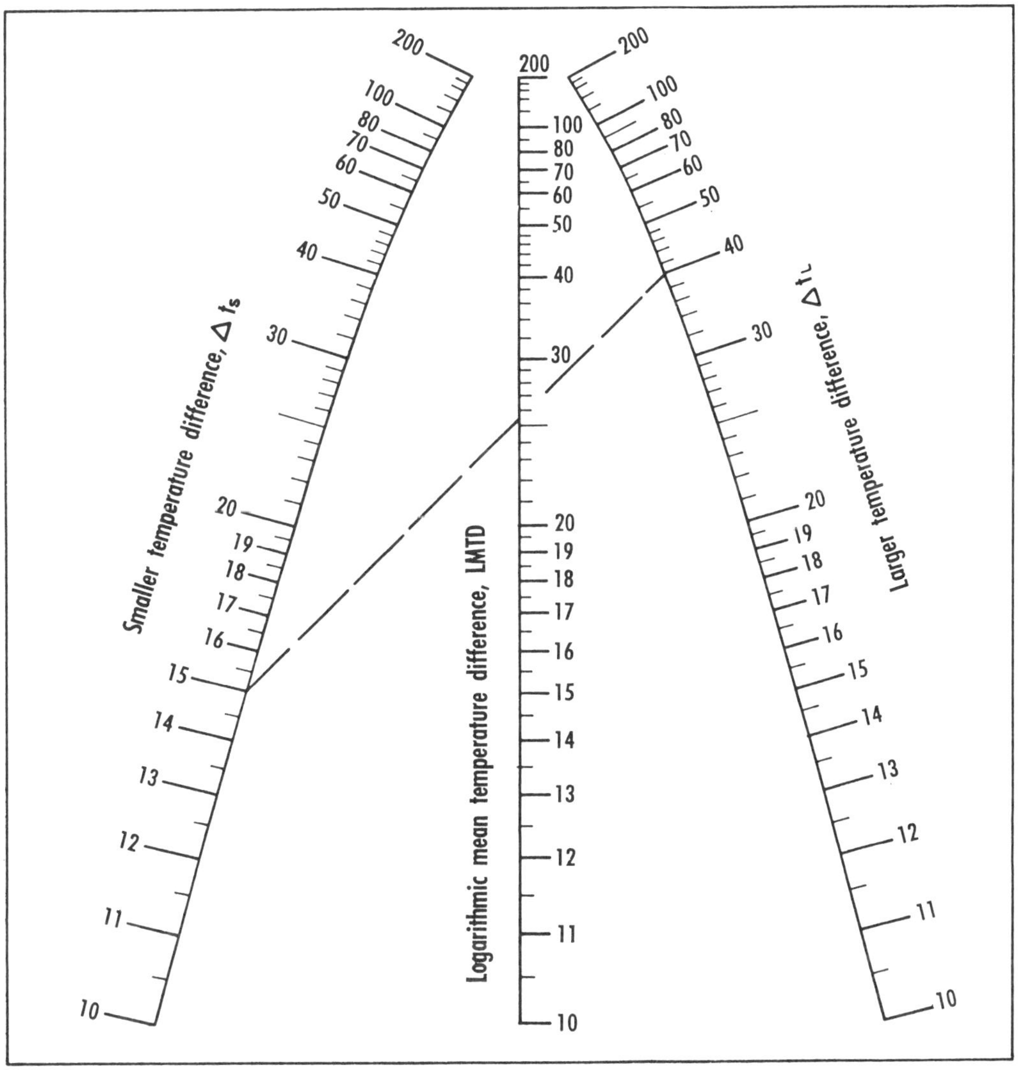

Nomogram tells condenser LMTD

F. Caplan
Kaiser Engineers

This nomogram helps you quickly find the logarithmic mean temperature difference for steam condensers. The LMTD for steam condensers is determined by this equation:

$$LMTD = TR/\log_e [TD/(TD - TR)]$$

Where:

TD = Initial temperature difference between steam and inlet cooling water

TR = Temperature rise of cooling water

See nomogram for a diagram of these two values.

Example. What is the LMTD for a steam condenser if the temperature rise of the cooling water is 12.8° F. and the initial temperature difference between the steam and inlet water is 21.2° F. Align TR = 12.8 with TD = 21.2 and read LMTD = 13.7° F. END

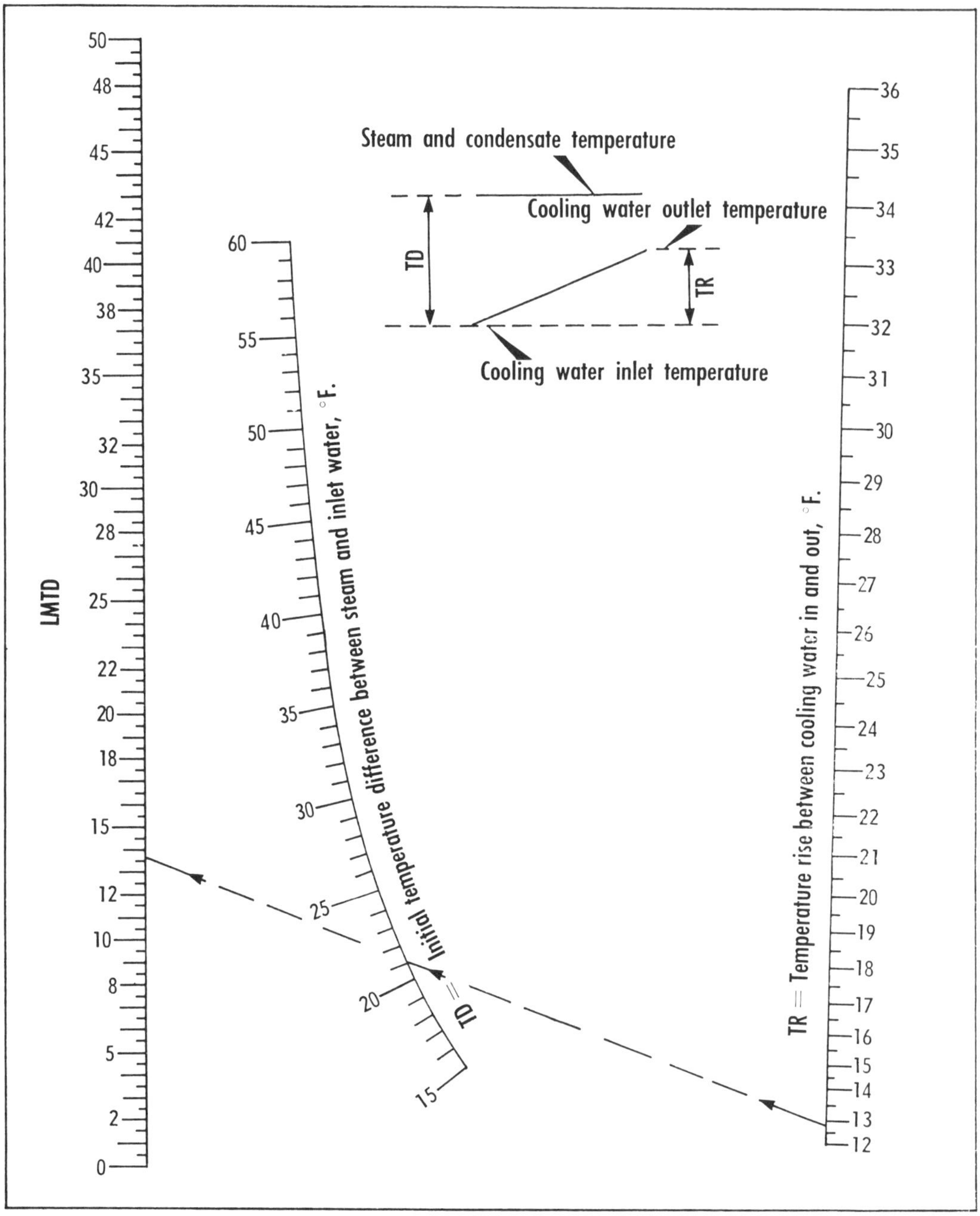

F. Caplan
Kaiser Engineers

Nomogram finds fouling-factor effect

THIS nomogram helps find the effect of the fouling factor on heat-exchanger size. Depending on the fluids handled and their velocities, and on the heat-exchanger material, scale may form which reduces the overall heat-transfer coefficient, U. The required heat exchanger surface,

$$A = Q/U\,(LMTD) \qquad (1)$$

Where:

A = required heat exchanger surface, sq ft
Q = heat load, BTU/hr
LMTD = Log mean temperature difference

The required heat-exchanger surface is inversely proportional to the heat-transfer coefficient, U. It can be shown that:

$$1/U_D = 1/U_c + r_t \qquad (2)$$

Where:

U_D = "dirty" overall coefficient
U_c = clean overall coefficient
r_t = fouling factor

Values of r_t can be found in the standards of the Tubular Exchanger Manufacturers Association (TEMA). The nomogram presented here permits a rapid solution of Equation 2.

Example. What is the effect on heat-transfer coefficient U if the fouling factor is 0.005 and the clean heat-transfer coefficient U_c is 200? Align these two values with a straight edge and read the value of U_D = 100.

END

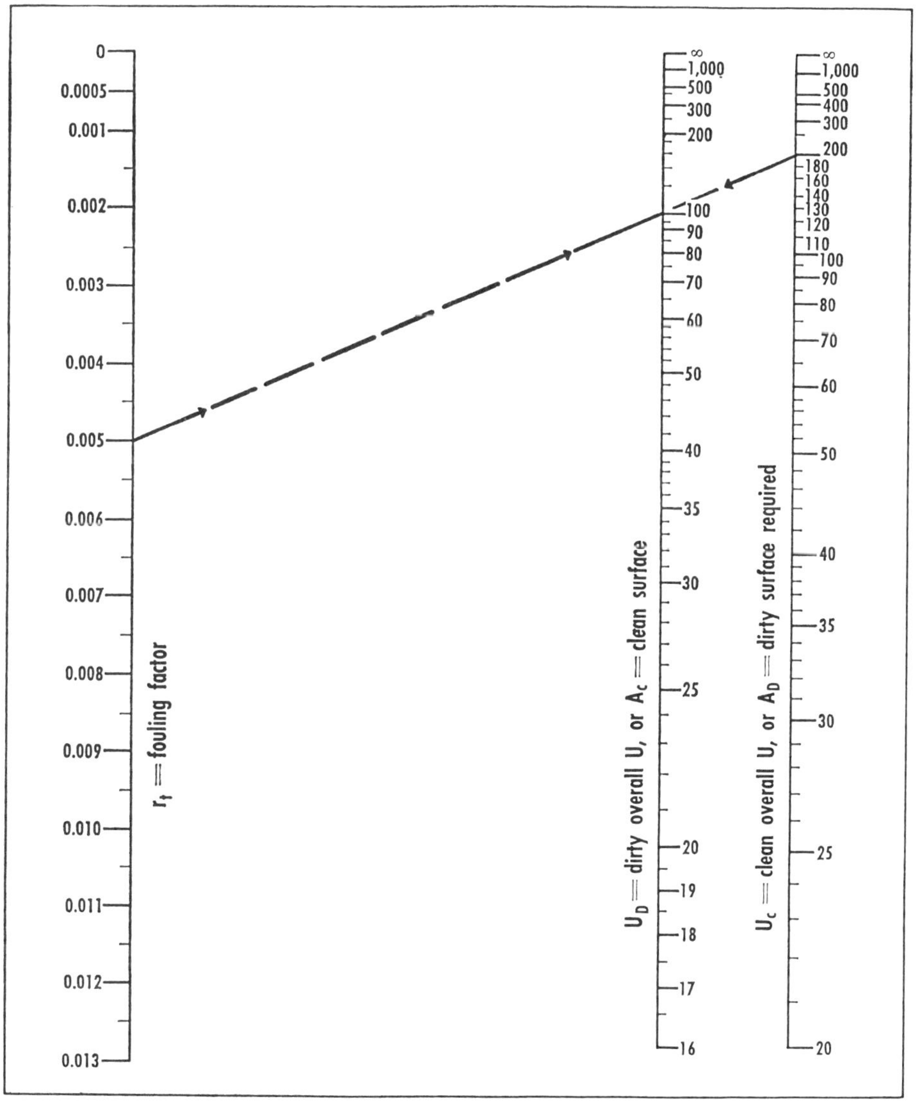

Nomogram tells heat transferred

Bill Sisson
Superintendent of Operations
Nipak, Inc.
Pryor, Okla.

WHEN you need to determine the heat transferred and the condensing rate of constant-steam-pressure, continuous-flow heat exchangers, this nomogram offers a fast solution. It is based on these formulas:

$Q = A\ U\ D_m$

$C = Q/L$

Where:

Q = quantity of heat transferred, BTU/hr
A = area of outside surface of coil, sq ft
U = overall heat transfer coefficient, BTU/hr/sq ft/°F. (See Table 1 for typical values)
C = condensate, lb/hr
L = latent heat of steam at given pressure
$D_m = (D_1 - D_2)/\log_e (D_1/D_2)$
D_1 = greatest temperature difference
D_2 = least temperature difference

Example. Determine the heat transferred, BTU/hr, and the condensing rate, lb/hr of condensate, for a heat exchanger with a coil having an outside surface area of 20 sq ft; overall heat-transfer coefficient of 500; inlet-water temperature of 40° F., outlet-water temperature of 150° F., steam pressure of 125 psig (353° F.); $D_1 = 353 - 40 = 313$; and $D_2 = 353 - 150 = 203$.

TABLE 1 — VALUES OF OVERALL HEAT-TRANSFER COEFFICIENT, U, BTU/hr/sq ft/°F.

Type of service	Value of U, BTU/hr/sq ft/°F. Natural Circulation	Forced Circulation
Pipe coils:		
Steam to water	50 - 200	150 - 1200
1.5-in. tube heaters	180	450
0.75-in. tube heaters	200	500
Steam to oil	10 - 30	50 - 150
Steam to boiling liquid	300 - 800	...
Steam to boiling oil	50 - 150	...
Embossed coils:		
Steam to watery solutions	100 - 200	150 - 275
Steam to light oil	40 - 45	60 - 110
Steam to medium oil	20 - 40	50 - 100
Steam to Bunker C fuel	15 - 30	40 - 80
Steam to tar asphalt	15 - 25	18 - 60
Steam to molten sulfur	25 - 35	35 - 45
Steam to molten paraffin	25 - 35	40 - 50
Dowtherm to asphalt	15 - 30	50 - 60

Align 203° F. on D_2 scale with 313° F. on D_1 scale and extend line to Pivot Line 1 (Step 1). From this point align with 20 sq ft on A scale and extend to Pivot Line 2 (Step 2). From point on Pivot Line 2 align with 500 BTU/hr/sq ft/°F. on U scale, extend line to Q scale and read 2.52 million BTU/hr transferred (Step 3).

To find condensing rate, align the point found on the Q scale with 125 psig on P scale and read 2,900 lb/hr of condensate on the C scale (Step 4).

END

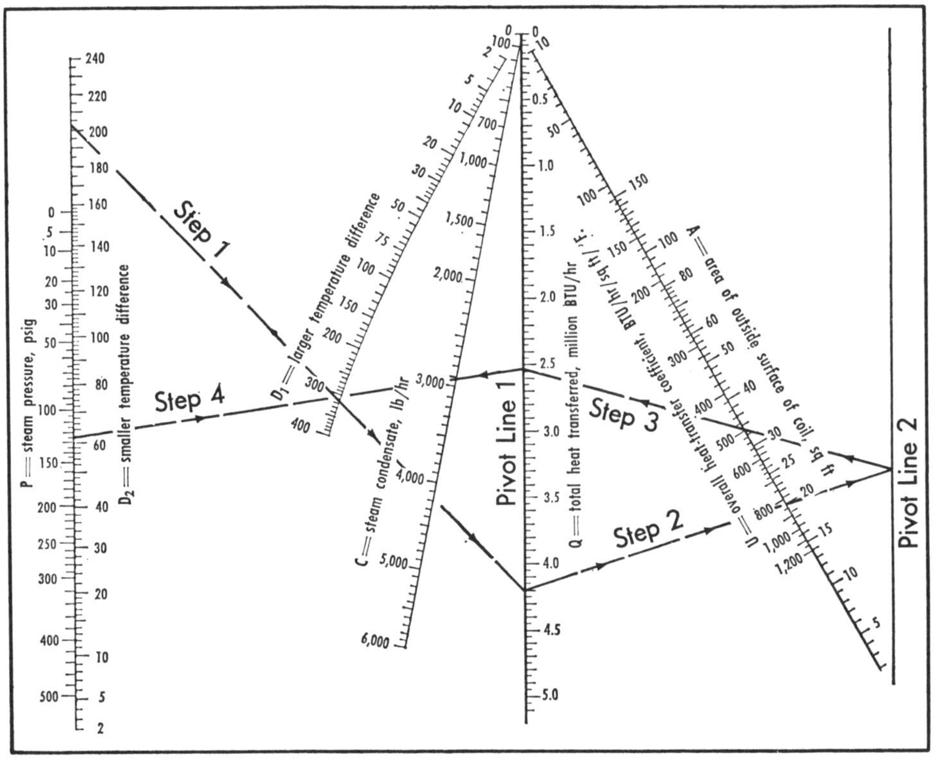

Nomogram gives steam-pipe heat loss

Alvaro Franco
Editor, Petroleo Internacional

WITH the help of a graph plotter coupled to an IBM-1130 computer, the Mexican Petroleum Institute (IMP) has developed a useful nomogram to calculate heat losses in steam pipes, such as between steam generators and steam-injection wellheads. The nomogram, developed by two IMP engineers—S. Casas L. and D. Noguerón T.—solves the following heat-transfer equation for an ambient temperature of 27° C. (81° F.):

$$Q = \frac{328\,\pi\,(T_v - T_a)\,1.8}{(\frac{1}{2}K_a)\ \ln[(D_e + 2E)/D_e] + 12/[h_{cr}(D_e + 2E)]}$$

Where:

Q = heat loss, 1,000 kilocalories/hr/100 meters (328 ft) of pipe

T_v = steam temperature, °C.

T_a = ambient temperature, °C.

k_a = heat conductivity of insulating material, kilocalories/hr/meter²/°C./meter

D_e = pipe OD, in.

E = insulation thickness, in.

h_{cr} = convection and radiation heat-transfer coefficient, kilocalories/hr/meter²/°C.

Example. Determine the heat loss per 100 meter of pipe, if $k_a = 0.061$, $h_{cr} = 9.00$, $D_e = 3$ in., $E = 2$ in., and $T_v = 310°$ C.

Solution. From k_a scale at left (Scale A), trace a line to $h_{cr} = 9.00$ on Scale B, and extend it to Pivot Line X (Step 1). From the intersection with Pivot Line X, trace a line to intercept the crossing of $E_a = 2$ (Scale F) and $D_e = 3$ (Scale C) in the gride-like scale on extreme right. Extend the line to Pivot Line Z (Step 2). From there, trace a line to intercept Scale D at 310° C. and Pivot Line Y (Step 3). From the intersection with Pivot Line Y, draw a final line to $k_a = 0.061$ (Step 4). Read the answer on Scale E which is a heat loss of 13,000 kilocalories/hr/100 meters of pipe.

END

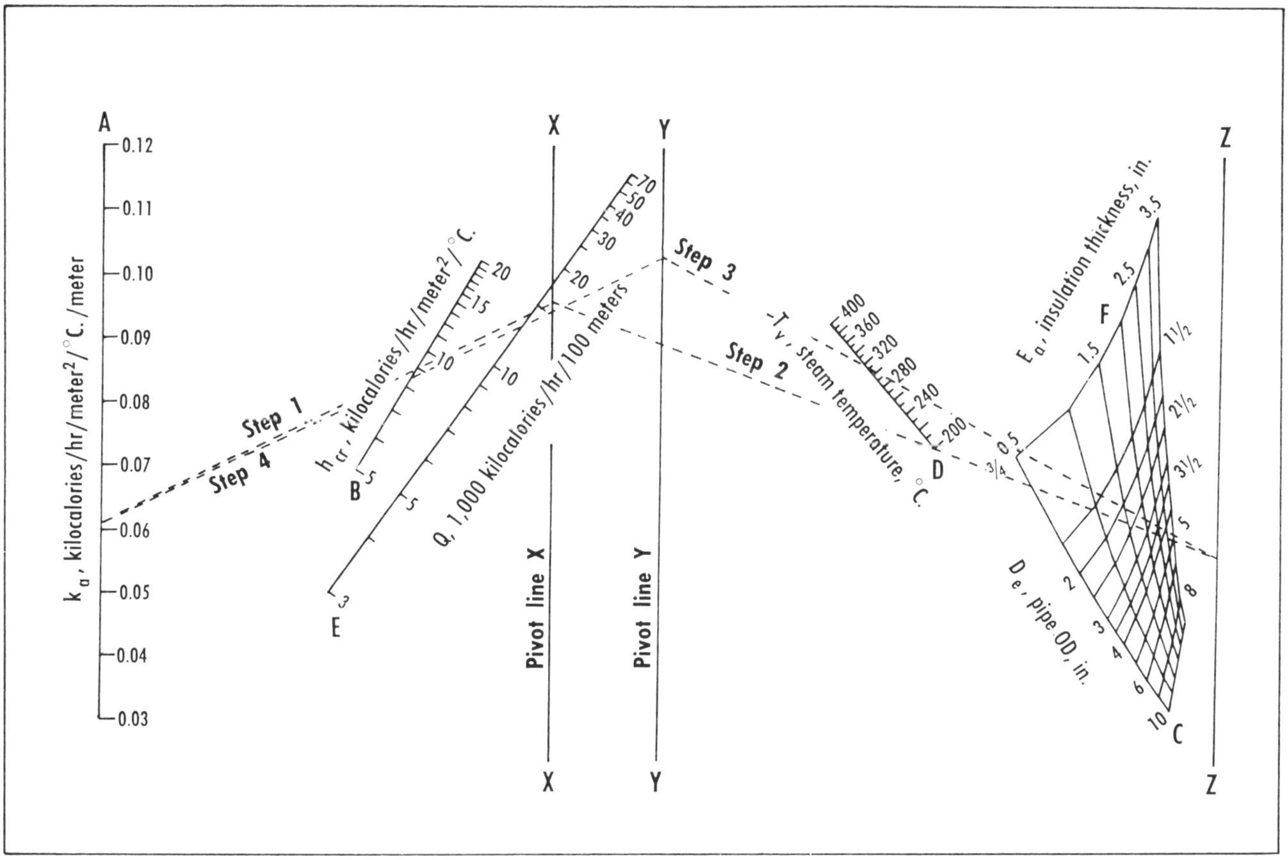

Use this nomogram to determine boiler capacity

Bill Sisson
Superintendent of Operations
Nipak Inc.
Pryor, Okla.

MODERN steam-generating units are often rated on their guaranteed output in thousands of pounds of steam per hour at the given steam conditions. Since steam output may vary in temperature and pressure, boiler capacity is more completely expressed as the heat transferred in BTU/hr. Boiler capacity is usually expressed as kilo BTU (kB)/hr, which is 1,000 BTU/hr, or mega BTU (mB)/hr which is 1,000,000 BTU/hr. Thus:

$$\text{Boiler capacity} = \frac{W(h_g - h_f)}{1{,}000} \text{ kilo BTU/hr}$$

Where h_g = total heat of steam, BTU/lb; h_f = heat of liquid, BTU/lb; and W = steam capacity, lb.

When steam boilers were employed principally to drive steam engines, it was convenient to rate them in horsepower. The term "boiler horsepower" is now obsolete but is employed to a limited extent in referring to certain conventional boilers of small capacity. The unit called "boiler horsepower" may be expressed as:

$$\text{Boiler horsepower} = \frac{W(h_g - h_f)}{970.3 \times 34.5}$$

That is, 1 boiler horsepower is equivalent to 34.5 lb of water evaporated per hour at standard atmospheric pressure and a temperature of 212° F. (970.3 BTU/lb latent heat of evaporation at 14.696 psia and 212° F.).

The nomogram shown here, based on the formulas above, will determine the boiler capacity in kilo BTU/hr or boiler horsepower when the operating pressure and pounds per hour are known.

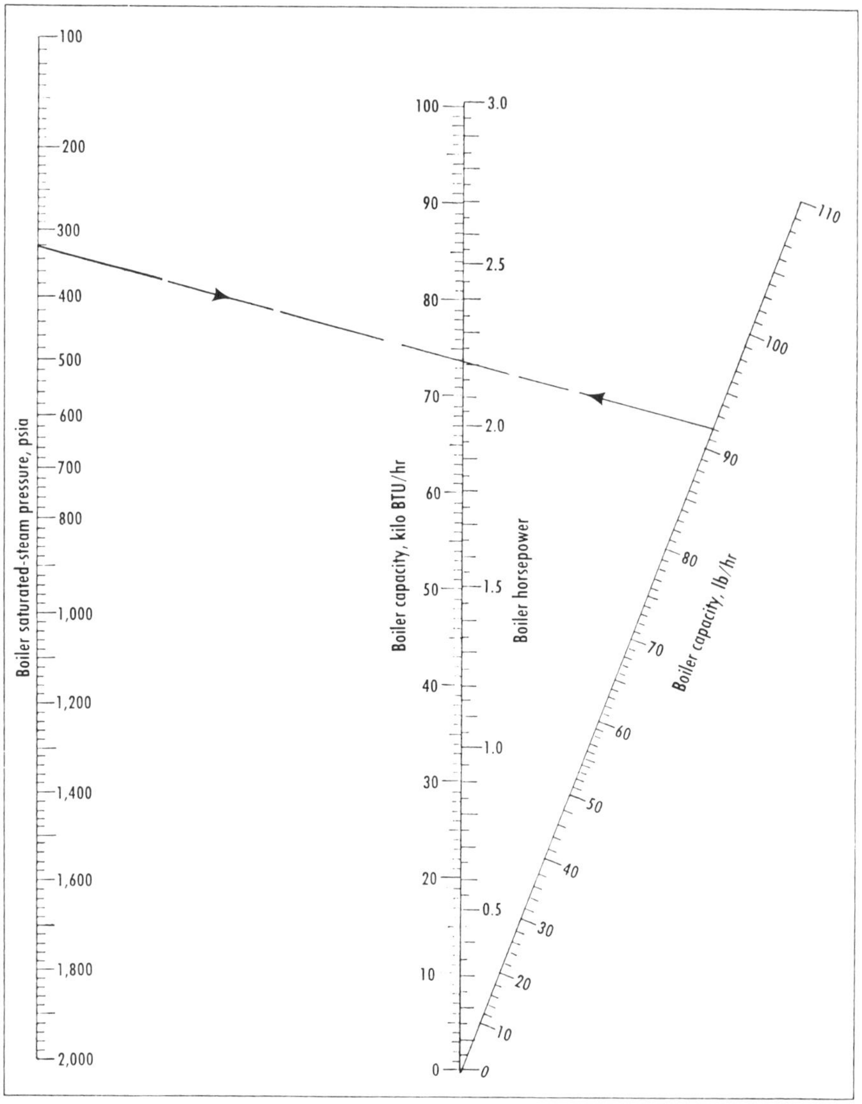

The steam rate, kilo BTU/hr, and boiler horsepower are extended by multiplying by 10, 100, 1,000, etc.

Example. If a steam boiler is operating at 320 psia saturated steam and has a maximum capacity of 92,000 lb/hr, what is the boiler capacity in both kilo BTU/hr and in boiler horsepower?

Solution. Connect 320 on the boiler pressure scale with 92 (92 × 1,000 = 92,000) on the boiler capacity—lb/hr scale, and read 73.88 (73.88 × 1,000 = 73,880 kilo BTU/hr) and 2.207 (2.207 × 1,000 = 2,207 boiler horsepower) where the line crosses the center scale. END

Nomogram determines heat needed to produce steam

Bill Sisson
Superintendent of Operations
Nipak Inc.
Pryor, Okla.

THE application of heat is an established method for recovery of crude oil. One method is steam flooding. It makes use of the fact that it is possible to add energy due to the heat content of the injected steam as well as the mechanical energy due to pressure differentials.

This nomogram provides a quick method for calculating the heat required to produce 1 lb of steam. If the total heat load has been established, division of this load by net enthalpy in BTU/lb as found with the nomogram gives the amount of steam generation in lb/hr of the selected steam quality. The nomogram also includes a scale for determining output in lb/hr/boiler horsepower.

Example. Determine heat requirement to produce 1 lb of 500 psia (467° F.) saturated steam of 70% quality from 182° F. feed water. What is actual evaporation in lb/hr/boiler horsepower?

Solution. Step 1, connect 70% on Steam-Quality scale with 500 psia on Pressure scale and read 978 BTU/lb where line intersects Total Enthalpy of Produced Steam scale. Step 2, align this value with 182° F. on Feed-Water Temperature scale, extend to Heat Required to Produce 1 lb of Steam scale, and read 828 BTU/lb. On opposite scale read 40.4 lb/hr/boiler horsepower of actual evaporation. END

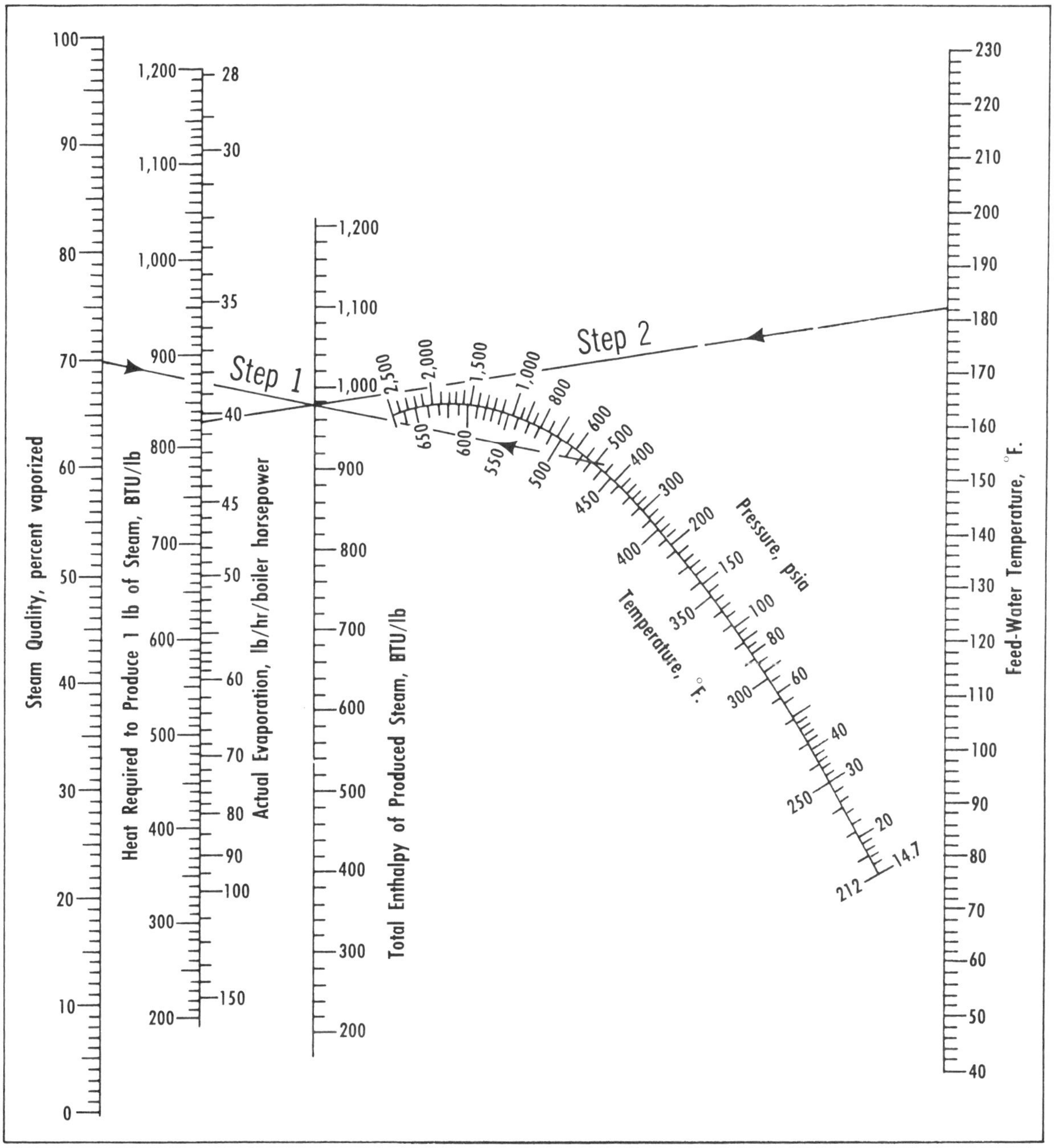

Effect of air on steam temperature

Bill Sisson
Superintendent of Operations
Nipak Inc.
Pryor, Okla.

SINCE air is a very good insulator, it has considerable effect on heat transfer. Steam condensing in a heat-transfer unit moves air to the heat-transfer surface where it collects or "plates out" to form an effective insulation.

Air can be present in steam at any time. It may be in boiler feedwater, and it is always present on start up. When a chamber contains steam and air, only the heat of the partial pressure of the steam is delivered, not the total pressure.

This month's nomogram will determine the temperature reduction caused by air as shown by the following examples.

Example 1. If the pressure in a steam chamber (or pipe) is 100 psia (85.3 psig), and the steam contains 10% air, what is the temperature of the mixture and the reduction in temperature caused by the air?

Solution. Align 100 psig on P scale with 10% (by volume) on A scale, extend to TR scale, and read temperature of steam mixed with air at 320.3° F. where the line crosses the TM scale and temperature reduction as 7.5° F. where the line intersects the TR scale.

Example 2. If a pressure gauge reads 15 psig, and the temperature indicator shows 246° F., what is the percent air in the steam, and what is the reduction in temperature caused by the presence of this air?

Solution. Connect 15 psig on the P scale with 246° F. on the TM scale and extend the line to the A scale and the TR scale. Read the percent air in steam as 6.5% on the A scale and reduction in temperature from that of 15 psig saturated steam as 3.7° F. on the TR scale. END

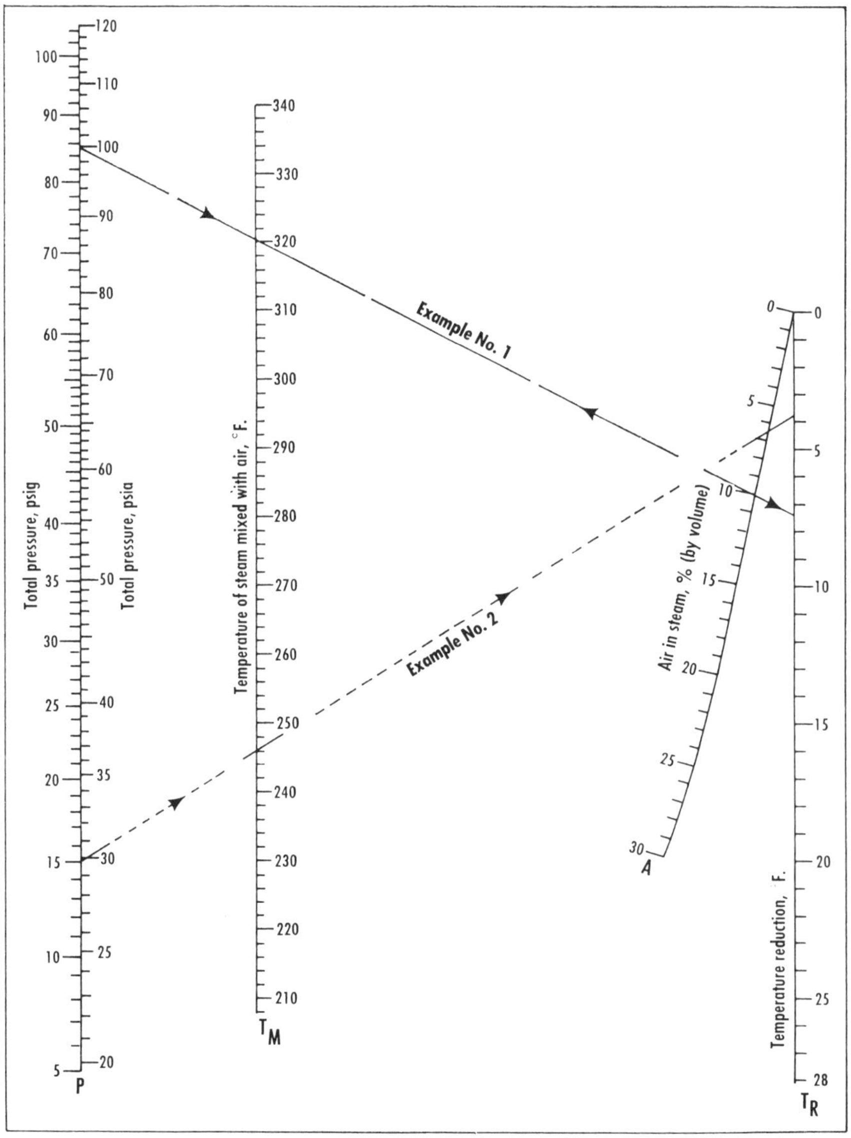

Nomogram determines . . . flash steam, its heat content

Bill Sisson
Superintendent of Operation
Nipak Inc.
Pryor, Okla.

ARE you wasting valuable energy by losing the heat that could be recovered from the hot condensate discharged from your steam traps? If the condensate formed in each part of a process is calculated, and the pressures and temperatures at the inlet and outlet of the traps are determined, the heat content and the volume of steam being flashed can be found with this nomogram.

The nomogram is based on condensate at or near steam temperatures. If the condensate is cooler, the actual temperature on the opposite side of the steam-pressure scale must be used.

Example. Forty steam traps discharge a total of 70,000 lb/hr of condensate from equipment operating at 150 psig. The condensate flows to a vented receiver that operates at 5 psig. How much heat is available and how much flash steam is formed, if the condensate is discharged very close to steam temperature?

Solution. Step 1, connect 150 psig on SP scale with 5 psig on FP scale, and mark at intersection with Pivot Line. Step 2, line this marked point with 70,000 lb/hr on C scale, extend to H scale, and read heat available as 9,945,000 BTU/hr. Step 3, connect 9,945,000 BTU/hr on H scale with 5 psig on FP scale, and read the quantity of flash steam formed as 10,360 lb/hr where line crosses F scale.

Example 2. If, at a pressure of 150 psig, the temperature of condensate at a trap is only 350°F. (instead of the 366°F., the 150-psig saturation temperature), what amount of heat is available, and how much flash steam is formed?

Solution. Step 4, connect 350°F. on TS scale with 5 psig on FP scale, and mark at intersect with Pivot Line. Step 5, line point on Pivot Line with 70,000 lb/hr on C scale, extend to H scale, and read heat available as 8,840,000 BTU/hr. Step 6, connect this BTU/hr on H scale with 5 psig on FP scale, and read quantity of flash steam as 9,200 lb/hr where line crosses F scale.

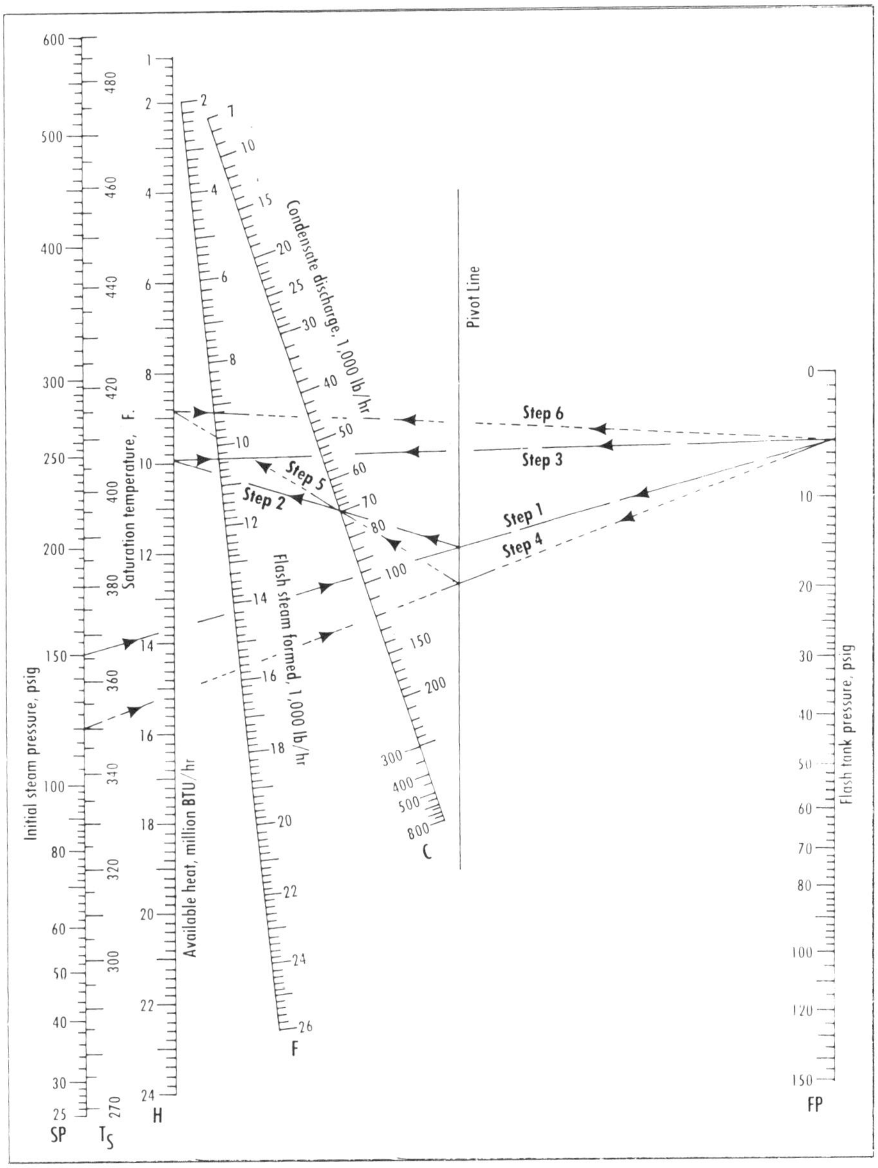

Now assume that water is heated from 40° to 140° F. with the heat energy from Example 1 in a water heater having an overall efficiency of 85%. How much water can be heated? Taking the specific heat of water as 1,

9,945,000 (0.85)/(100 - 40) =
84,532 lb/hr water heated
84,532/8.3 = 10,184 gph heated

Further saving is realized by returning the condensate from the water heater to an atmospheric-pressure receiver and using it again in the boiler.

END

Flash steam recovered from blowdown

Bill Sisson
Superintendent of Operations
Nipak, Inc.
Pryor, Okla.

THE NOMOGRAM presented here can be used to quickly determine the amount of steam recovered from continuous-blowdown systems. It can be a valuable time saver as it eliminates the need to make time-consuming calculations and does away with the need to refer to steam tables. It can also greatly reduce the chance for errors in calculations.

The nomogram is based on the equation,

$$S_F = (H_B - H_F)B_D/V_f$$

Where:

S_F = flash steam recovered, lb/hr
B_D = amount of blowdown, lb/hr
H_B = heat of liquid at boiler pressure, BTU/lb
H_F = heat of liquid at flash pressure, BTU/lb
V_F = latent heat of vaporization at flash pressure, BTU/lb

Example. If a boiler operates at 1,000 psig, with a continuous blowdown of 16,000 lb/hr, how much of the blowdown can be recovered as 50-psig flash steam?

Solution. Align 50 psig flash-steam pressure on Scale B with 1,000 psig boiler operating pressure on Scale E (Step 1), and mark where the line crosses Pivot Line D. From this marked point on Pivot Line D align with 16,000 lb/hr on Scale C, extend the line to Scale A, and read 4,870 lb/hr of flash steam recovered on Scale A (Step 2). END

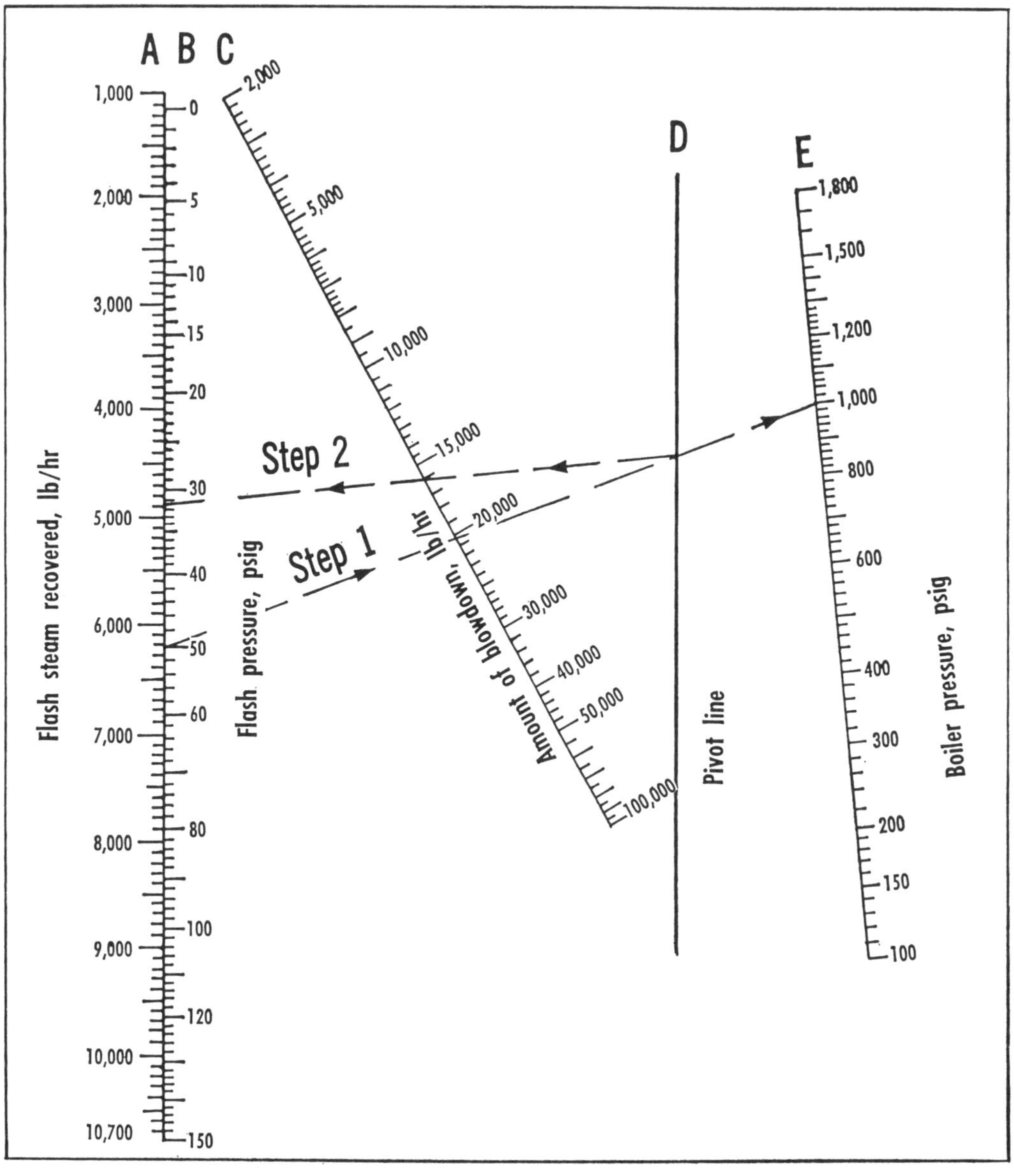

F. Caplan
Kaiser Engineers

Finding flash steam percentage

WHEN water near its saturation temperature, as in a boiler or upstream of a steam trap, enters a lower pressure zone, part of the water flashes into steam. If the receiver is above atmospheric pressure, this steam can be recovered for heating purposes. This nomogram permits a rapid determination of the amount of steam formed.

The nomogram is based on the equation:

$$\% \text{ water flashed} = 100\,(h_b - h_f)/h_{fg}$$

Where:

h_b = enthalpy of saturated water at higher pressure

h_f = enthalpy of saturated water at lower (flash) pressure

h_{fg} = latent heat of vaporization at flash pressure

Example. If 1,000 lb/hr of boiler water at 100 psig is blown down to a tank at atmospheric pressure, how much steam is formed? Using a straight edge, align 100 psig on gauge-pressure-of-water scale with 0 psig on the gauge-pressure-of-flash tank scale and read 13.3% on the %-of-water scale, or 133 lb/hour of steam formed (1,000 lb/hr × 0.133).

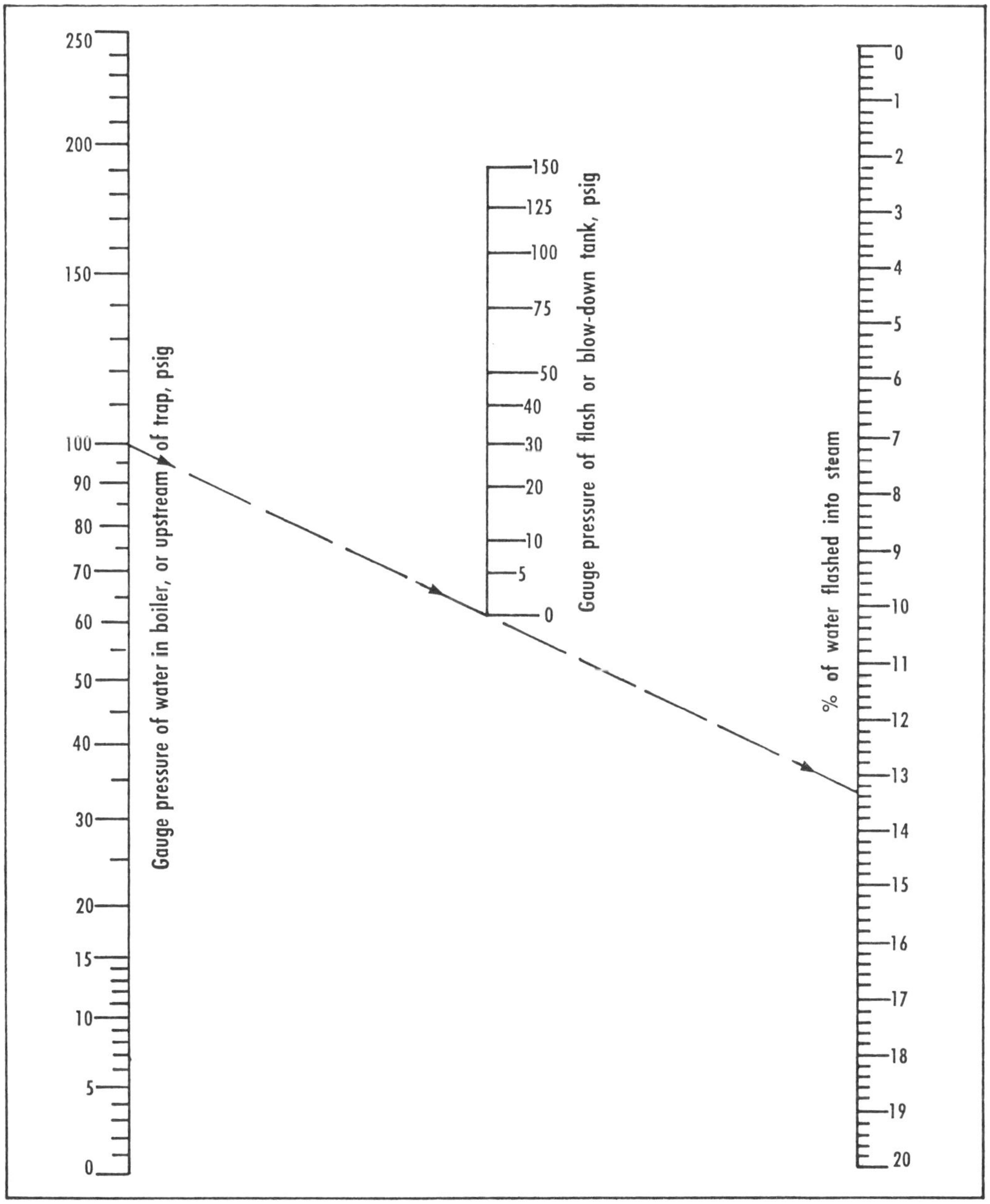

Nomogram gives heat loss for continuous boiler blowdown

Bill Sisson
Superintendent of Operations
Nipak, Inc.
Pryor, Okla.

THE HEAT LOSS caused by boiler blowdown goes up quite rapidly as boiler blowdown is increased. Considerable savings can be realized if the boiler blowdown rate is regulated to maintain the boiler-water concentration at the optimum value. The heat loss can be found with this formula:

$$L = (h_{fB} - h_{fW})\,([W/(1.00 - B_d)] - W)$$

Where h_{fB} = enthalpy of liquid at boiler pressure, BTU/lb; h_{fW} = enthalpy of make-up water, BTU/lb; W = boiler load, lb/hr; and B_d = blowdown rate, percent (as decimal).

Example. Determine heat loss with 6% and 4% blowdown rates for a boiler operating at 650 psia with a load of 700,000 lb/hr and makeup water temperature of 72° F.

Solution. Connect 650 psia on P scale (Step 1) with 72° F. on T_w scale, and mark at intersection with Pivot line 1. Line marked point with 6% on B_d scale (Step 2) and extend to Pivot line 2. From this point, line with 700,000 lb/hr on W scale (Step 3), extend to L scale, and read heat loss as about 19,725,000 BTU/hr for 6% blowdown. To find loss for 4% blowdown, connect point (Step 4) found on Pivot line 1 in Step 1 with 4% on B_d scale, and extend to Pivot line 2. From this point, line with 700,000 lb/hr on W scale (Step 5), extend to L scale, and read heat loss as 12,900,000 BTU/hr. Thus a reduction in blowdown rate from 6% to 4% would save about 6,825,000 BTU/hr. END

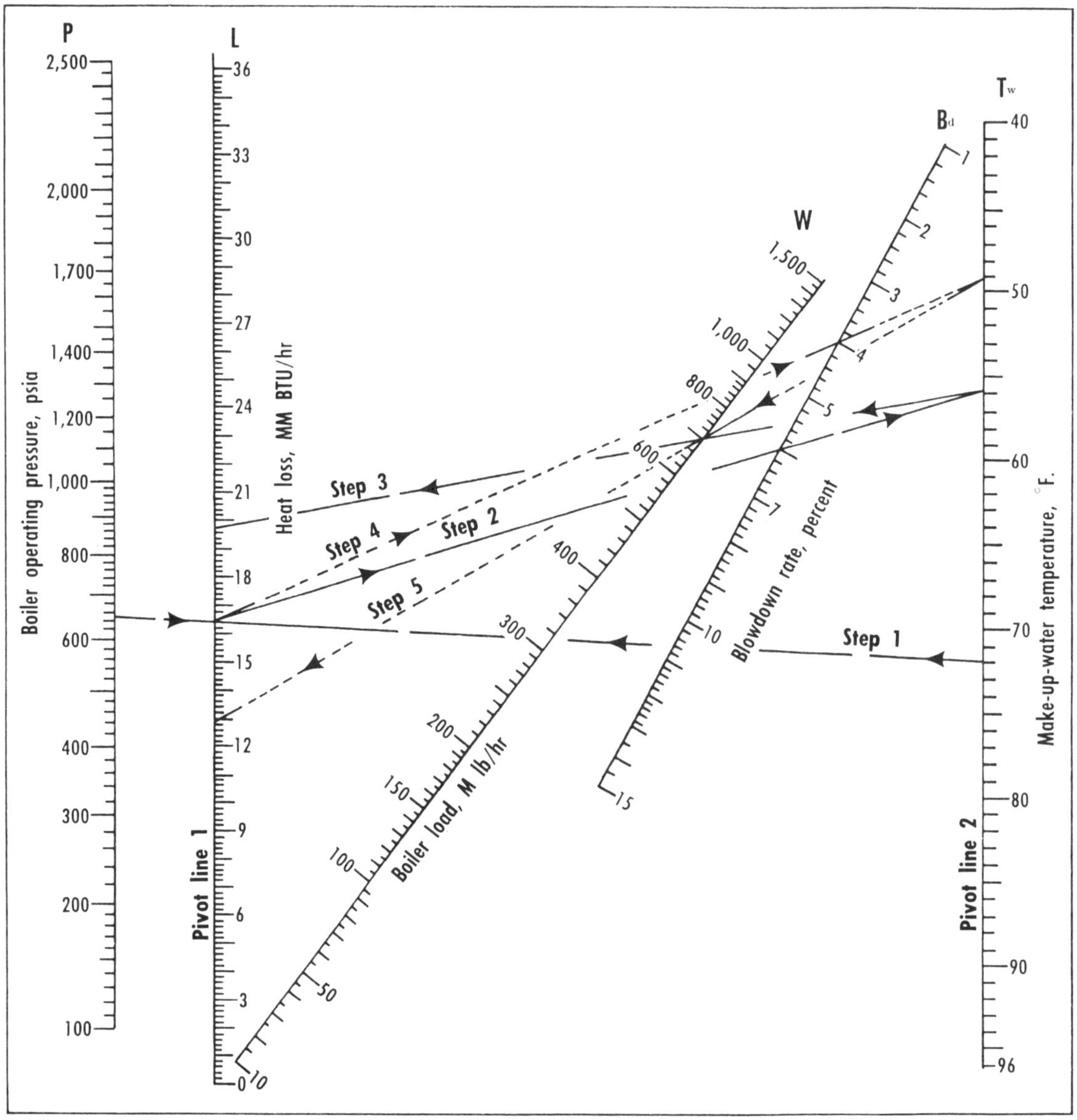

Nomogram sizes flash-tank discharge pipe

Bill Sisson
Superintendent of Operations
Nipak Inc.
Pryor, Okla.

FLASH tanks are used to collect condensate and the steam formed when hot condensate under pressure is released to a lower pressure. For economy, condensate is normally returned to the boiler-feed water, and the flash steam is used in low-pressure heating systems, etc.

If you wish to use the flash steam by discharging it into a low-pressure main, it is necessary to properly size the line that connects the flash tank to the low-pressure main. To do this, you must determine the amount of condensate flashed to steam. This is found from the formula:

$$\text{lb steam flashed} = \frac{W\,(H_B - H_F)}{L_F}$$

Where:
H_B = enthalpy of liquid at steam-line pressure, BTU/lb
H_F = enthalpy of liquid at flash-tank pressure, BTU/lb
L_F = enthalpy of evaporation at flash-tank pressure, BTU/lb
W = condensate load, lb/hr

Then you must decide what steam velocity in the line will be acceptable. If noise level is important, and it usually is, a relatively low velocity must be selected. Generally, 4,000 to 6,000 fpm is considered a reasonable velocity. If noise level is not a consideration, a velocity of 12,000 fpm or higher may be used. Velocity can be determined with the formula:

$$V = 3.06\,W\,\mathbf{V}/d^2$$

Where:
W = rate of flow, lb/hr
$\mathbf{V}$ = specific volume of steam, cu ft/lb
d^2 = internal diameter of pipe, in.

The above formulas are not complicated, but it is necessary to refer to steam and pipe tables for enthalpy, specific volume, and pipe sizes before they can be used.

The nomogram shown is based on the above formulas. It will determine the required pipe sizes and eliminate the tiresome chore of reference to tables.

Example. If the condensate discharged to a flash tank from a 125-psig steam system is 15,000 lb/hr and the flash tank is piped into a 10-psig low-pressure heating system, what size pipe should be used to connect the flash tank to the steam main? Assume that a low noise level is important.

Solution. Line 10 psig on B-1 scale with 125 psig on C scale (Step 1) and extend to D scale. From this point, connect with 15,000 lb/hr on E scale (Step 2), extend to F scale, and read 1,830 lb/hr of flashed steam. From 1,830 lb/hr on F scale, line with 10 psig on B-2 scale (Step 3), extend to D scale, and mark. As noise level is important, the velocity should be 4,000 to 6,000 fpm. Thus a line from the marked point on the D scale through the 4-in. pipe size on the G scale and extended to the A scale (Step 4) shows a velocity of 5,650 fpm. A 4-in. pipe would be satisfactory. **END**

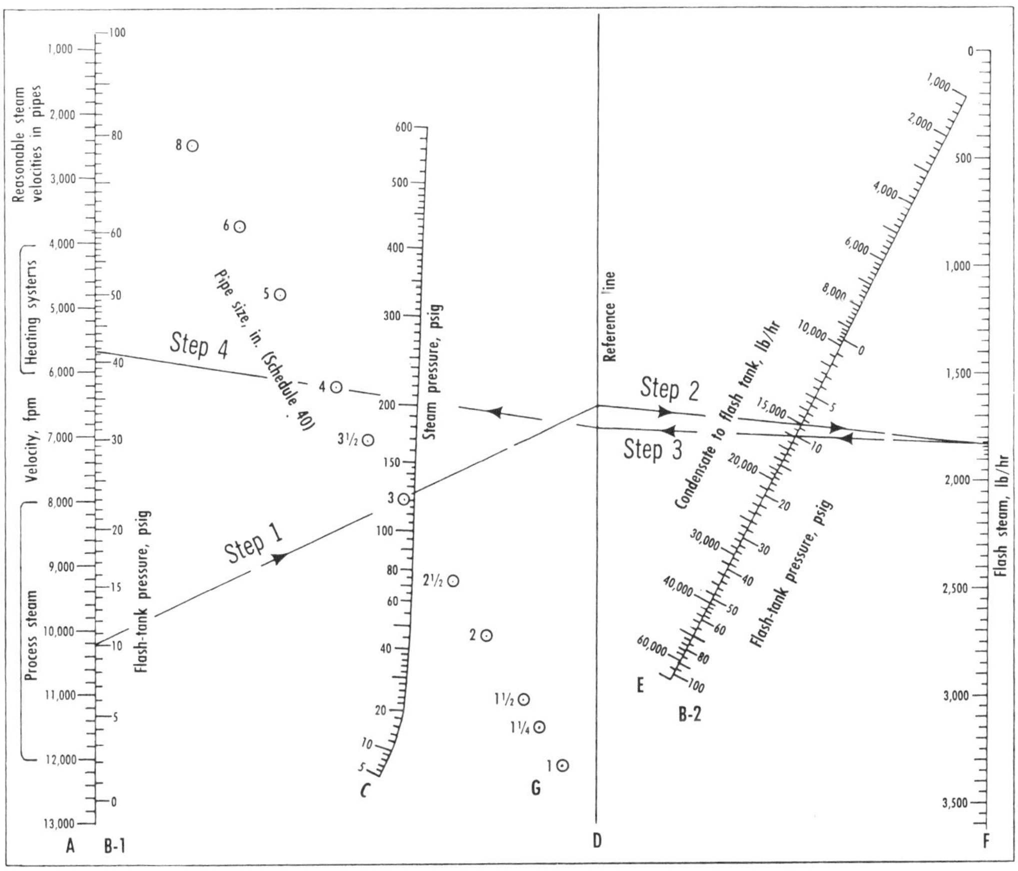

F. Caplan
Kaiser Engineers

Saturated steam-flow rate in pipe

HERE is a nomogram that gives you a fast way to find the friction pressure drop of saturated steam in a steel pipe of known size, or vice versa, to select a pipe size to limit the pressure drop to a desired value. It's based on the equation:

$$\Delta P_{100} = 6.44 \times 10^{-3}\ W^{1.83}/d^{4.83}\ P^{0.193}$$

Where:

ΔP_{100} = pressure drop, psi, per 100 ft length of pipe
W = flow rate, lb/hr
d = actual inside diameter of pipe, in.
P = average steam pressure, absolute, psia

Example. What is the pressure drop of 3,250 lb/hr of 100-psig steam in a 3-in. standard weight pipe? Using a straight edge, align d = 3 in. with w = 3,250 lb/hr and mark the intersection with the pivot line (Step 1). Next align p = 100 psig with the mark on the pivot line and read pressure drop = 1.0 psi/100 ft length of pipe (Step 2).

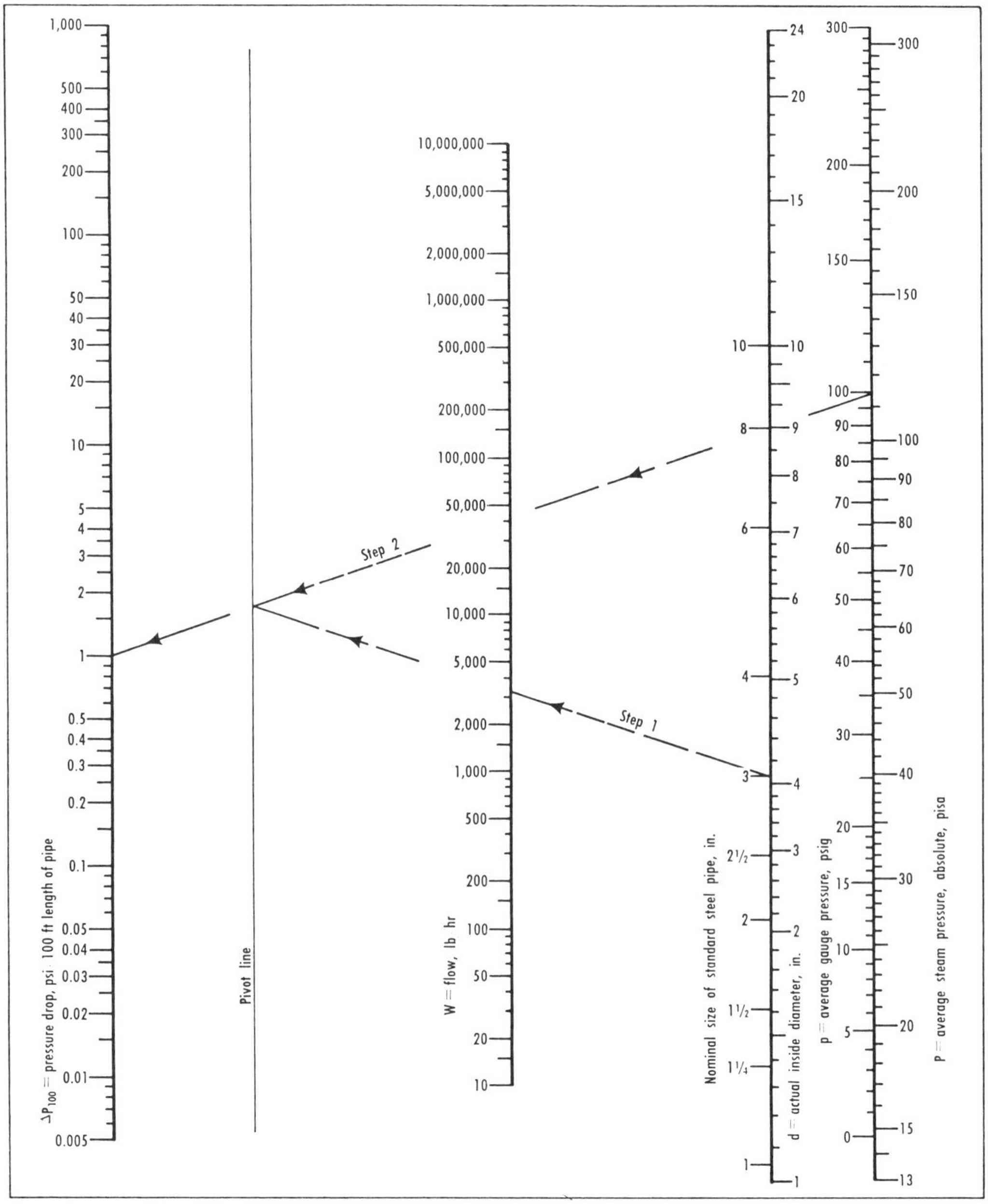

F. Caplan
Kaiser Engineers

Use this nomogram to find hot-water flow rate in steel pipe

THIS nomogram provides a rapid means of determining the friction pressure drop of hot water flowing turbulently in a steel pipe of known size, or, vice versa, the selection of a pipe size to limit the pressure drop to a desired value. It's based on the equation:

$$\Delta h_{100} = 0.312\, Q^{1.83}/t^{0.185}\, d^{4.83}$$

Where:

Δh = friction pressure drop, ft of hot water, per 100 ft length of pipe.

Q = flow, gpm

t = water temperature, °F.

d = actual inside diameter of pipe, in.

Example. What is the pressure drop of 5.7 gpm of 400° F. water flowing in a 1-in. standard weight pipe? Using a straight edge, align t = 400° F. with Q = 5.7 gpm and mark the intersection with the pivot line (Step 1). Next align this pivot point with d = 1 in. and read friction loss as 2 ft of hot water/100 ft of pipe length (Step 2).

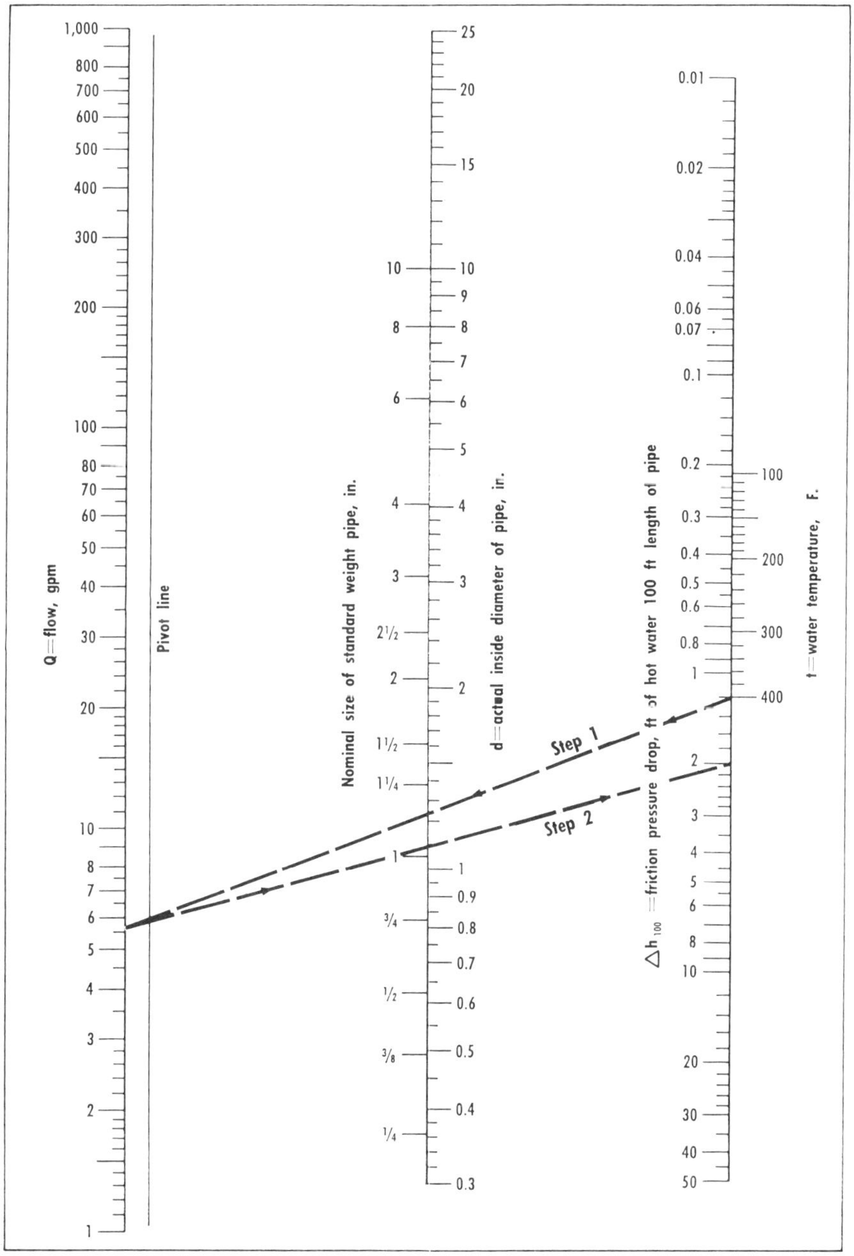

Nomogram determines steam flow, pressure drop

Bill Sisson
Superintendent of Operations
Nipak Inc.
Pryor, Okla.

THE friction of fluid flowing in a pipeline causes a drop in initial pressure. This drop is approximately proportional to the square of the velocity and directly proportional to the effective length of the line.

Under certain conditions, sufficient accuracy in pressure drop calculations is obtained with simplified equations which allow a direct graph of flow against pressure drop.

Flow or pressure drop for either saturated or superheated steam may be found with this nomogram as shown by the following examples.

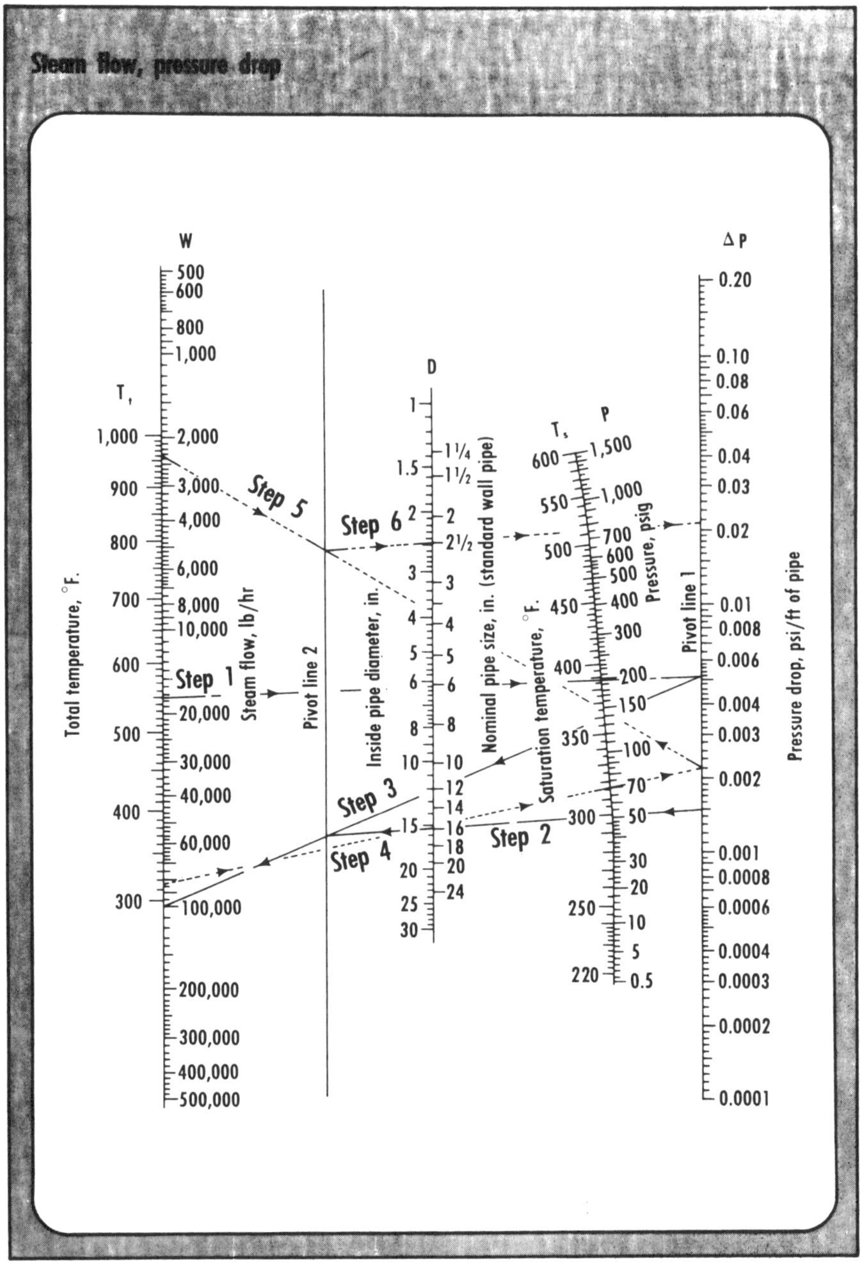

Example 1. Given a steam pressure of 200 psig at 550° F., a pipe size of 16 in. (standard wall), and a pressure drop of 0.0015 psi/foot of pipe, determine superheat (°F.) and flow rate.

Solution. On T_s scale, note that temperature of 200-psig saturated steam is 388° F. Then superheat = 550° − 388° = 162° F. On the nomogram, Step 1, line 550° F. on T_t scale with 200 psig on P scale, extend to Pivot line No. 1 and mark. Step 2, line 0.0015 psi/ft of pipe on Δ P scale with 16-in. pipe on D scale, extend to Pivot line 2 and mark. Step 3, connect marked points on pivot lines, extend to W scale, and read steam flow as 100,000 lb/hr.

Example 2. Given 70-psig saturated steam, 2,300 lb/hr steam flow, and 2½-in. standard-wall pipe, find pressure drop.

Solution. Note on the P and T_s scales that the temperature of 70-psig saturated steam is 316° F. Step 4, connect 70 psig (316° F.) on P scale with 316° F. on T_t scale and extend to Pivot line 1. Step 5, from point found on Pivot line 1, connect with 2,300 lb/hr on W scale, and mark at intersection with Pivot line 2. Step 6, line this marked point with 2½-in. pipe on D scale, extend to Δ P scale, and read pressure drop as 0.0216 psi/ft of pipe.

Note that the inside-diameter scale may be used to determine values for pipe other than standard-wall pipe. END

Nomogram determines air actually used in fuel oil combustion

Bill Sisson
Superintendent of Operations
Nipak Inc.
Pryor, Okla.

EFFICIENCY of combustion is based on the relation between the amount of air theoretically required for perfect combustion and the amount actually supplied. The difference between the two is excess air. When the nitrogen in the fuel is negligible and there is no carbon monoxide or hydrogen in the flue gas, it's percentage may be computed from the equation:

$$\text{Percent excess air} = O_2/(0.264N_2 - O_2)\ 100$$

Where O_2 and N_2 are the volumetric percentages of these compounds in the dry flue gas.

Theoretical air required may be calculated either from the ultimate analysis of the fuel, which is the most accurate method, or from the heating value of the fuel. This nomogram is based on the formula shown and the theological air requirements (Table 1) for combustion of fuel oils.

Example. If a fuel being burned is Commercial Standard No. 3 fuel oil and an Orsat analysis of the flue gas shows 4.3% oxygen and 14.3% carbon dioxide, what is the percent excess air and how much air is actually used in cu. ft./gal of fuel?

Solution. Connect 4.3% on O_2 scale with 14.3% on CO_2 scale (Step 1) and read excess air as 25% where line intersects Excess-aîr scale. From 25% on Excess air scale (Step 2), connect with No. 3 fuel oil reference point and read 1,743 cu ft air/gal of oil where line crosses air actually used scale. END

Table 1—Air required for combustion of fuel oil

Fuel oils	BTU/gal	Air required for combustion, cu ft air/gal fuel
Commercial Standard No. 1	136,000	1,342
Commercial Standard No. 2	138,500	1,370
Commercial Standard No. 3	141,000	1,395
Commercial Standard No. 4	145,000	1,431
Commercial Standard No. 5	148,500	1,468
Commercial Standard No. 6	152,000	1,498

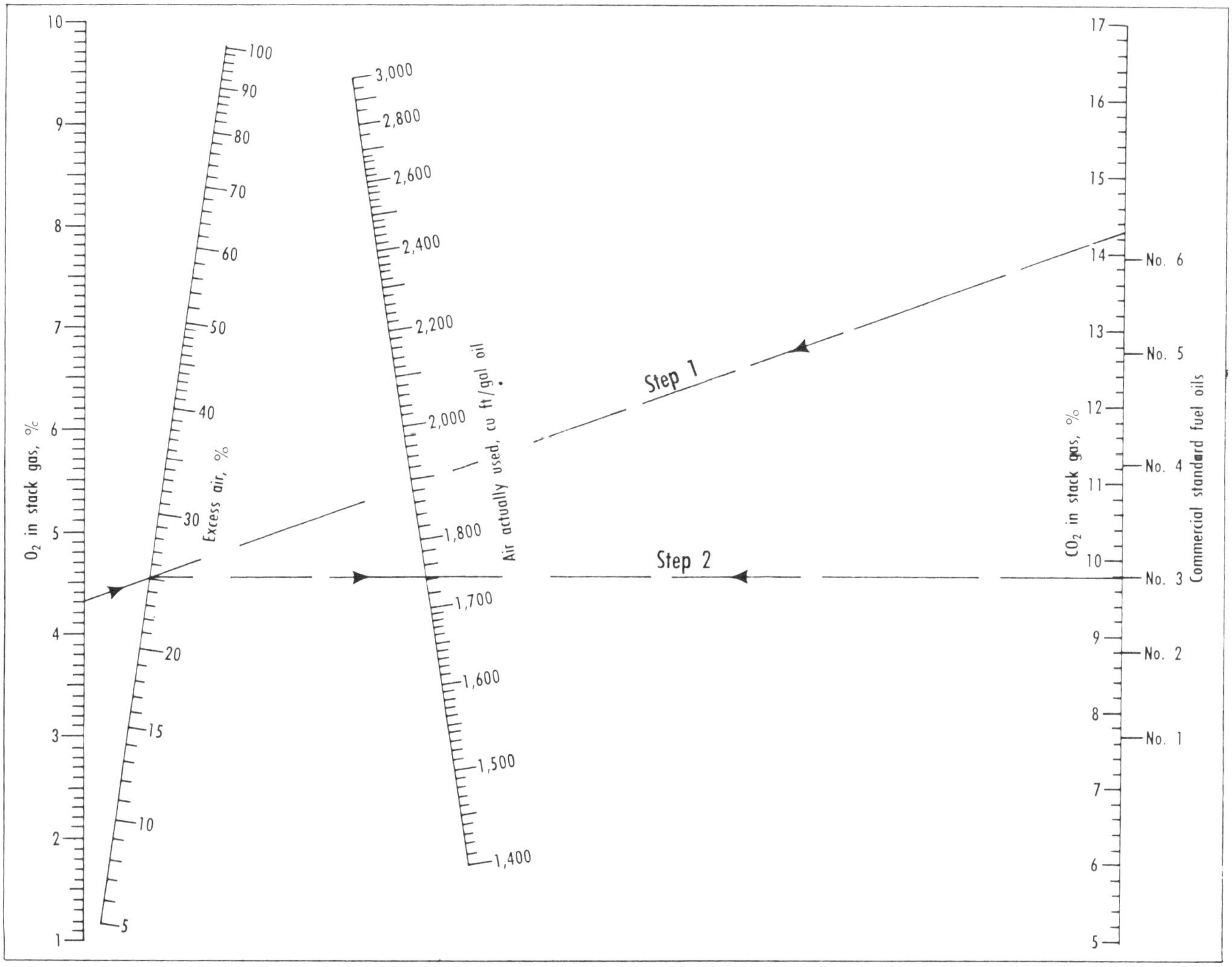

Nomogram finds weight of dry flue gas

Bill Sisson
Superintendent of operations
Nipak, Inc.
Pryor, Okla.

WHEN the pounds of carbon burned per pound of fuel and the flue-gas analysis are known, the accompanying nomogram provides a convenient means of determining the weight of dry flue gas per pound of fuel burned. As knowledge of the weight of dry flue gas is required in a number of formulas concerning boilers, furnaces, heaters, etc., this nomogram eliminates the necessity of making this tiresome calculation. It is based on the formula:

$$W = C\ (44\ CO_2 + 28\ CO + 28\ N_2 + 32\ O_2/12\ (CO_2 + CO)$$

Where:
W = weight of dry flue gases, lb/lb fuel
C = carbon burned, %

Example. Assume the ultimate analysis of a fuel and the combustible in the refuse shows 0.70 lb of carbon per lb of fuel was burned, or 70% of the fuel weight. Orsat analysis of the flue gas shows $O_2 = 5.5\%$, $CO_2 = 11.6\%$, $CO = 0.4\%$, and $N_2 = 82.5\%$. What is the weight of dry flue gas per pound of fuel burned?

Solution. Draw a line from 5.5% on the O_2 scale through 12% on the $CO_2 + CO$ scale. Mark where it intersects the pivot line (Step 1). From this point, align with 70% on C scale, extend line to W scale, and read 14.61 lb of dry flue gas per lb of fuel (Step 2). END

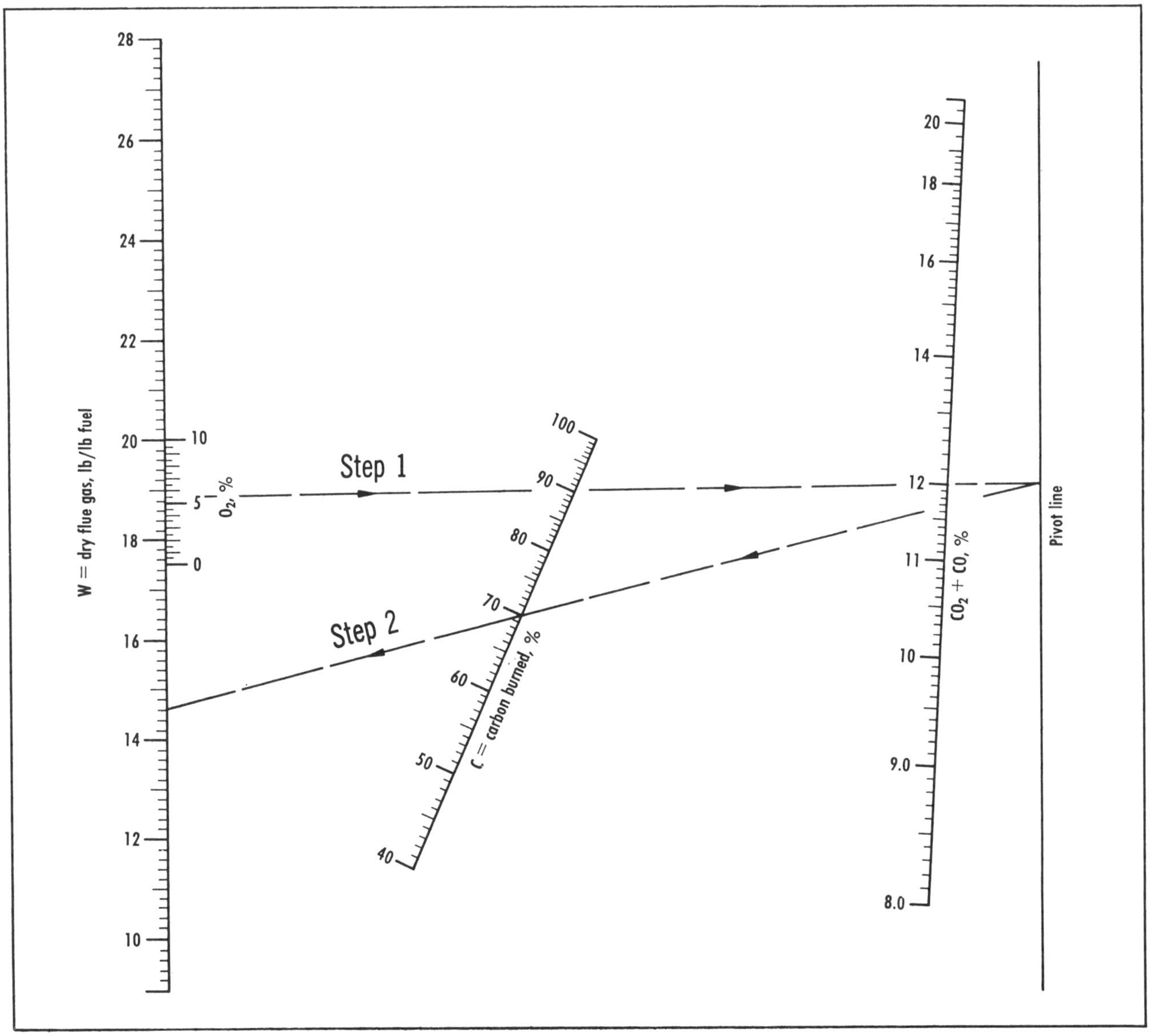

Nomogram determines flue-gas density

Bill Sisson
Superintendent of Operations
Nipak Inc.
Pryor, Okla.

THE DENSITY of a gas mixture is calculated from the molecular weight of the mixture, which is calculated from the composition by volume of the mixture.

The volumes occupied by 1 lb-mol of every ideal gas are identical for any stated conditions of temperature and pressure. For standard conditions of a gas, defined as 32° F. and 1 atmosphere of pressure (29.92 in. Hg), the molar volume of the gas is 359 cu ft. At any other temperature the molar volume, based upon the ideal gas laws, is:

$$V_m = 359\,[(t + 460)/492]\,(29.92/p)$$

Where V_m = molar volume, cu ft; t = temperature, °F.; and p = absolute pressure, in. Hg.

Most combustion gases deviate slightly from the ideal gas laws, but, for engineering calculations, no serious errors are caused by assuming them to be ideal.

The nomogram shown here is based on the formula:

$$D = (44\ CO_2 + 32\ O_2 + 28.16\ N_2)/V_m$$

Where D = flue gas density, lb/cu ft; and CO_2, O_2, and N_2 = volume percentages (as decimals) in flue gas.

The nomogram is based on 29.92 in. Hg pressure. To convert to other pressures, divide D by 29.92/p.

Example. If a flue gas contains 15% CO_2, 6% O_2, and 79% N_2, and the mixture is at 500° F. and 29.92 in. Hg, what is the density of the mixture?

Solution. Connect 15% on CO_2 scale with 79% on N_2 scale, and mark where line crosses Pivot line No. 1 (Step 1). From this point connect with 6% on O_2 scale, extend to Pivot line No. 2, and mark (Step 2). From point found on Pivot line No. 2, line with 500° F. on GT (Temperature) scale, extend to D (Density) scale, and read 0.0439 lb/cu ft (Step 3).

If you desire to determine the density of the flue gas at the mean temperature in the stack, the mean temperature of the gas may be estimated by allowing a drop in temperature of 1° F. for each foot of stack height. END

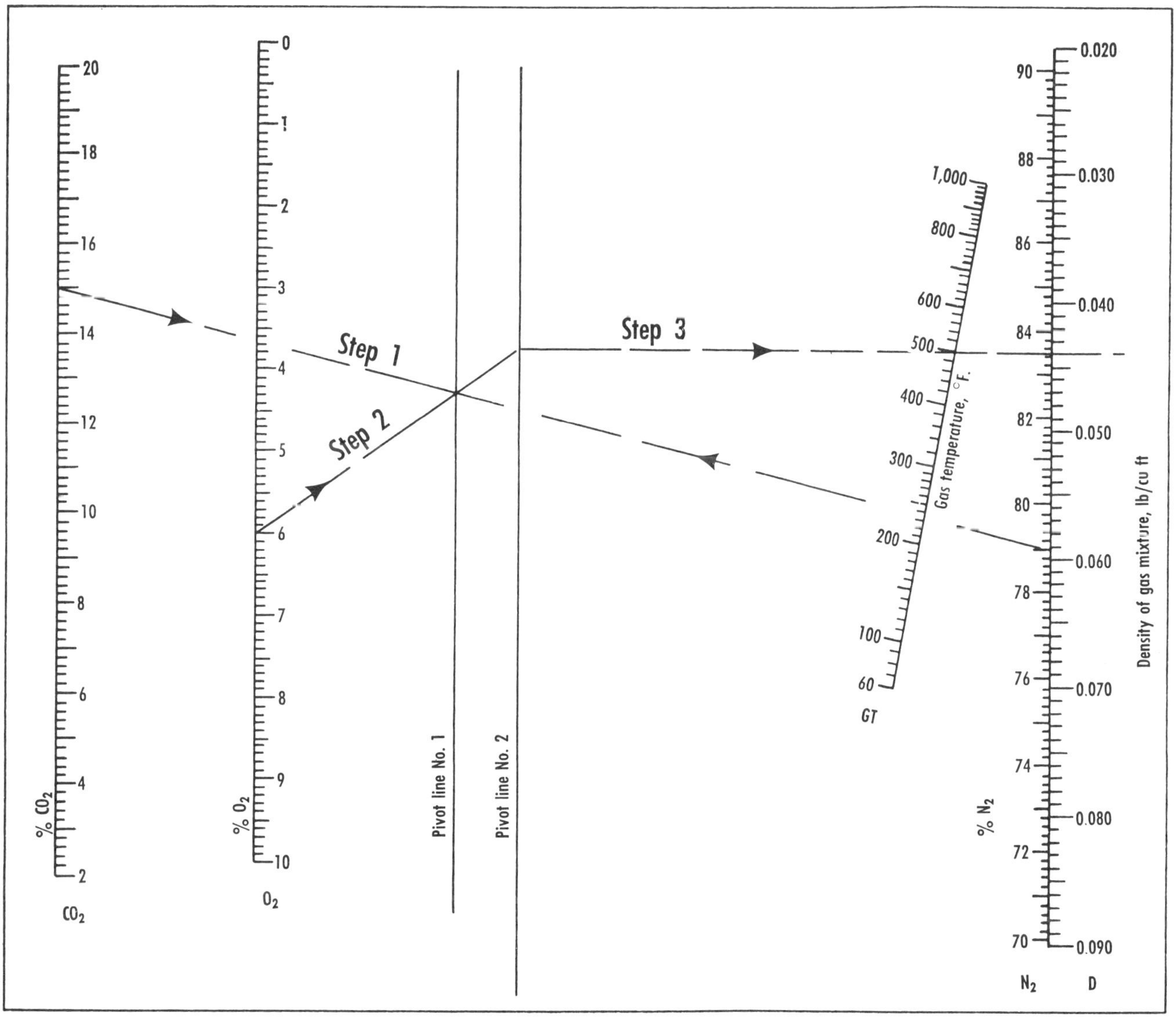

Nomogram helps find excess air

Bill Sisson
Superintendent of Operations
Nipak, Inc.
Pryor, Okla.

WHEN you operate boilers or furnaces with excess air, you are losing dividends that could be realized by establishing a definite schedule for checking boiler or furnace efficiency. Normal ranges of the amount of excess air for efficiently operated plants range from 10 to 25% for gas burners to 20 to 30% for oil, and 30 to 40% for coal.

The excess amount of air can be determined by Orsat flue-gas analysis and the nomogram presented here. In the Orsat analyzer, the flue gas is analyzed by measuring the volume decrease of the sample when the flue gas components, O_2, CO_2, and CO are absorbed, one by one, in various solutions. The nitrogen percentage is determined by finding the difference in volume resulting from removal of the other components. As the moisture condenses out, the analysis is on a dry percentage basis. The nomogram presented here is based on this equation:

% Excess air $= 100\ (O_2 - 0.5CO)/[0.264N_2 - (O_2 - 0.5CO)]$ where O_2, CO, and N_2 are volumetric percentages.

Example. How much excess air is present in flue gas if an Orsat analysis shows that CO is 1.5%, O_2 is 9.2%, and N_2 is 78%.

Step 1, from 1.5 on the CO scale, align with 9.2 on the O_2 scale, and extend the line to the reference line and mark the intersection. Step 2, from the marked point, align with 78 on the N_2 scale, and read 69.6% where the line crosses the excess-air scale. END

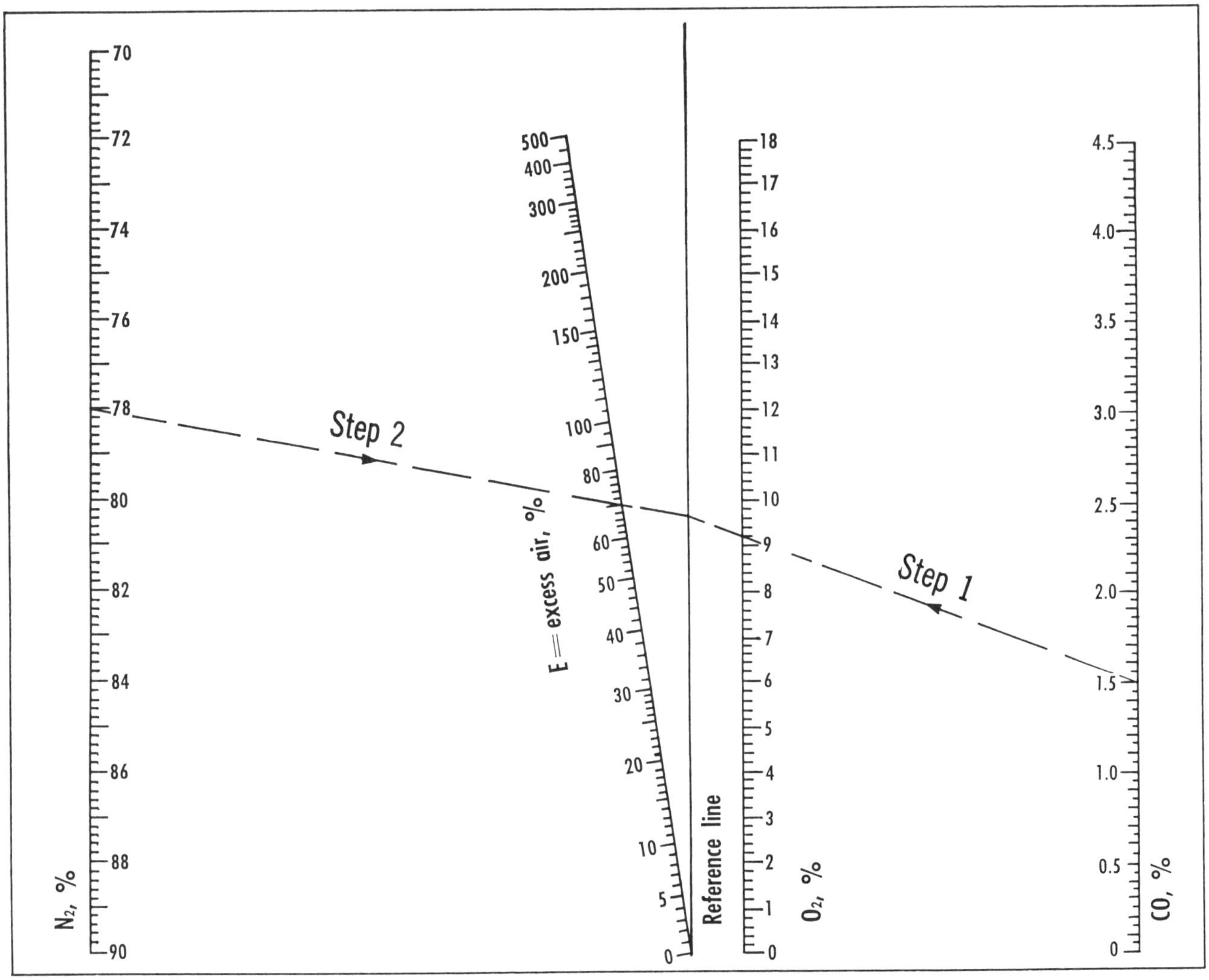

Determine excess air from orsat analysis

F. Caplan
Kaiser Engineers

THIS nomogram is designed to help determine the percent excess air from an orsat analysis of exhaust gas. It is based on the equation:

$$E = 100\,(O_2 - CO/2)/[0.264\,N_2 - (O_2 - CO/2)]$$

Where:

E = % excess air

O_2 = volumetric percentage of oxygen

CO = volumetric percentage of carbon monoxide

CO_2 = volumetric percentage of carbon dioxide

N_2 = volumetric percentage of nitrogen (obtained by difference: 100 − sum of $O_2 + CO + CO_2$)

Example: What is the percent excess air from an orsat analysis if volumetric percentage of carbon monoxide = 2.0%, oxygen = 8.0%, and nitrogen = 79.0%?

To solve this problem, align Step (1) with a straight-edge 2.0 on the carbon monoxide, CO, scale with 8.0 on the oxygen, O_2, scale and mark the pivot line at the intersection with the straight edge. Next align 79.0 on the nitrogen, N_2, scale with the mark on the pivot line, Step (2), and read 50.5% excess air, E, on the excess-air scale.

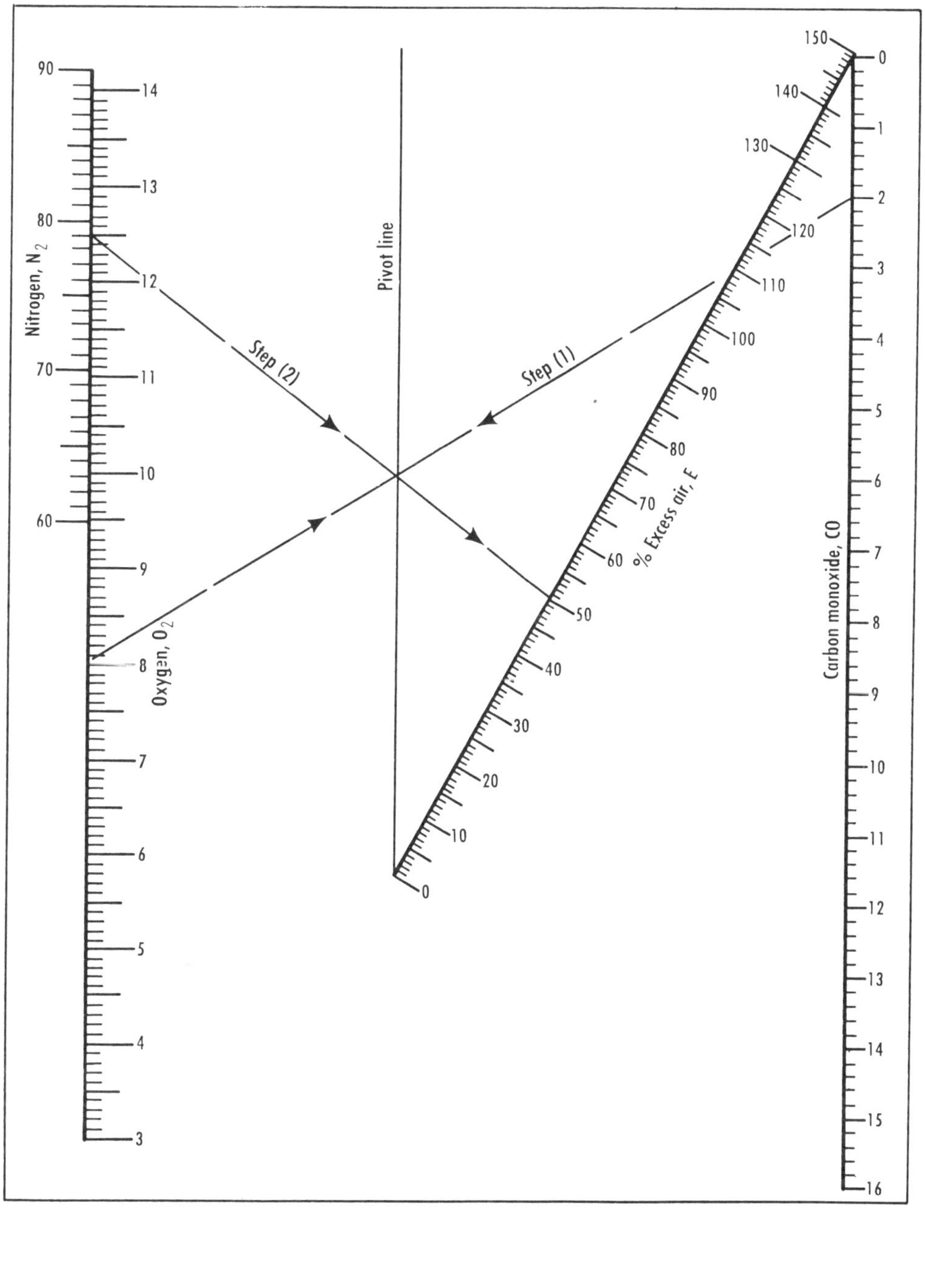

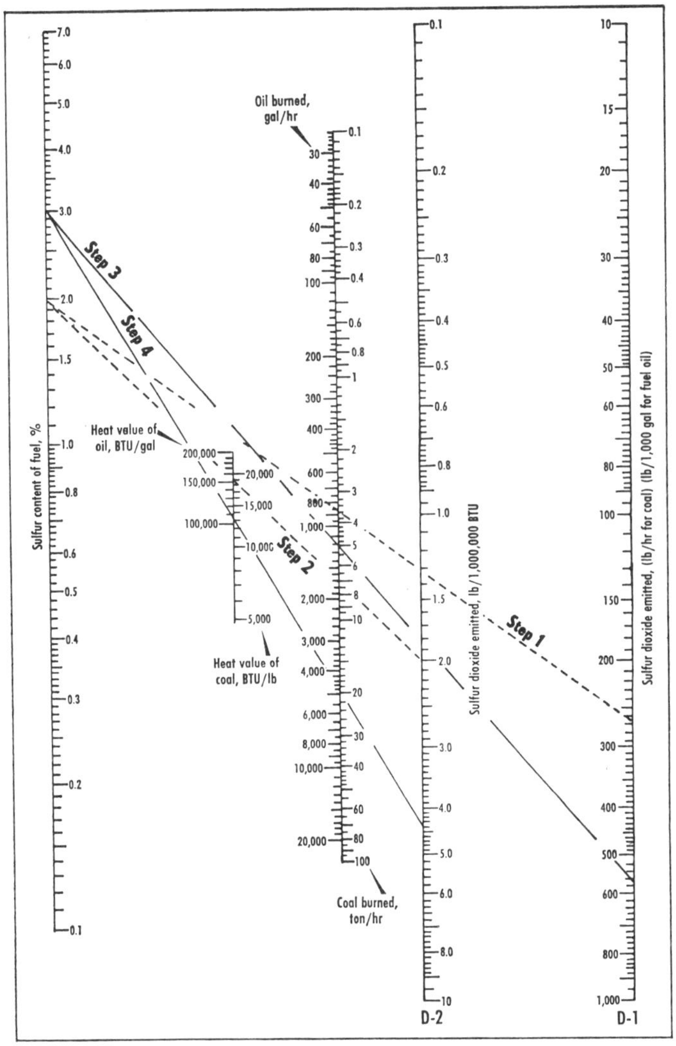

Nomogram estimates sulfur dioxide emission

BILL SISSON
Superintendent of Operations
Nipak Inc.
Pryor, Okla.

REGULATIONS controlling the emission of sulfur dioxide from combustion processes are being put into effect in many states and municipalities. The limits of SO_2 emission are expressed in several ways. These include pounds of SO_2 emitted per hour, pounds emitted per million BTU of heat input, and SO_2 concentration in parts per million.

Where instruments are not available for measuring actual concentration of SO_2 in the flue gas, the average concentration can be estimated if the sulfur content of the fuel and fuel rate are known. Where coal is used as fuel, about 95% of its sulfur content appears as a sulfur oxide in the flue gases. When fuel oil is used, 98% of the sulfur appears as sulfur oxide in flue gases.

A method for estimating the SO_2 emitted in pounds per hour for coal or fuel oil is provided by the nomogram presented here. Pounds of SO_2 emitted per million BTU can also be found if the heating value of the fuel is known.

Example 1. Assume 850 gal/hr of fuel oil, with a heat value of 154,000 BTU per gal, and a sulfur content of 2%, is burned. How many pounds of SO_2 are emitted per 1,000 gal of fuel oil burned? How many pounds are emitted per million BTU?

Solution: Connect 2% on Sulfur-content-of-fuel scale (Step 1, dashed line) with 850 gal/hr on Oil-burned scale, extend to D-1 scale, and read 267 lb of SO_2 emitted per 1,000 gal of fuel oil burned.

Connect 2% on Sulfur-content-of-fuel scale (Step 2, dashed line) with 154,000 BTU/gal on Heat-value-of-oil scale, extend to D-2 scale, and read 2 lb of SO_2 emitted per million BTU.

Example 2. Assume 5 ton per hour of 13,000 BTU/lb coal that contains 3% sulfur is burned. How many pounds of SO_2 are emitted per hour? How many pounds are emitted for each million BTU burned?

Solution. Connect 3% on Sulfur-content-of-fuel scale (Step 3, solid line) with 5 tons per hr on Coal-burned scale, extend to D-1 scale, and read 570 lb/hr of SO_2 emitted.

Connect 3% on Sulfur-content-of-fuel scale (Step 4, solid line) with 13,000 BTU/lb on Heat-value-of-coal scale, extend to D-2 scale, and read 4.4 lb SO_2 per million BTU.

END

Nomogram makes NO_X conversions

Bill Sisson
Superintendent of Operations
Nipak Inc.
Pryor, Okla.

NITROGEN oxides are present in the effluent gases from all combustion sources and in the waste gases generated from the production of nitric acid, nitration processes, etc. Since oxides of nitrogen play an important role in the formation of smog, they are important air pollutants. The most important oxide pollutants are nitric oxide (NO) and nitrogen dioxide (NO_2). These two gases are sometimes measured individually. But, in many cases, a measurement of total nitrogen oxides (NO_X) is sufficient.

Emission limits for nitrogen oxides are specified in many states as NO_X in lb/million BTU. Since sampling instruments are not calibrated to read in lb/million BTU, a conversion must be made. This nomogram provides a convenient means for converting NO_X in parts per million to the desired unit for average fuels.

Example. If a wet gas sample (10% moisture) from a boiler burning fuel oil with 18% excess air contains 400 ppm NO_X, what is the NO_X in lb per million BTU?

Solution. Step 1, connect the Oil reference point with 18% on Excess-air scale, extend to Pivot Line, and mark. Step 2, connect 400 ppm on NO_X, ppm, wet scale with 10% on moisture, %-by-weight scale, extend to NO_X, ppm, dry scale, and mark. Step 3, connect the two marked points and read 0.63 where line crosses the NO_X, lb/million BTU scale.

END

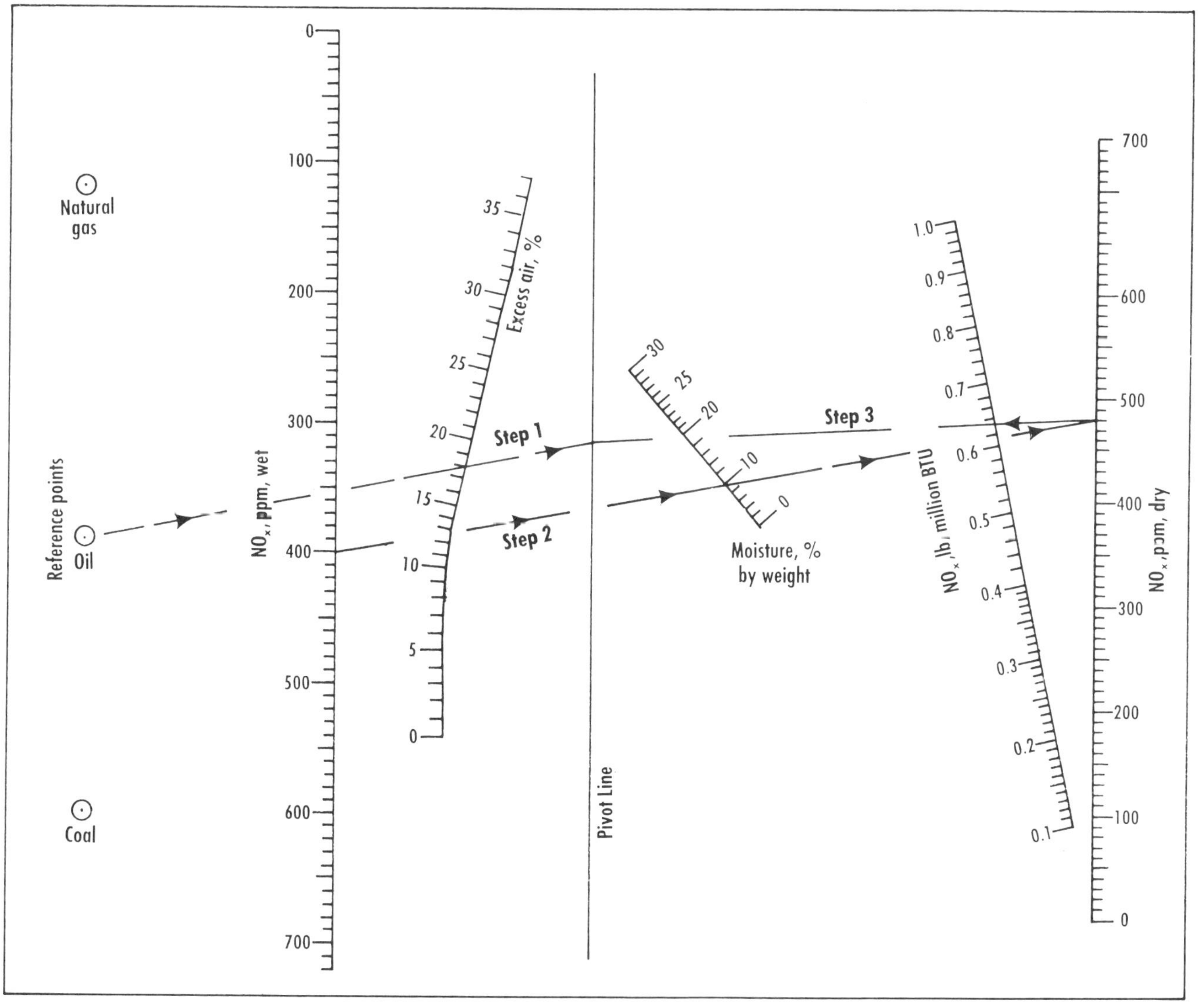

Nomogram estimates maximum ground-level concentrations of stack gases

Bill Sisson
Superintendent of Operations
Nipak Inc.
Pryor, Okla.

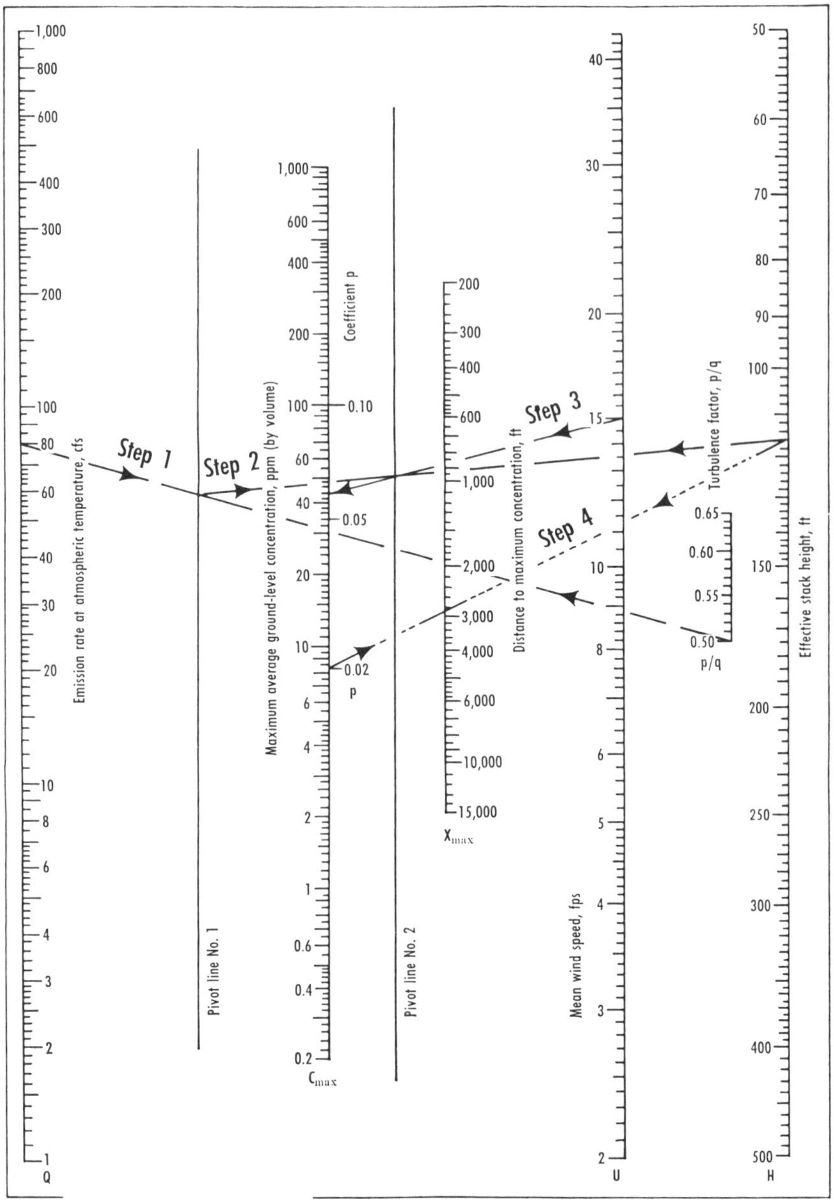

FOR a given situation, the severity of air pollution is determined largely by the weather. Although the total discharge of contaminants to the atmosphere in a given location remains constant, the degree of air pollution will vary widely. The Bosanquet and Pearson equation for the maximum ground-level gas concentration that will occur directly down wind can be stated as:

$$C_{max} = 2.15 \frac{Q\ 10^5}{U\ H^2} \left(\frac{p}{q}\right)$$

at a distance, $X_{max} = H/2p$.

Where:

- Q = emission rate for gas at atmospheric temperature, cfs
- U = mean wind speed, fps
- H = effective stack height, ft
- p = diffusion coefficient, dimensionless
- q = diffusion coefficient, dimensionless
- C_{max} = maximum average ground-level concentration for gas, ppm
- X_{max} = distance from stack to maximum ground-level concentration, ft

In calculations involving particulate matter, Q (cfs) may be replaced by M (kg/sec) so that calculation yields C (mg/cu ft) instead of C (ppm by volume).

The height H is often assumed to be equal to the stack height. Actually, it is the height at which the smoke levels off, a level which may be much higher than the stack height because, if stack gases are at an elevated temperature or have appreciable exit velocity, the effective height is greater than the physical stack height due to buoyancy and jetting action.

The values of p and q commonly used with the Bosanquet and Pearson equation are shown in Table 1.

The nomogram shown here, based on the above formulas, provides an easy method for estimating the maximum ground level concentration and distance from stack.

Example. If stack emission rate is 80 cfs, effective stack height is 116 ft, and mean wind speed is 15 fps (10.23 mph), what is the maximum average ground level concentration for little air turbulence, and what is the maximum distance from stack to point of maximum ground-level concentration?

Solution. Check Table 1 and note that the turbulence factor for little turbulence is $p/q = 0.5$, and $p = 0.02$.

Line from 80 cfs on Q scale to 0.50 on p/q scale and mark where line crosses Pivot line No. 1 (Step 1). From marked point line with 116 ft on H scale and mark where line crosses Pivot line No. 2 (Step 2).

Connect this point with 15 fps on U scale, extend to C_{max} scale, and read 42.6 ppm maximum average ground-level concentration (Step 3).

Table 1 — Bosanquet & Pearson Coefficients

Turbulence	p	q	p/q
Low	0.02	0.04	0.50
Average	0.05	0.08	0.63
Moderate	0.10	0.16	0.63

Connect 116 ft on H scale with 0.02 on p scale, and read 2,900 ft to point of maximum ground-level concentration where line crosses X_{max} scale (Step 4).

If we assume that gas discharging from a stack contains 0.5% by volume of SO_2, the maximum average ground-level concentration of SO_2 would be:

$42.6 \times 0.005 = 0.213$ ppm END

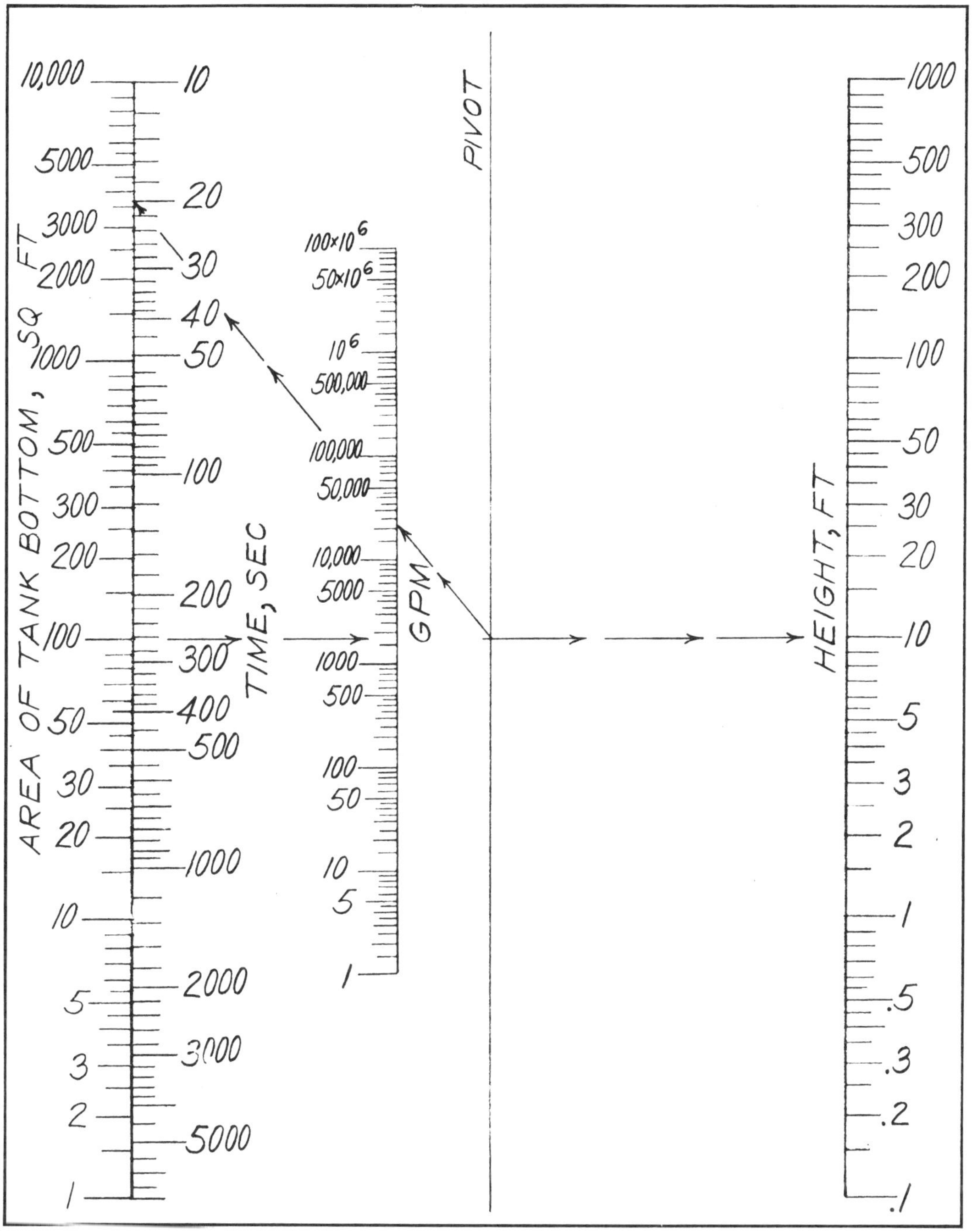

How to find pump capacity

IN TESTING A PUMP to determine its true capacity, engineers often use a tank on either the suction or discharge side of the pump. Water or other liquid is pumped out of or into the tank for a given period, using the pump being tested. This chart allows easy determination of pump capacity by this type of test when the area of the tank bottom, fall or rise of the liquid, and duration of the test are known.

Example: What is the capacity of a pump discharging to a tank with a bottom area of 100 sq. ft. if the liquid rises 10 ft. in 20 seconds?

Solution: Enter the chart on the left at the bottom area of 100 sq. ft. and draw a straight line through a liquid height rise of 10 ft. on the right. From the intersection of this line with the pivot, draw a straight line through the test time, 20 seconds, on the left. At the intersection with the capacity scale read the pump capacity as 22,400 g.p.m.

Note that this chart can be used for any type of pump—centrifugal, reciprocating, or rotary. It can also be used for any kind of tank—square, rectangular, circular, etc., provided the walls of the tank are vertical and the bottom horizontal. The height scale can be used for either a rise or fall in the liquid level, depending on the tank arrangement.

This chart is based on $Q = 448.8\ AH/T$, where Q = pump capacity, g.p.m.; A = area of tank bottom, sq. ft.; H = liquid rise or fall, ft.; T = duration of test, seconds.

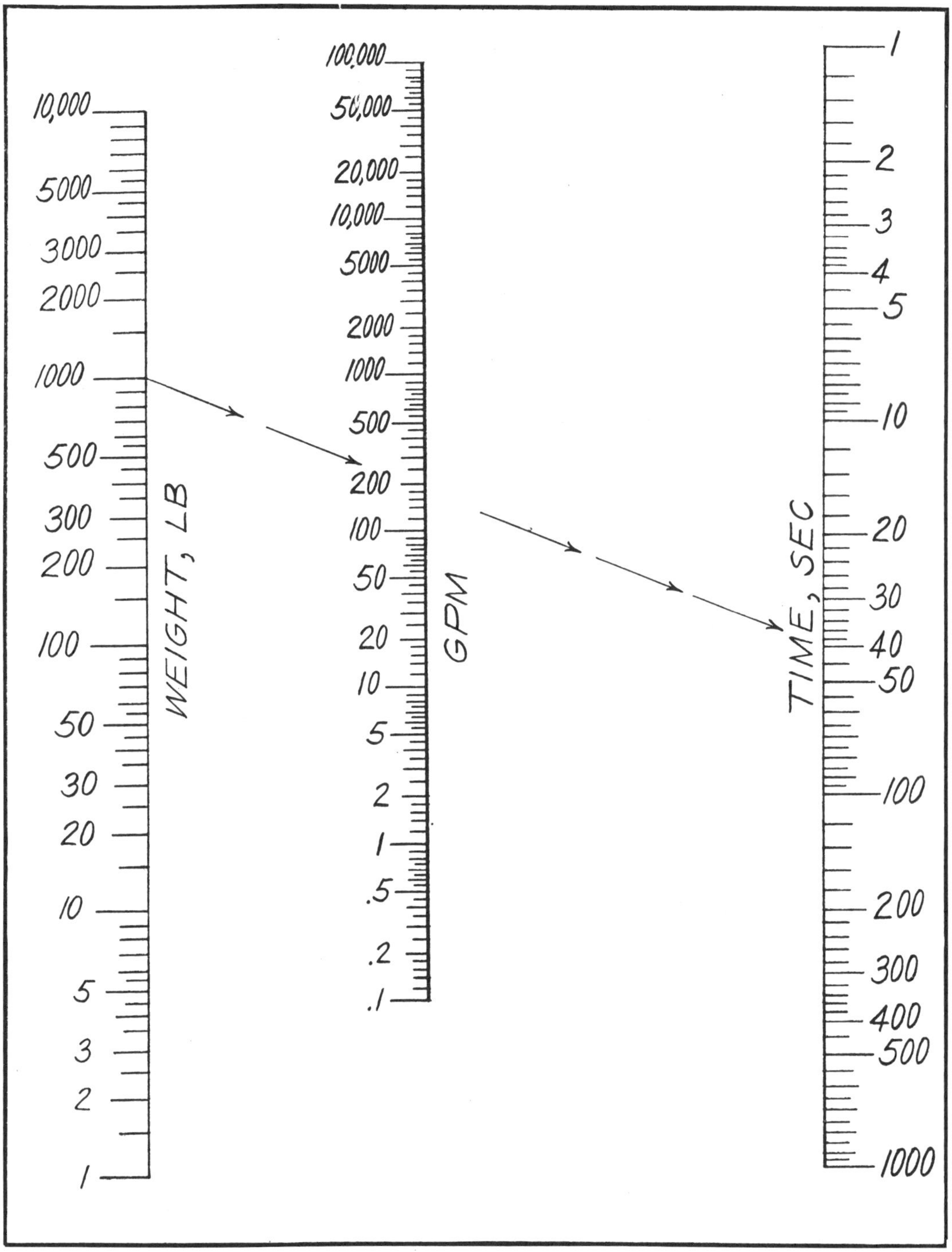

Pump capacity by weight test

TO CHECK the capacity of any kind of pump—centrifugal, reciprocating, or rotary—design engineers often weigh the liquid discharged during a given period. This is an important test in determining if a pump design meets its guarantee. The accompanying chart makes it easy to compute pump capacity from weight test results.

Example: During a weight test, a pump discharges 1,000 lb. of liquid in 40 seconds. What is the capacity of the pump?

Solution: Enter the chart on the left at a weight of 1,000 lb. and draw a straight line through 40 seconds on the right. At the intersection with the central scale read the pump capacity as 180 g.p.m.

Note that this chart can be used for any pump handling any liquid. This chart is based on $Q = 7.2\ W/T$, where Q = pump capacity, gal. per minute = g.p.m.; W = weight of liquid, lb., discharged during time T; T = time of test, seconds. **End.**

Minimum pump flow, orifice size

F. Caplan
Kaiser Engineers

WHEN a pump operates near its shutoff point (low flow), overheating and pump damage can result, unless a minimum flow is maintained through the pump. It is common practice to maintain this minimum flow by means of an orifice in a line off the pump discharge line.

This nomogram permits a rapid approximation of this minimum flow and orifice size from the known normal operating (rated) flow and head. It is based on these equations:

$$Q_m = 0.000061\ Q_d\ H_d$$

$$d_o = Q_m/3.55\ H_d^{1/4}$$

Where:
Q_m = minimum required flow, gpm
Q_d = pump flow at normal operating (design) point, gpm
H_d = pump head at normal operating (design) point, ft
d_o = diameter of sharp-edged orifice, in.

Example. Determine the approximate minimum flow and orifice size for a pump with a rated flow of 1,000 gpm and a head of 164 ft. Using a straightedge, align Q_d = 1,000 with H_d = 164. Then read minimum flow, Q_m = 10 gpm and orifice diameter, d_o = 0.250 in. in.

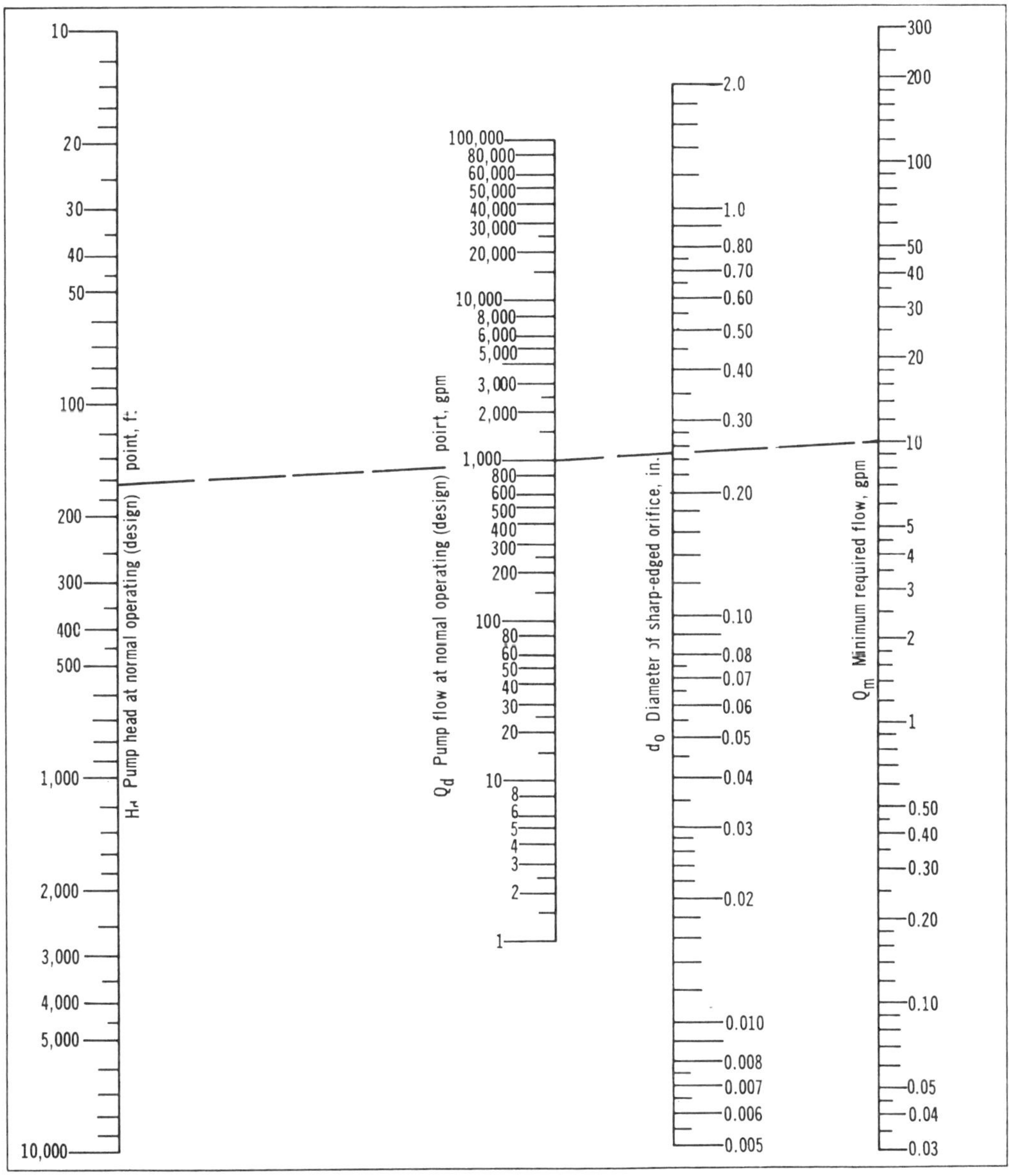

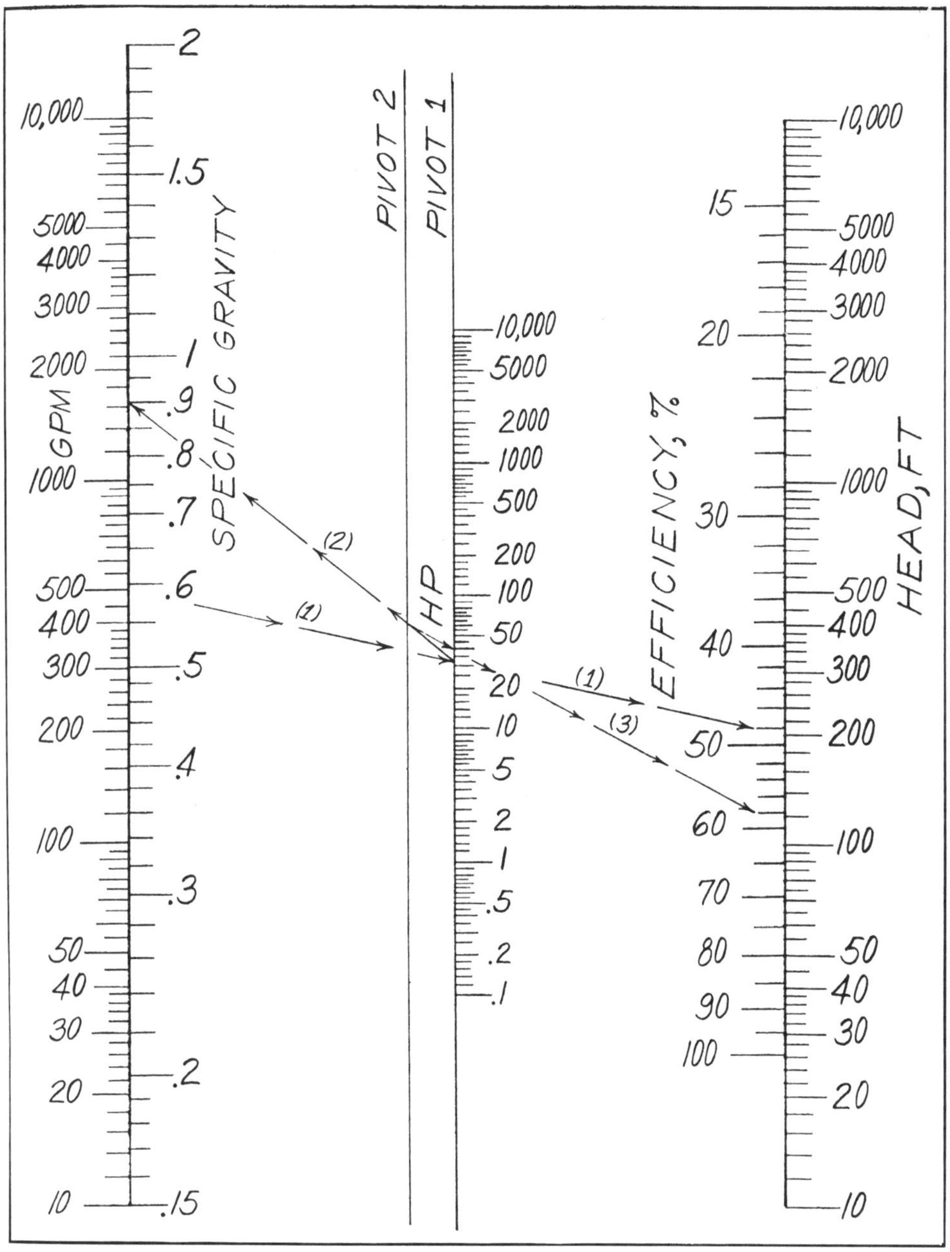

Find power needed for pumping

THIS CHART makes it easy to quickly compute the power required to pump any liquid met in engineering practice today. The chart takes into consideration all factors affecting the power input to a pump—liquid flow rate, specific gravity, pumping head, and efficiency.

Example: How much power is required to drive a pump handling 500 g.p.m. of 0.9-specific-gravity oil against a total head of 200 ft., if the pump has an efficiency of 60%?

Solution: Enter the chart on the left at the capacity of 500 g.p.m. and draw a straight line (1) through the total head of 200 ft. on the right. From the intersection of this line with pivot 1 draw a straight line (2) through the specific gravity of the liquid, 0.9, on the left. Then from the intersection of this line with pivot 2 draw a straight line through the pump efficiency, 60%, on the right. At the intersection with the power scale read the required input as 38 hp.

This chart can be used for any kind of pump — centrifugal, reciprocating, or rotary. Note that pivot 1 corresponds with the horsepower scale.

The chart is based on $H = GHS/3960E$, where H = horsepower input; G = gallons per minute of liquid actually handled by the pump; H = total head against which the pump operates, ft. of liquid handled; S = specific gravity of liquid handled; E = pump efficiency, expressed as a decimal.

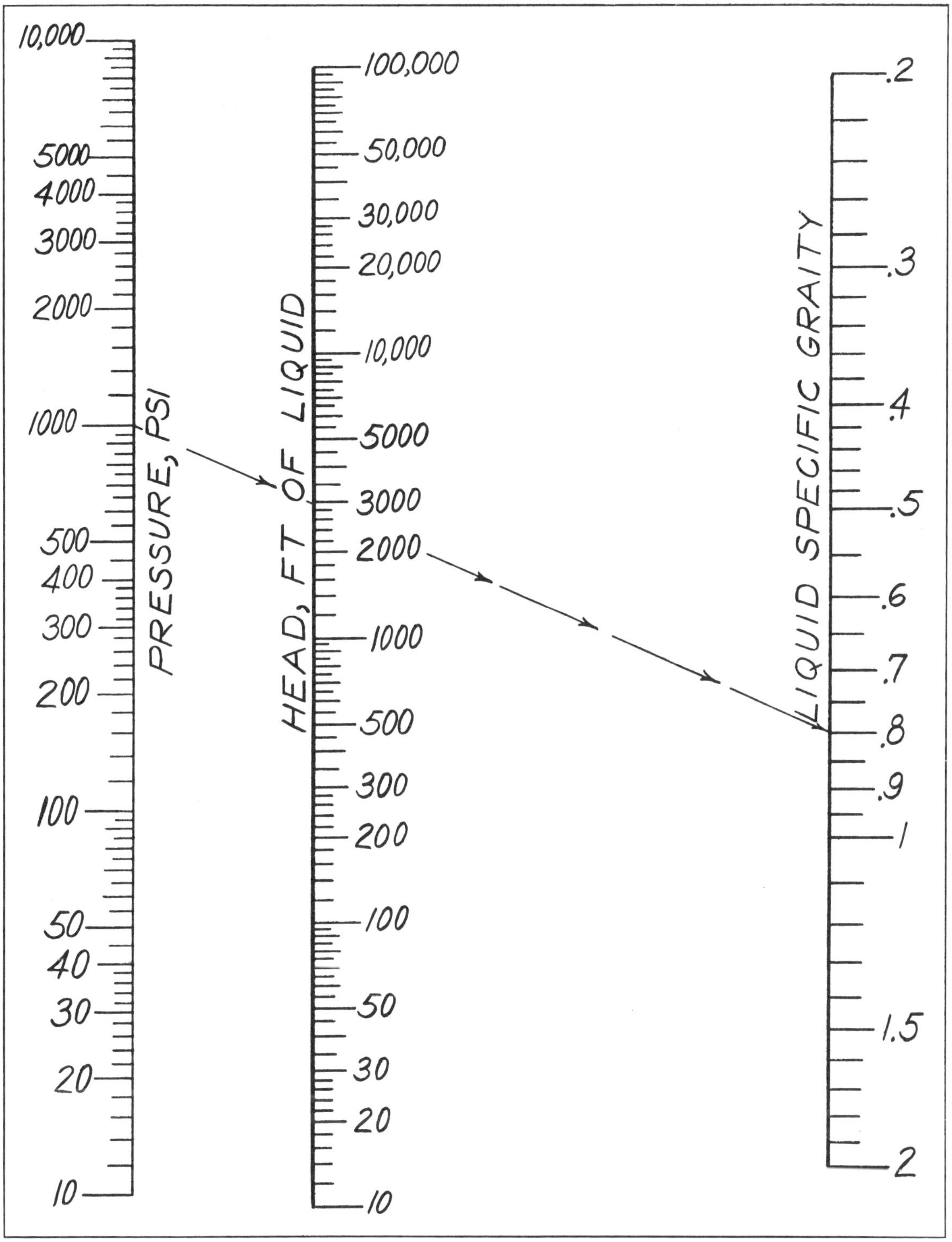

Pressure conversions for pumping

IN ALL pumping calculations, you will frequently find it necessary to convert from pressure in pounds per square inch to head in feet of liquid, or vice versa. The accompanying chart takes the tediousness out of this calculation, particularly where the specific gravity is not 1.0.

Example: What is the head in feet of 0.8 specific gravity oil when its pressure is 1,000 psi.?

Solution: Enter the chart on the left at the pressure of 1,000 psi. and draw a straight line through the specific gravity of 0.8 on the right. At the intersection with the central scale read the head as 2,900 ft. of liquid.

This chart is based on head, ft. of liquid = 2.31 p/s, where p = pressure, pounds per square inch; s = specific gravity of liquid.

Determine head of centrifugal pumps

Bill Sisson
Superintendent of Operations
Nipak Inc.
Pryor, Okla.

A CENTRIFUGAL pump imparts velocity to a liquid. This velocity energy is then transformed largely into pressure energy as the liquid leaves the pump. Thus, the head developed approximately equals the velocity energy at the periphery of the impeller. This relationship is expressed by the formula, $H = V^2/2g$, where H = total head developed, ft; V = peripheral velocity of impeller, fps; and $g = 32.16$ ft/sec^2 (acceleration of gravity).

The approximate head of any centrifugal pump can be predicted by calculating the peripheral velocity of the impeller and substituting into the above formula. A convenient formula for peripheral velocity is $V = ND/229$, where N = speed, rpm, and D = impeller diameter, in. Always think in terms of feet of liquid rather than pressure when working with centrifugal pumps as a given pump with a given impeller diameter and speed will raise a liquid to a certain height regardless of the weight of the liquid. The nomogram, based on the formulas above, will predict the approximate head of a centrifugal pump as shown by this example.

Example. If the impeller diameter of a centrifugal pump is 8 in. and the speed is 3,550 rpm what is the peripheral velocity of the impeller and the total head developed?

Solution. Connect 3,550 rpm on N scale with 8 in. on D scale, and read peripheral velocity as 124 fps and total head as 239 ft where line intersects V and H scales. END

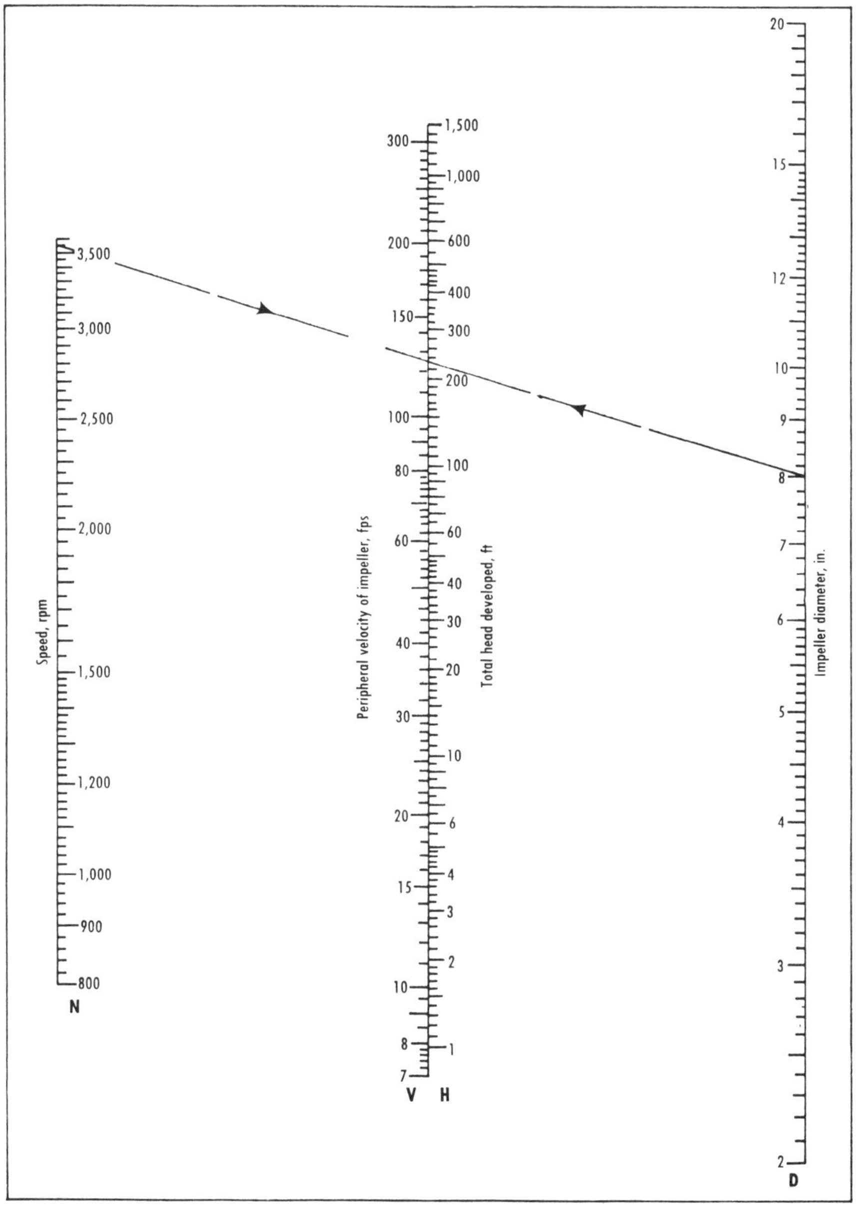

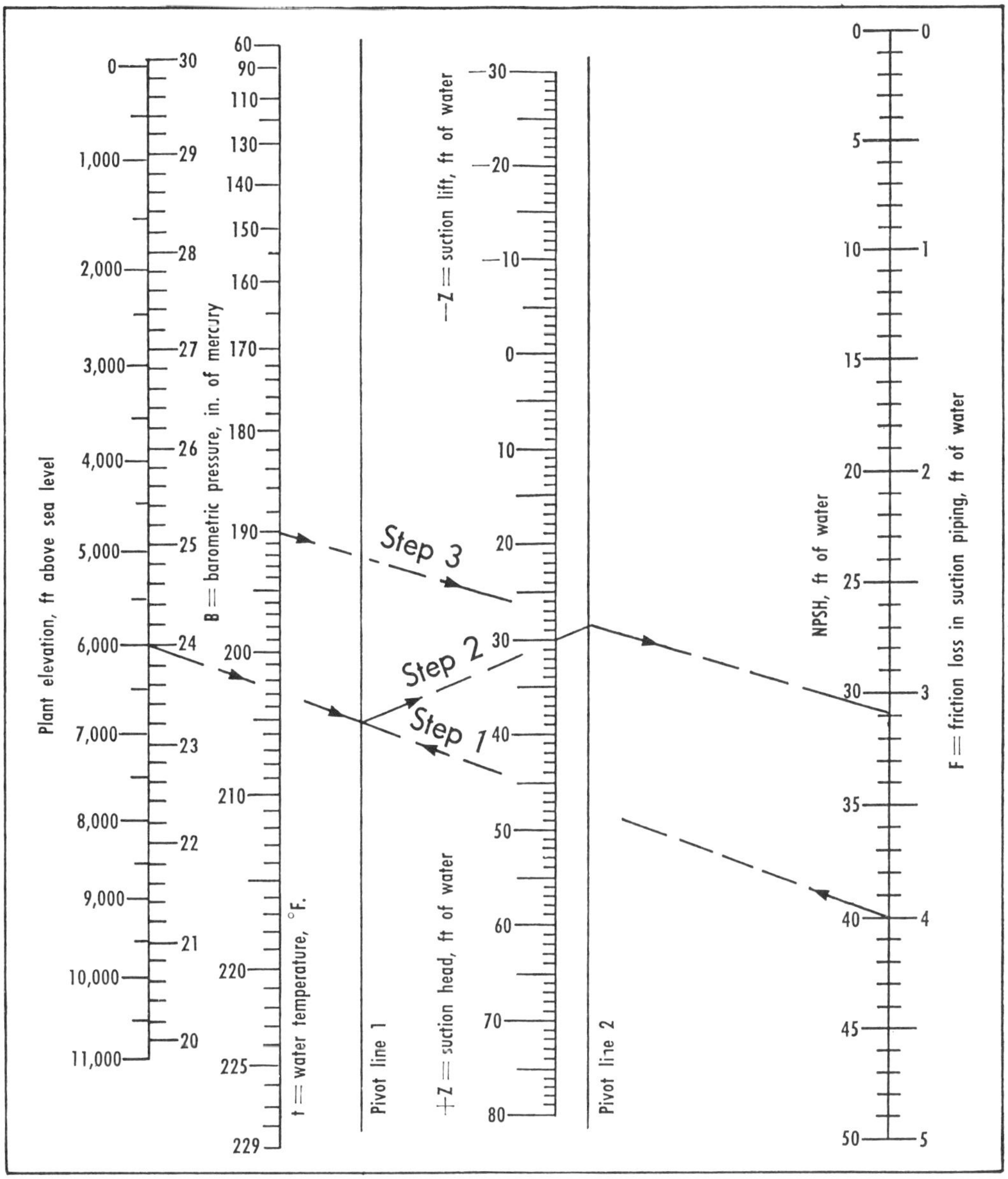

F. Caplan
Kaiser Engineers

Finding available pump NPSH for hot water

To PREVENT cavitation, the available net positive suction head (NPSH) at pump inlet must exceed the required NPSH of the pump. This nomogram permits a quick determination of the pump available NPSH at the inlet for hot water at atmospheric pressure (for water under pressure, convert the gauge pressure to feet of water and add it to the chart value).

The nomogram is based on the equation:

$$\text{Available NPSH} = A \pm Z - V - F$$

Where:

A = atmospheric pressure, ft of water

Z = suction head (positive) or suction lift (negative), ft of water

V = vapor pressure of water at temperature t, ft of water

F = friction loss in suction piping, ft of water

Note that all of these variables should be expressed in ft of water at temperature t.

Example. What is the available NPSH of a pump at 6,000-ft elevation handling water at 190° F. with 30-ft suction head, if the suction piping loss is 4 ft. Using a straight edge, align elevation = 6,000 ft with friction loss F = 4 ft and mark the intersection with Pivot Line 1 (Step 1). Next align this mark with suction head Z = 30 ft and mark the intersection with Pivot Line 2 (Step 2). Then align this mark with temperature t = 190° F. and read NPSH = 30.9 ft of water (Step 3).

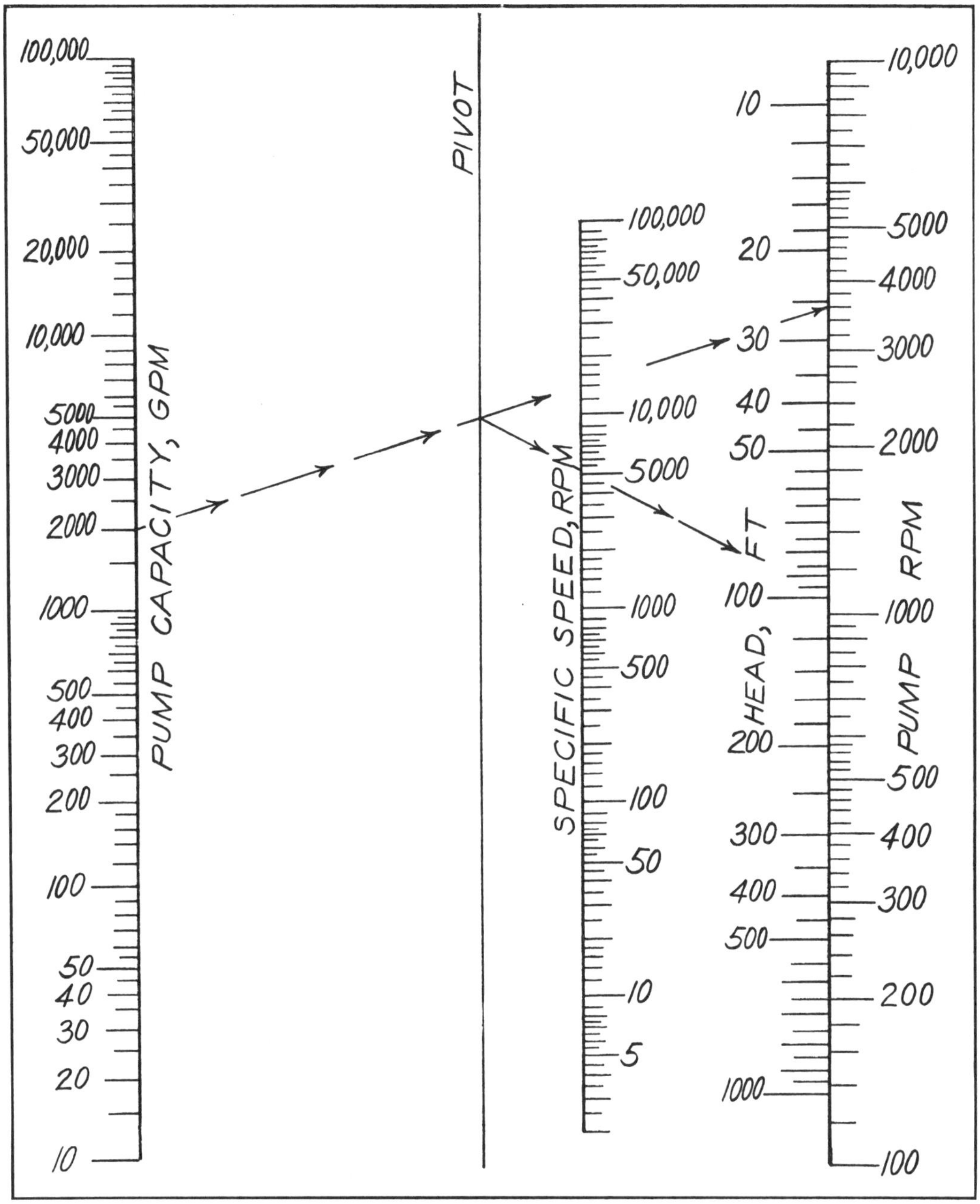

Find specific speed of centrifugal pumps

ONE OF THE MOST IMPORTANT measures of centrifugal-pump performance is the specific speed. This quantity is widely used in the selection and application of all types of centrifugal pumps. But since computation of the specific speed of a pump involves taking the square root of the flow rate, and raising the head to the 0.75 power, the calculation can be lengthy, particularly when a number of pumps are involved. This chart simplifies all specific-speed calculations.

Example: What is the specific speed of centrifugal pump handling 2,000 g.p.m. at 3 600 r.p.m. if the head is 100 ft. of liquid?

Solution: Enter the chart on the left at 2,000 g.p.m. and draw a straight line through 3,600 r.p.m. on the right. From the intersection of this line with the pivot, draw a straight line through the head, 100 ft., on the right. At the intersection with the specific-speed scale, read the result as 5,040 r.p.m.

This chart is based on $N = G^{1/2} r/H^{0.75}$, where N = specific speed, revolutions per minute; G = pump capacity, gallons per minute; r = pump speed, revolutions per minute; H = head developed by pump, feet of liquid. **End.**

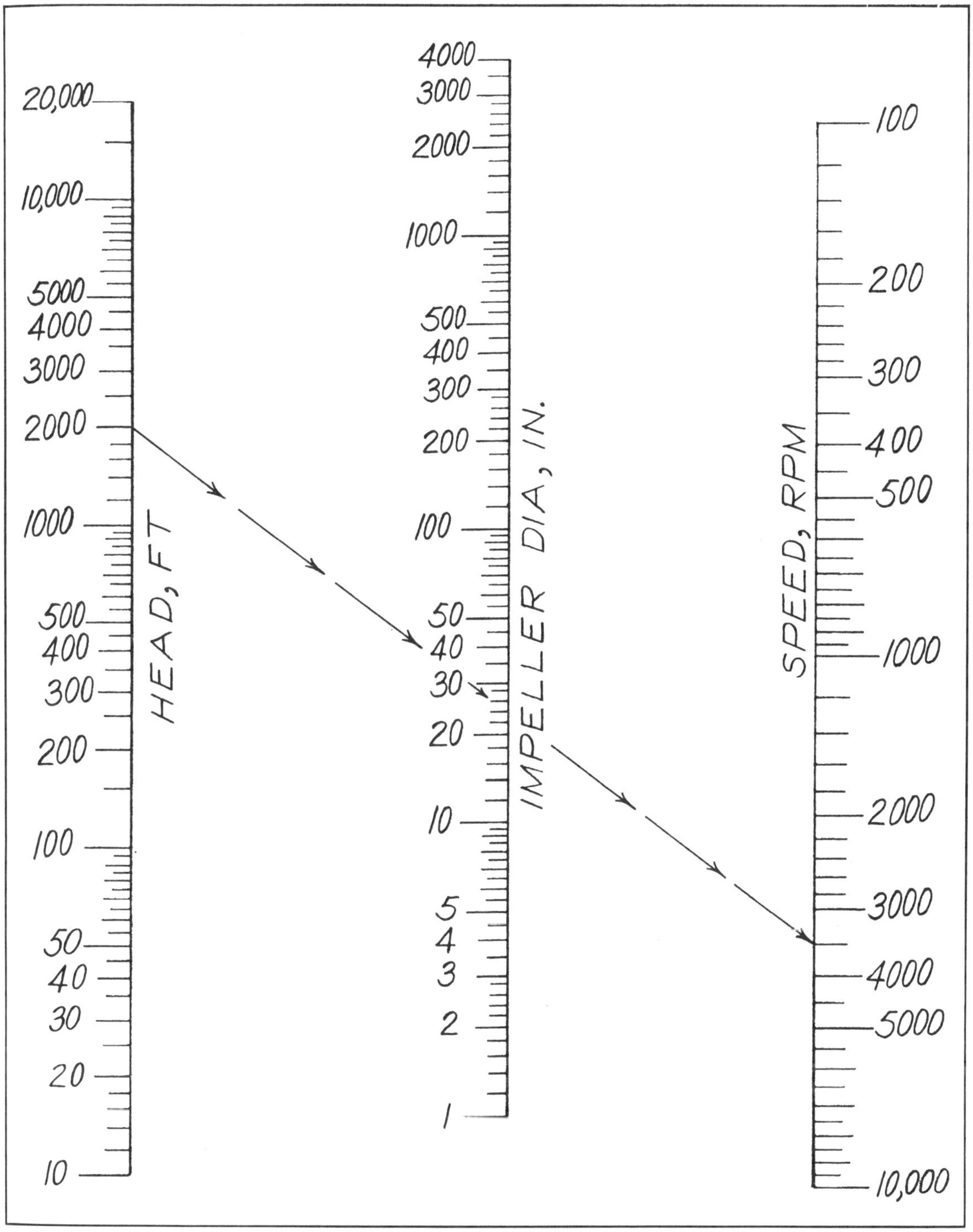

Impeller diameter for centrifugal pumps

HERE IS A QUICK WAY to determine the diameter of an impeller for a centrifugal pump to develop a given head at a specified speed. It is valid for all types of centrifugal pumps if the developed head at zero capacity is used when entering the chart.

Example: What diameter of impeller can be used in a centrifugal pump that is to develop a shutoff head of 2,000 ft. when the impeller speed is 3,500 r.p.m.?

Solution: Enter the chart on the left at the shutoff head of 2,000 ft. and draw a straight line through the impeller speed of 3,500 r.p.m. on the right. At the intersection with the central scale read the impeller diameter as 23.5 in.

This chart is based on $D = 1{,}840K\,(H)^{1/2}/\text{r.p.m.}$, where D = impeller diameter, in.; K = a constant = 1.00 when H is the shutoff head; H = shutoff head, ft. of liquid; r.p.m. = impeller speed, revolutions per minute.

Nomogram gives temperature rise in centrifugal pumps

Bill Sisson
Superintendent of Operations
Nipak Inc.
Pryor, Okla.

WHEN a centrifugal pump is operated at an extremely reduced flow, a thermodynamic problem is caused by the heating up of the liquid handled by the pump. The difference between the brake horsepower consumed and the water horsepower developed represents the power losses within the pump itself. These power losses are converted into heat and transferred to the liquid flowing through the pump.

If liquid is flowing through the pump, conditions become stabilized, and the amount by which the temperature at the discharge exceeds that at the suction can be calculated for any given flow. This rise is determined by multiplying the difference between the brake horsepower and the water horsepower by 2,545 (BTU equivalent to 1 hp-hr) and thus determining the total heat imparted to the liquid in BTU. Assuming that the liquid is water (specific heat = 1 BTU/lb/°F.) this total heat is divided by the flow in pounds per hour to yield the temperature rise in °F. A more readily used formula, based on total head and pump efficiency is this:

$$T_R = (H/778[(1/E)-1]$$

Where T_R = temperature rise, °F./min; H = Total head, ft; E = pump efficiency at capacity involved (expressed as a decimal); and 778 = BTU equivalent to 1 hp.

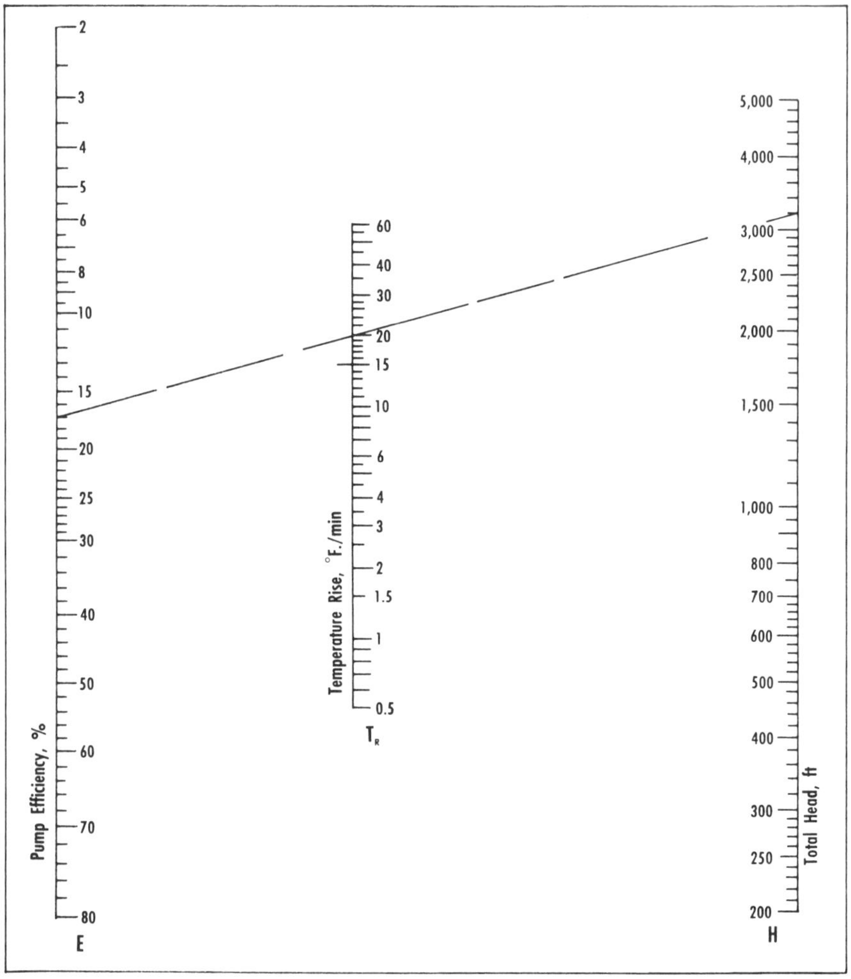

Since temperature rise varies with pump capacity, the minimum permissible capacity from the thermodynamic standpoint depends on the maximum permissible temperature rise. With boiler feed-water and other hot-water pumps, temperature rise is generally limited to 15° F.

This nomogram provides a graphical solution to the formula given above and permits determination of the minimum permissible operating capacity when the maximum permissible temperature rise has been determined.

Example. Assume a centrifugal pump handles 220°F. water and the NPSH (net positive suction head) is 18.8 ft. The shutoff head is 3,200 ft, and the allowable temperature rise is determined to be 20°F. What is the efficiency for minimum safe flow.

Solution. Connect 3,200 ft on the H scale with 20° F. on the T_R scale, extend to the E scale, and read 17% efficiency for minimum safe flow. END

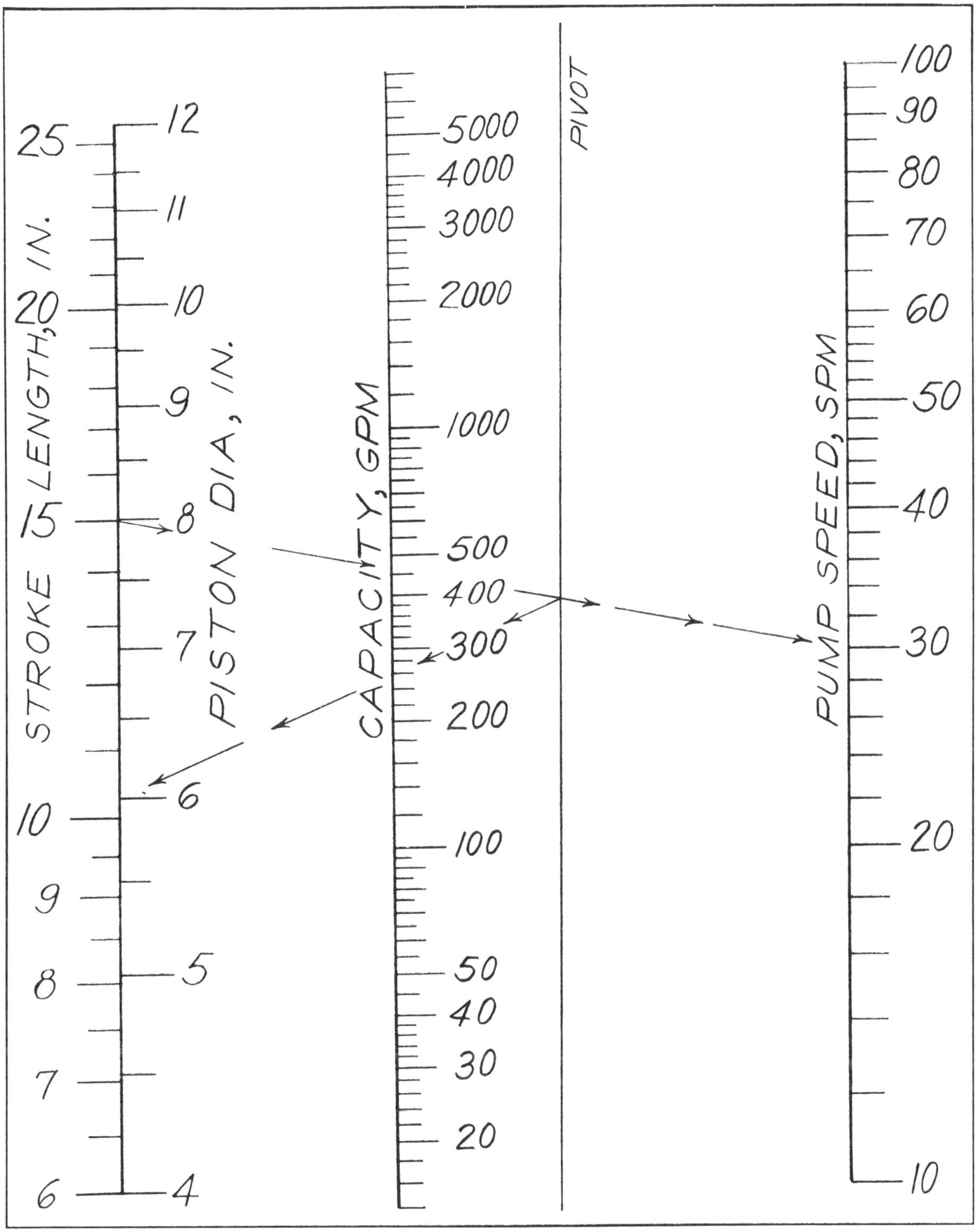

Reciprocating pump capacity

THIS CHART permits rapid computation of the liquid capacity of duplex double-acting reciprocating pumps. It saves much of the tedious labor involved in this computation.

Example: What is the liquid capacity of a duplex double-acting reciprocating pump having a stroke length of 15 in., a speed of 30 strokes per minute, and a liquid piston diameter of 6 in.?

Solution: Enter the chart on the left at the stroke length of 15 in. and draw a straight line through the pump speed of 30 strokes per minute on the right. From the intersection with the pivot draw a straight line through the piston diameter of 6 in. on the left. At the intersection with the central scale read the capacity as 260 g.p.m.

This chart is based on $Q = 0.01602\ (SND^2)$, where Q = pump capacity, gallons per minute; S = stroke length, inches; N = number of strokes per minute; D = liquid piston diameter, inches. **End.**

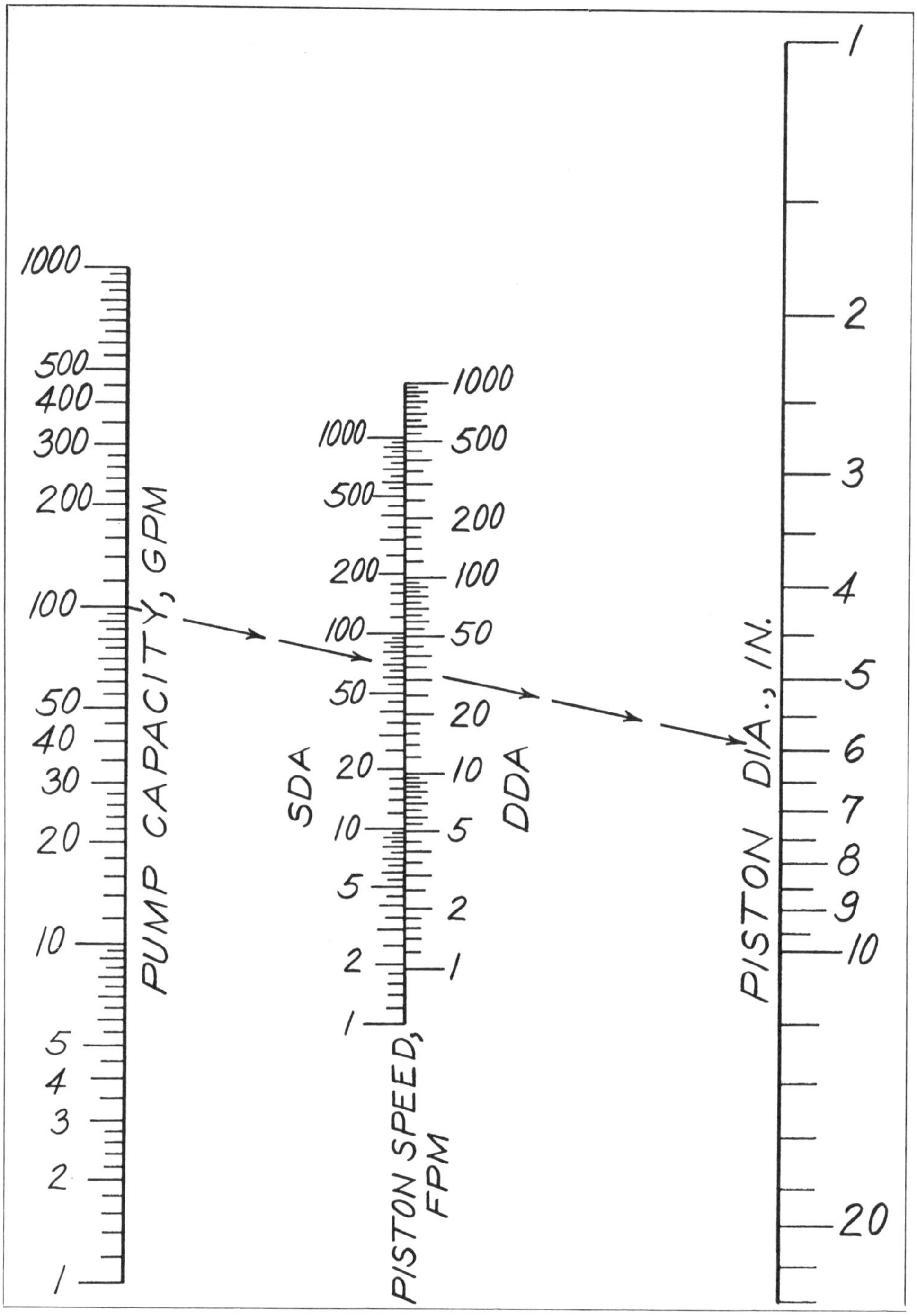

Piston speed of reciprocating pumps

SIMPLEX and duplex double-acting pumps are widely used throughout industry for a variety of pumping jobs. The accompanying chart allows easy computation of the piston speed of simplex double-acting (SDA) and duplex double-acting (DDA) pumps for a wide range of capacities and liquid-piston diameters. Piston speed is important in net positive suction head and similar calculations.

Example: What is the piston speed of SDA and DDA reciprocating pumps handling 100 g.p.m. with 6-in.-diameter liquid pistons?

Solution: Enter the chart on the left at 100 g.p.m. and draw a straight line through the liquid-piston diameter of 6 in. on the right. At the intersection with the central scale, read the piston speeds as 68 ft. per minute (SDA) and 34 ft. per minute (DDA).

This chart is based on $S_s = 24.5\ G/D^2$ and $S_d = 12.25\ G/D^2$, where S_s and S_d = piston speed of simplex and duplex double-acting pumps, feet per minute, respectively; G = pump capacity, gallons per minute; D = liquid-piston diameter, in.

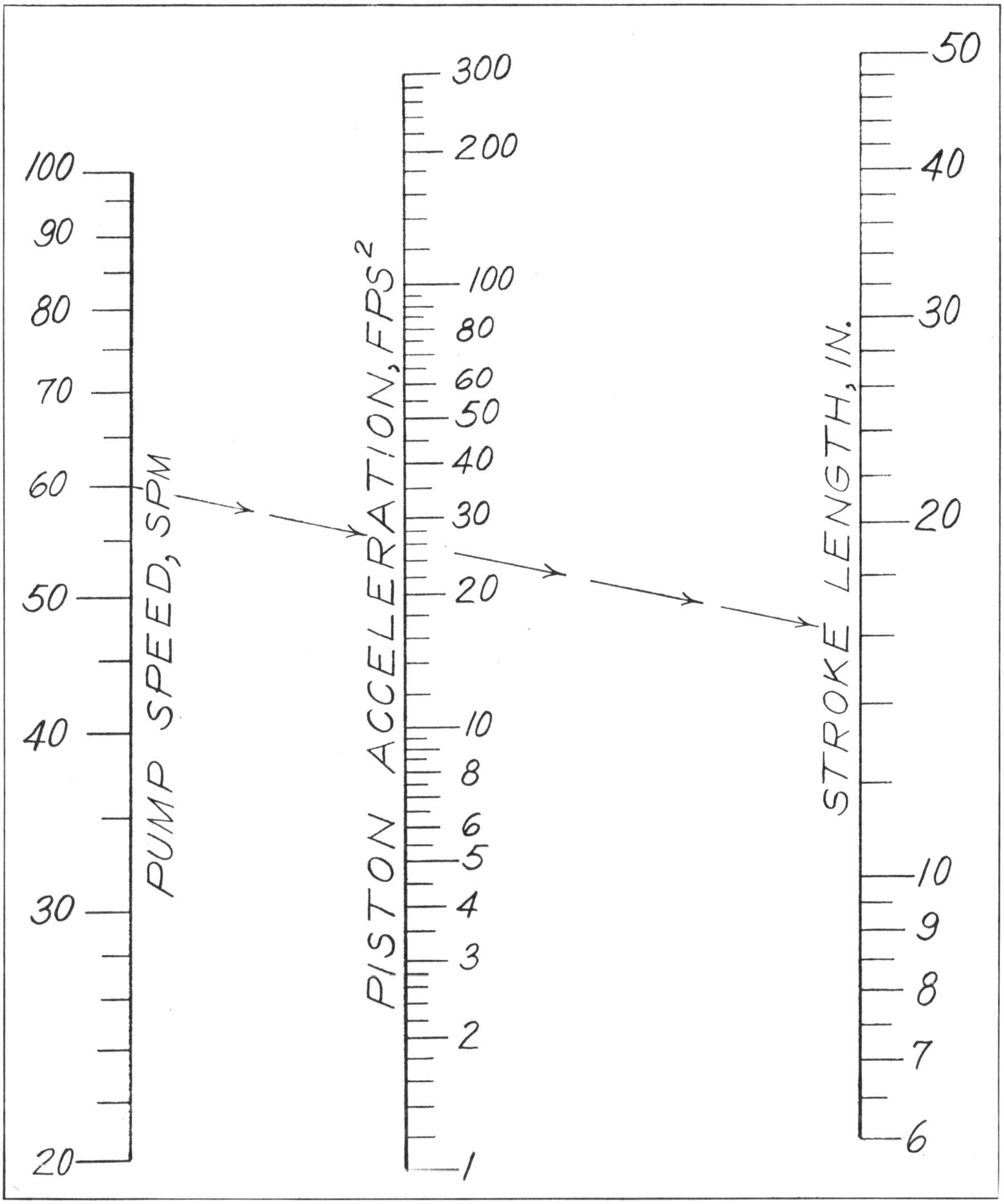

Piston acceleration in reciprocating pumps

WHEN YOU COMPUTE the required net positive suction head and the capacity of reciprocating pumps, often you may find it necessary to determine the liquid piston acceleration. This chart allows quick and simple computation of piston acceleration.

Example: What is the piston acceleration in a reciprocating pump having a speed of 60 s.p.m., if the stroke length is 24 in.?

Solution: Enter the chart on the left at the pump speed of 60 s.p.m. and draw a straight line through the stroke length, 24 in., on the right. At the intersection with the central scale, read the piston acceleration as 26.3 ft. per second per second.

This chart is based on $A = 0.000456\ N^2S$, where A = piston acceleration, feet per second per second = fps^2; N = number of pump strokes per minute; S = stroke length, in. **End.**

Effect of liquid viscosity and water temperature on reciprocating pumps

Bill Sisson
Superintendent of Operations
Nipak Inc.
Pryor, Okla.

DIRECT-ACTING or power type of reciprocating pumps are usually rated for maximum delivery when handling 250-SSU liquids or 70° F. water. At higher liquid viscosities or water temperatures, the speed is reduced. Since pump capacity is a function of piston speed, it is directly proportional to the number of strokes per minute it makes or to its rpm.

The Standards of the Hydraulic Institute give typical recommended speed-correction factors for reciprocating pumps for various liquid viscosities and water temperatures. This nomogram is based on these factors.

Example. If a direct-acting steam-driven pump delivers 120 gpm of 70° F. water when operating at 50 strokes per minute, how much 2,000-SSU crude oil will this pump deliver? How much 125° F. water will this pump deliver?

Solution. Step 1, align 2,000 SSU on Liquid-Viscosity scale with 120 on Pump Delivery Rate at 70° F. scale, extend to Corrected Pump Delivery Rate scale, and read 96 gpm. Step 2, when handling 125° F. water, align 125° F. on Water-Temperature scale with 120 on Pump Delivery Rate at 70° F. scale, extend line to Corrected Pump Delivery Rate scale, and read 90 gpm for 125° F. water. Note that the two Pump-Delivery-Rate scales may be simultaneously increased or decreased by multiples of 10. END

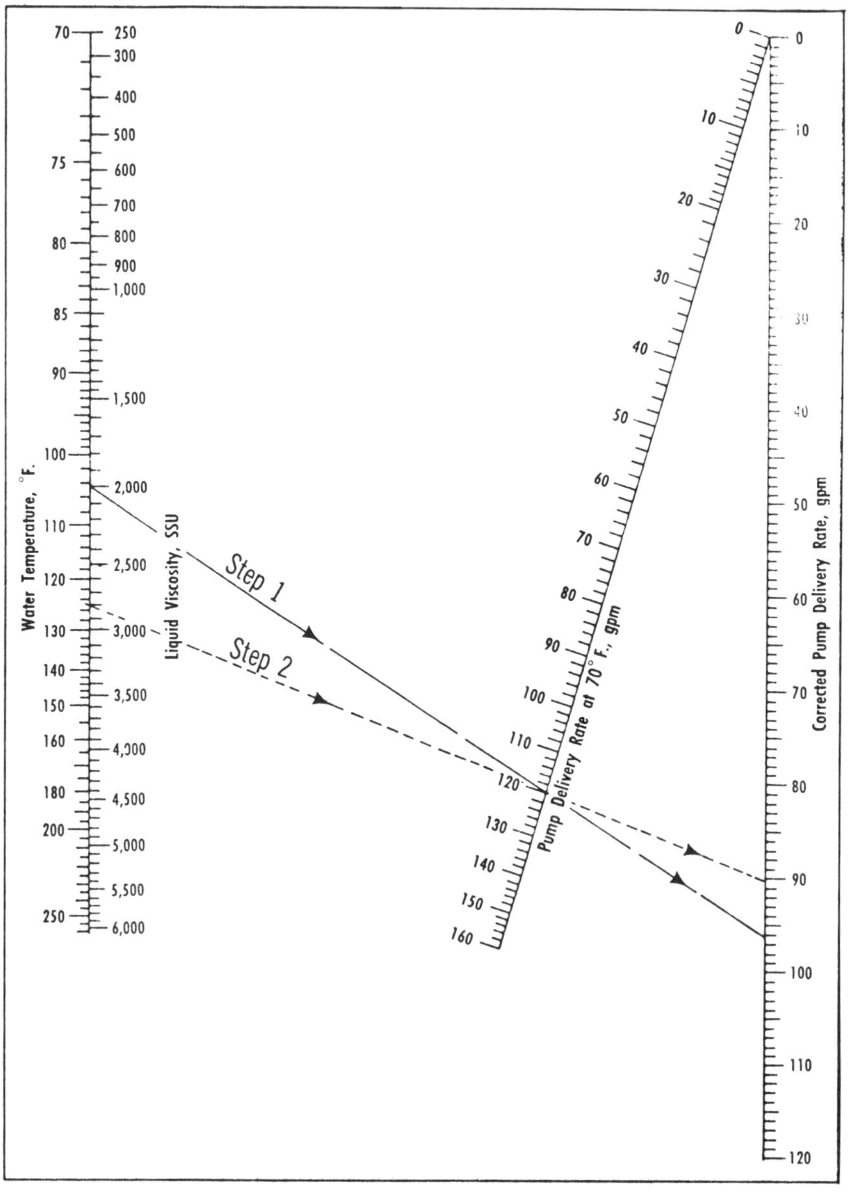

F. Caplan
Kaiser Engineers

flow-nozzle permanent pressure loss

THIS nomogram is designed to help calculate the permanent pressure loss for a flow nozzle mounted inside of a pipe. It solves the equation:

$$\Delta P = \Delta h\,(D^2 - d^2)/(D^2 + d^2)$$

Where:

ΔP = flow-nozzle permanent pressure loss, upstream pressure P_1 − downstream pressure P_4, psi or inches water (use same units as those for measured pressure differential, Δh)

Δh = measured pressure differential, upstream pressure P_1 − downstream nozzle pressure P_2, any units

D = pipe inside diameter, any units

d = flow-nozzle diameter (use same units as pipe diameter)

Example. What is the permanent loss ΔP of a 6.0-in.-diameter flow nozzle mounted in a 10.0-in.-id pipe if the measured differential pressure Δh is 30 in. water?

Align 10.0 in. on pipe-inside-diameter scale with 6-in. on flow-nozzle-diameter scale, Step (1), and mark intersection of straight edge with pivot line. Then align this point with 30-in. water on measured-pressure-differential scale, Step (2), and read ΔP = 14.1 in. water on permanent-pressure-loss scale. Note that the Δh and ΔP scales and the D and d scales can be simultaneously multiplied by such numbers as 0.1, 2, 10, 100, etc. to extend the scale ranges.

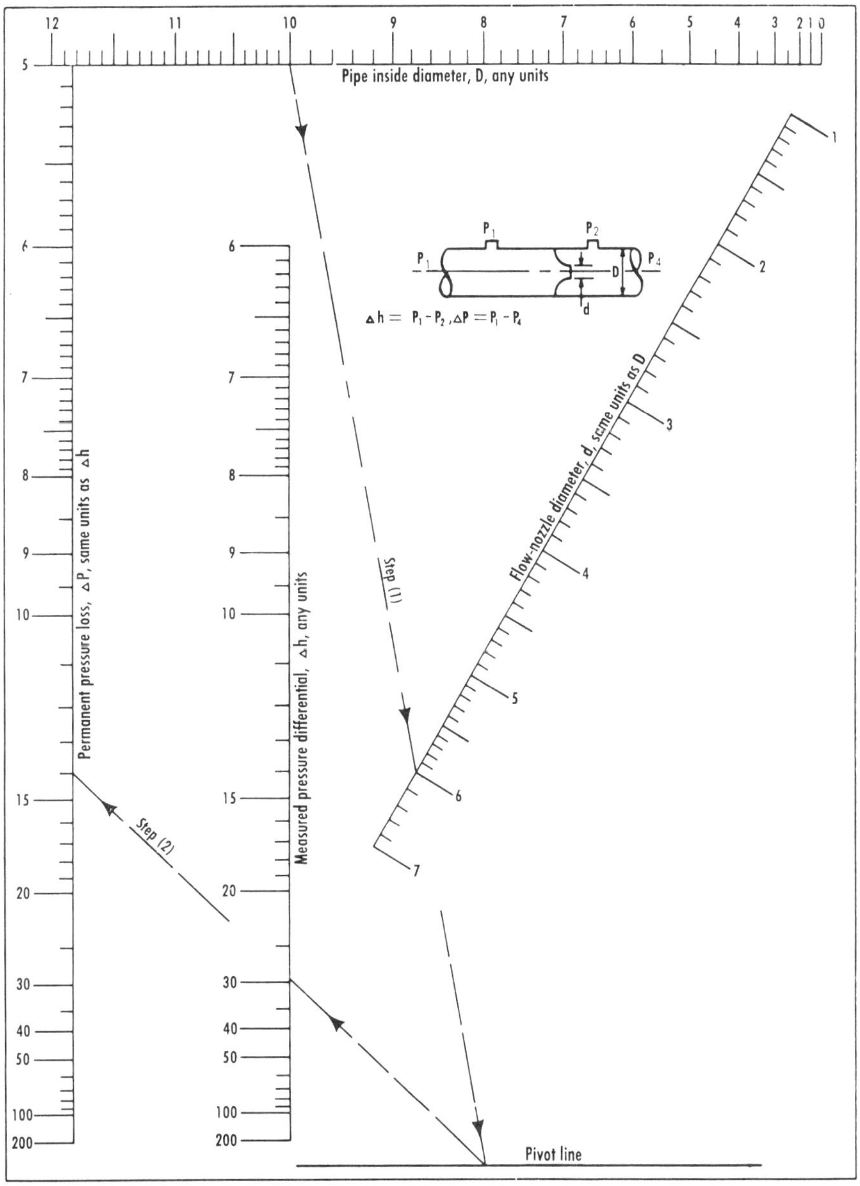

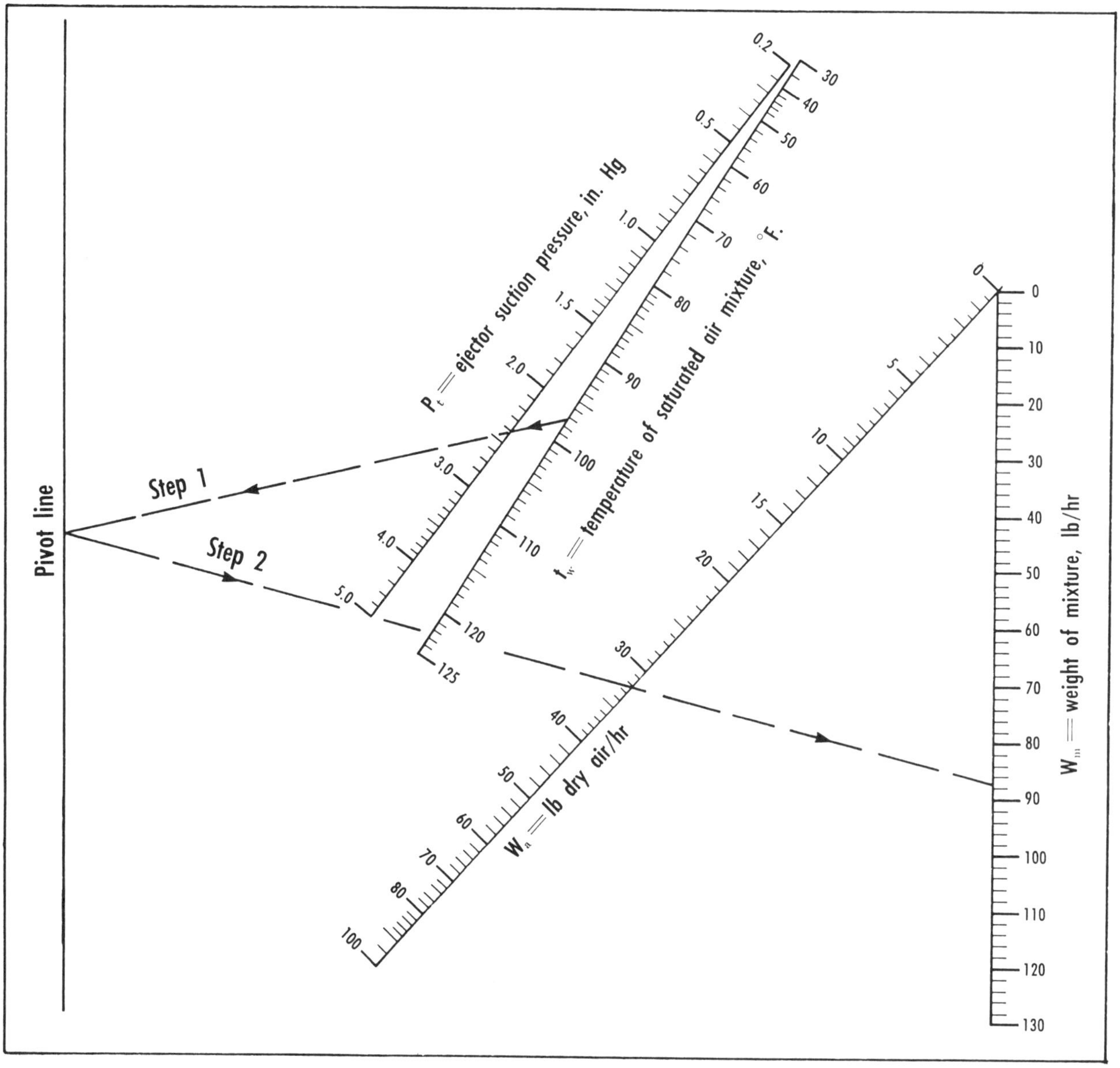

Air-water vapor mix for steam-jet ejectors

Bill Sisson
Superintendent of Operations
Nipak, Inc.
Pryor, Okla.

STEAM-JET air ejectors are widely used as a vacuum pump for removing air, gases, or vapors from condensers and vacuum equipment. When handling saturated air, their capacity is expressed in terms of total pounds of mixture handled. This is the sum of dry air plus water vapor. The air-water vapor mixture weight can be determined with the nomogram presented here which is based on the following formulas:

$$W_w = 0.622\ P_w/(P_t - P_w)$$
$$W_m = W_a\ (1 + W_w)$$

Where:

W_w = lb water vapor/lb dry air

P_w = saturated vapor pressure of water at temperature of mixture (t_W)

P_t = ejector suction pressure, in. Hg (absolute total pressure)

W_m = weight of mixture handled, lb/hr

W_a = dry air handled, lb/hr

Example. How much mixture is handled by an ejector if suction pressure is 2.4 in. Hg, moving 32 lb/hr dry air at 97° F.?

Solution. Connect 97° F. on t_w scale with 2.4 in. Hg on P_t scale, extend to pivot line and mark (Step 1). From this point, line with 32 lb/hr on W_a scale, extend line to W_m scale, and read 87.3 lb/hr handled (Step 2). END

Valve coefficient for control valve

F. Caplan
Kaiser Engineers

This nomogram is designed to help size control valves handling liquids. It is designed to solve the equation:

$$C_v = F\,(s)^{1/2}/(\Delta P)^{1/2}$$

Where:

C_v = valve coefficient
F = flow rate, gpm
s = specific gravity of liquid (water = 1)
ΔP = pressure drop, psi

Example 1. What size valve is required if the allowable pressure drop is 4 psi, flow rate is 100 gpm, and specific gravity is 1? To solve this, align 4 psi on the pressure-drop scale with 100 gpm on the flow-rate scale, Step (1). Mark the intersection of your straightedge with the pivot line. Next align the mark on the pivot line with 1 on the specific-gravity scale, Step (2). Read C_v = 50 on the flow-coefficient scale.

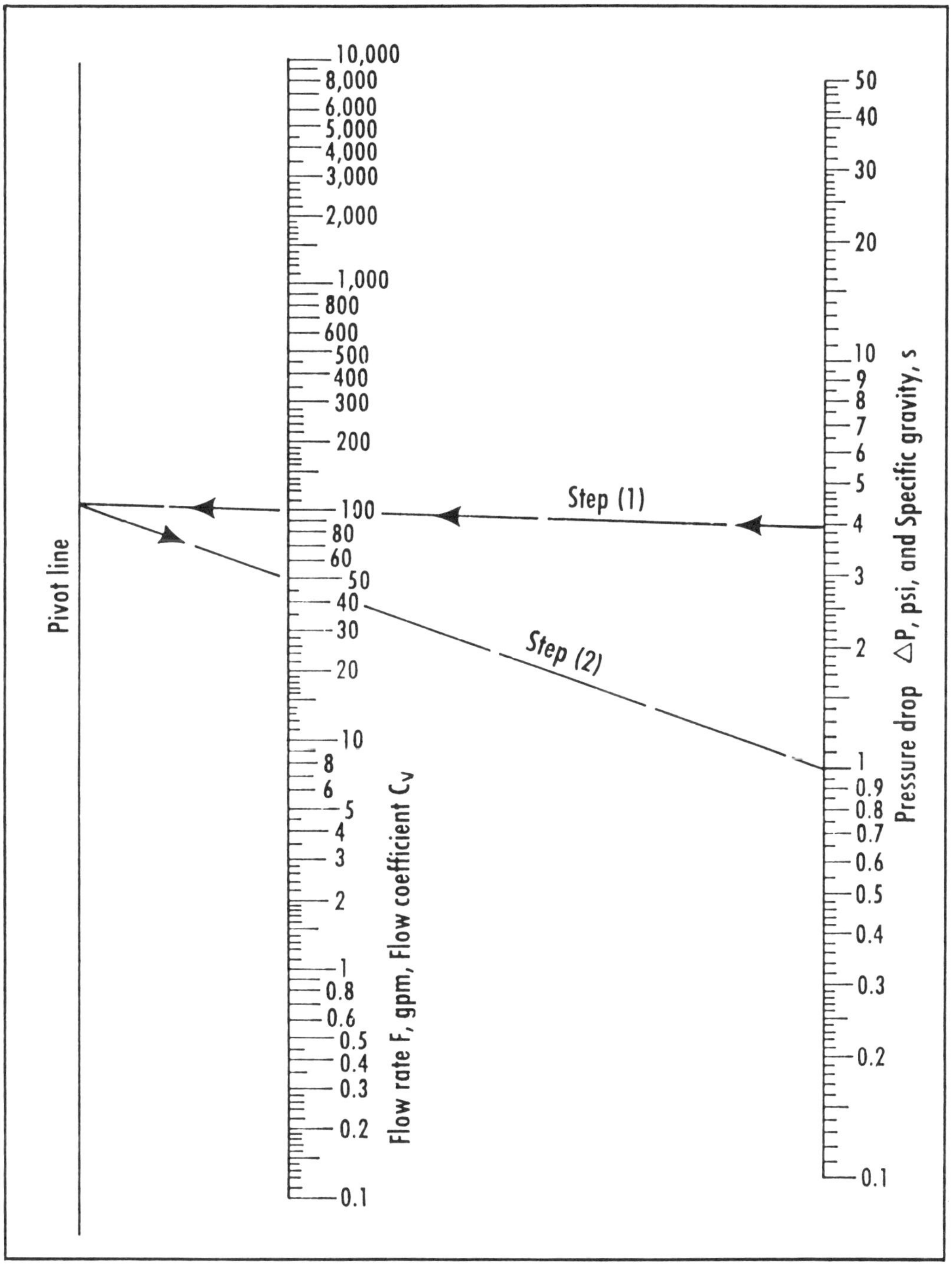

Nomogram determines time required to pressure air receivers

Bill Sisson
Superintendent of Operations
Nipak Inc.
Pryor, Okla.

THE nomogram presented here will provide a convenient way to quickly estimate the time required to pressure-up air receivers of different capacities, pressures, and altitudes with compressors of various volumetric efficiencies. Construction of the nomogram is based on the formula,

$$T=(p_1-p_2)V/BDE$$

Where T = time required to pressure receiver, min; V = volume of receiver, cu ft; D = compressor piston displacement, cfm; p_1 = final receiver pressure, psia; p_2 = initial receiver pressure, psia; E = average volumetric efficiency of compressor, % (as a decimal); B = barometric pressure, psia.

Example. Assume an air compressor installed at 6,000 ft elevation with an average volumetric efficiency of 80% and a piston displacement of 100 cfm pumping to a receiver with a volume of 800 cu ft. If the initial receiver pressure is 20 psig, how much time is required to raise the pressure to 130 psig?

Solution. First, determine the pressure rise: $p_1-p_2=130-20=110$ psi. Align from 800 cu ft on V scale (Step 1) with 110 psi on p scale and extend line to Pivot line 1 and mark. From the marked point, line with 100 cfm on D scale (Step 2) and extend line to Pivot line 2. From point found on Pivot line 2, align with 80% on E scale (Step 3) and extend to Pivot line 1. Line this point with 6,000 ft on A scale (Step 4), extend line to T scale, and read 93.3 min required to raise receiver pressure to 130 psig.

END

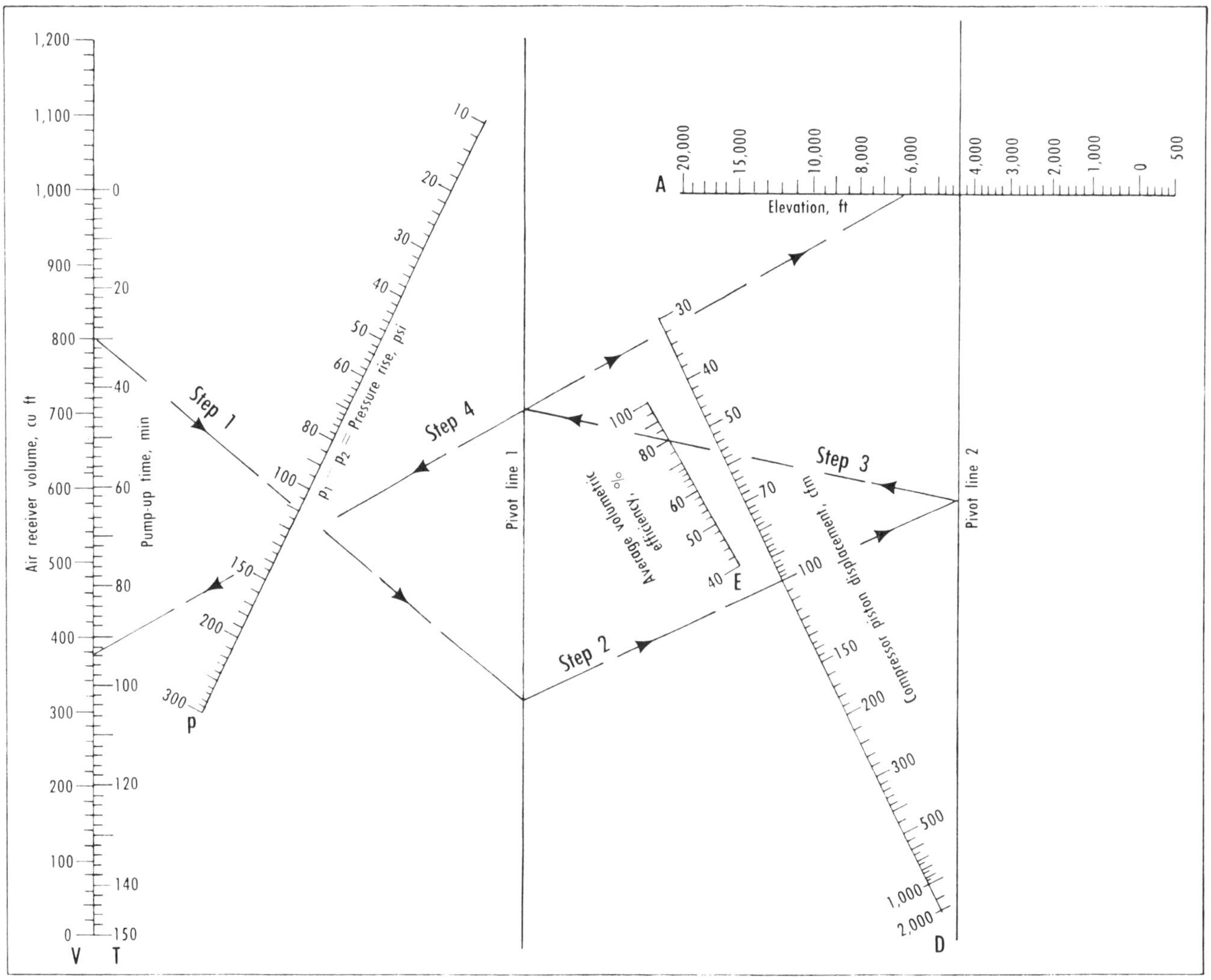

How to find pressure drop in air lines

F. Caplan
Kaiser Engineers

HERE'S a fast way to size an air line. The nomogram is based on the equation:

$$\Delta P_{100} = \frac{5.88 \times 10^{-5}\, Q^{1.83}\, T^{1.129}}{P\, d^{4.83}}$$

Where:

ΔP_{100} = pressure drop per 100 ft of air-line pipe, psi
P = average air pressure, absolute, psia (p = psig)
Q = flow rate, scfm
d = pipe ID, in.
T = air temperature, absolute, °R. (2 = °F.)

Note that you can solve for any variable in the equation, if you know all the others.

Example. What is the pressure drop for a flow rate of 400 scfm of air at 60° F. and 100 psig in a 2-in. Schedule 40 steel pipe? To find the answer, use a straightedge and align t = 60° F. with p = 100 psig. Mark the intersection of the straightedge with pivot line 1 (Step 1).

Next align the mark on pivot line 1 with Q = 400 scfm and mark the intersection of the straightedge with pivot line 2 (Step 2). Now align the mark on pivot line 2 with d = 2 in. (Step 3). Read pressure drop, ΔP_{100} = 1.0 psi per 100 ft of pipe.

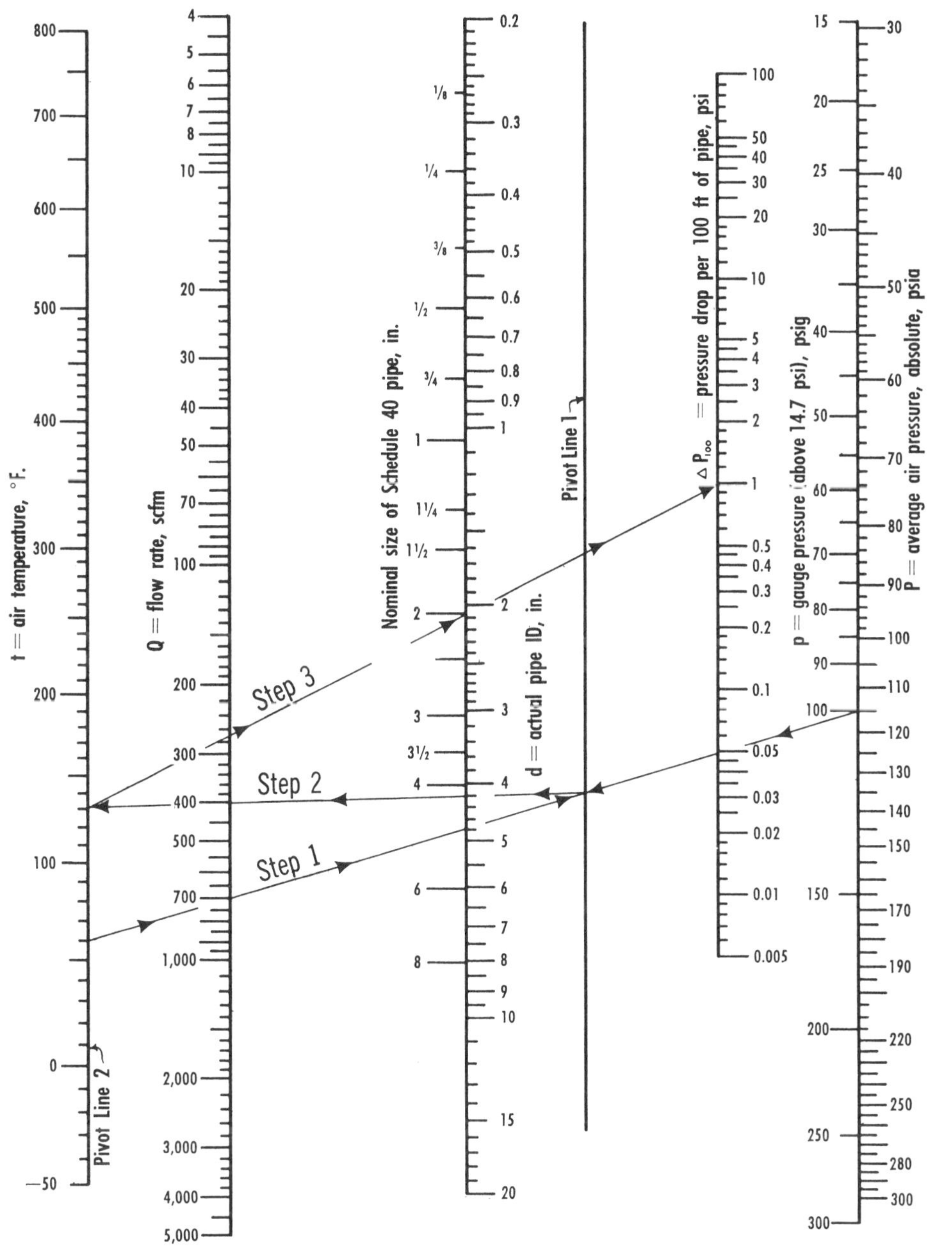

Nomogram gives water content, dew point of compressed air

Bill Sisson
Superintendent of Operations
Nipak Inc.
Pryor, Okla.

THE WATER-VAPOR content of air is often expressed in terms of dew point. But there is no standard way to express moisture concentration. Each industry uses its own units. This nomogram correlates dew points of compressed air with parts per million (ppm) of water vapor on both a volume and a weight basis.

Example 1. If air at 50 psig has a dew point of −75° F., what is its dew point when expanded down to atmospheric (0 psig) pressure?

Solution 1. Connect 50 psig on P scale (Step 1) with −75° F. on DP scale and extend to ppm scale. From point found on ppm scale (Step 2), connect with 0 psig on P scale, and read dew point at 0 psig as −94° F.

Example 2. What is the water content of air compressed to 200 psig, if it has a dew point of −52° F.?

Solution 2. Line 200 psig on P scale (Step 3) with −52° F. on DP scale, extend to ppm scale, and read water content as 2.5 ppm by weight or 4 ppm by volume. END

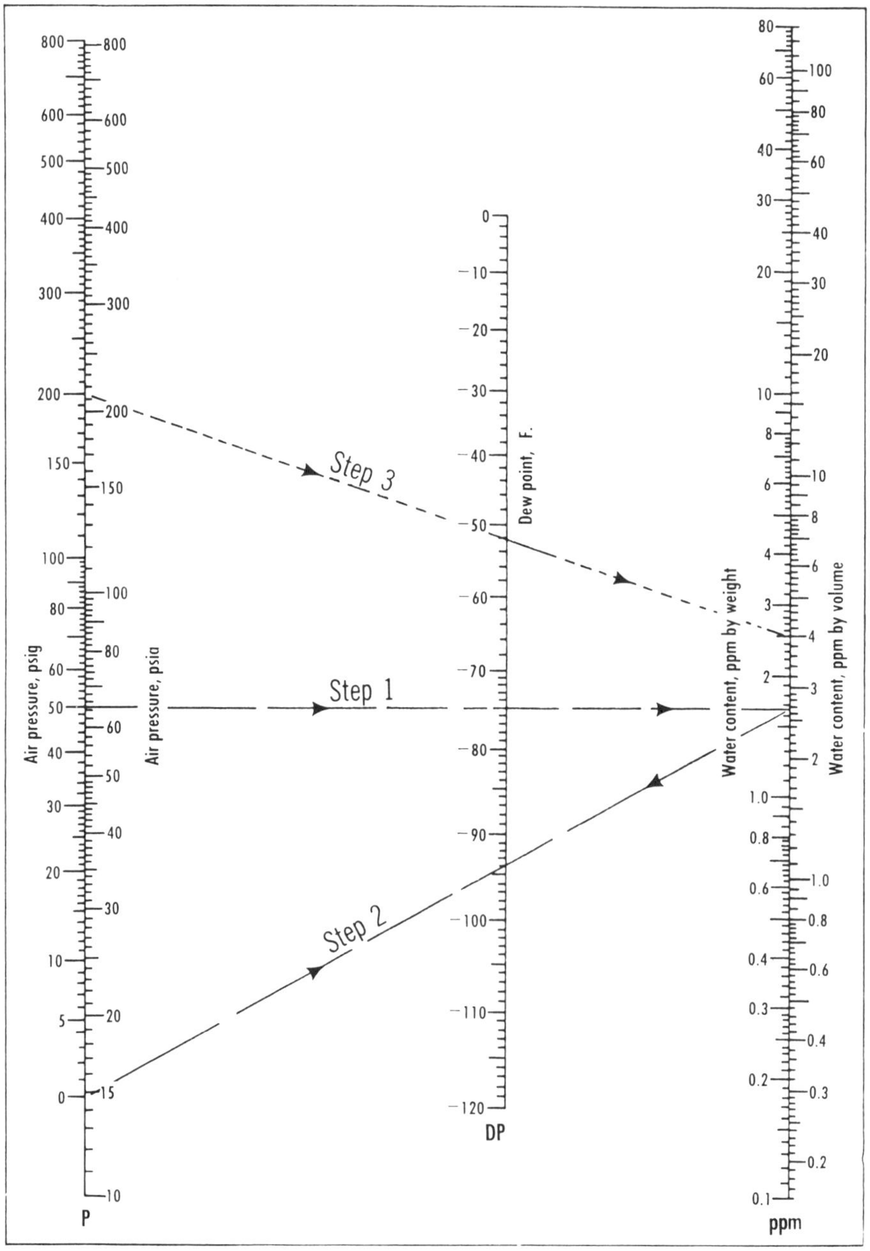

Nomogram gives dewpoint, specific humidity of air-water vapor mixtures

Bill Sisson
Superintendent of Operations
Nipak Inc.
Pryor, Okla.

IF YOU have to determine the dewpoint and specific humidity of air-water vapor mixtures, this nomogram will provide a quick, no-calculation, no-reference-to-tables method for determining these values when the temperature, relative humidity, and altitude are known. Construction of the nomogram is based on the following formulas:

$$R = p_v/p_d$$
$$W = p_v/1.608\,(B - p_v)$$

Where R = relative humidity, percent (as decimal); p_v = actual vapor pressure, in. Hg (corresponds with dew point, t_c); p_d = vapor pressure, in. Hg (corresponds with dry-bulb temperature, t_d); W = specific humidity (humidity ratio), lb water vapor/lb dry air; and 1.608 = equivalent molecular weight of air/molecular weight of water = 28.97/18.

Example. If the dry-bulb temperature is 90° F. and the relative humidity is 70%, what is the dewpoint and specific humidity of air at 2,500 ft altitude?

Solution: Connect 90° F. on t_d scale (Step 1) with 70% on R scale, and read dew point as 79° F. where line intersects t_c scale. Next line 79° F. on t_c scale (Step 2) with 2,500 ft (27.3 in. Hg) on A scale, and read specific humidity as 0.0235 lb water vapor/lb of dry air where line crosses W scale. To convert to grains: 0.0235 x 7,000 = 164.5 grains water vapor/lb dry air.

END

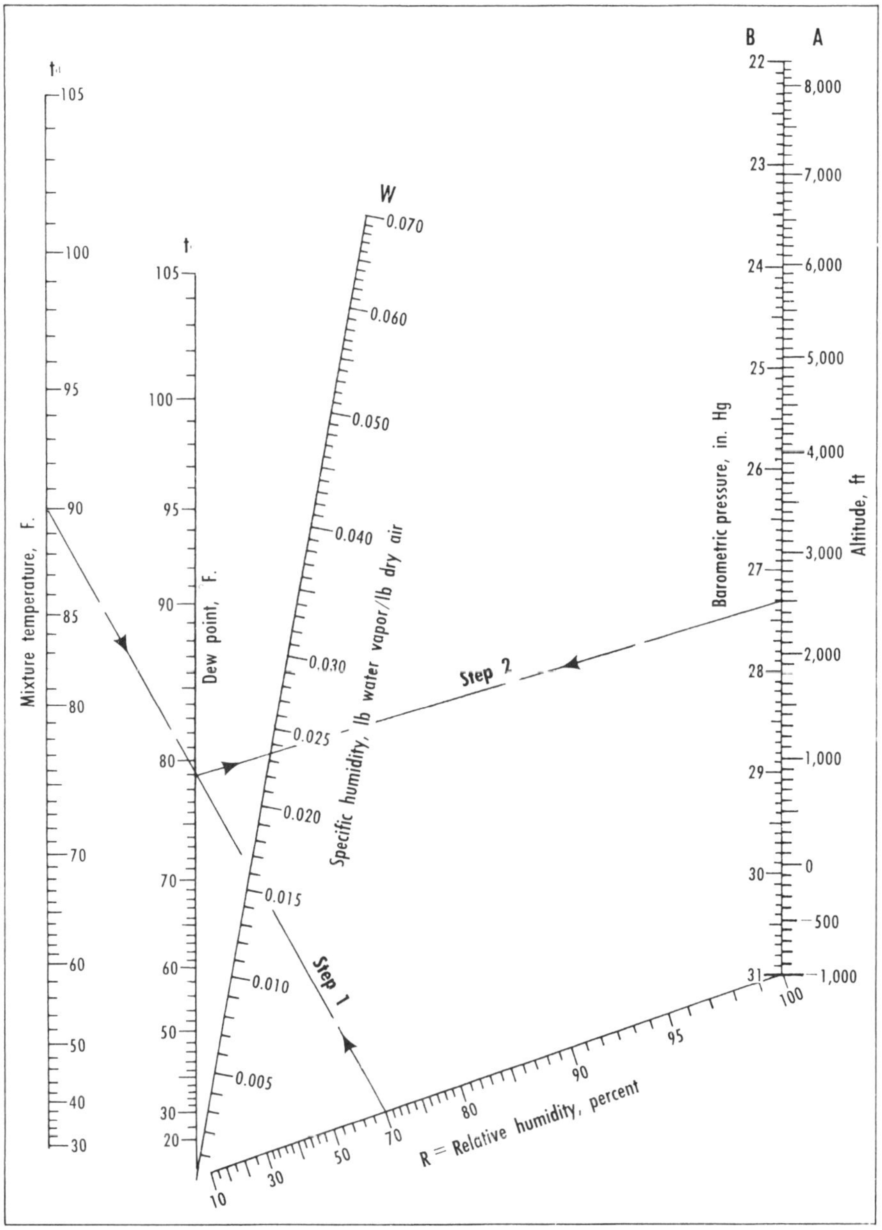

Enthalpy of saturated air

Bill Sisson
Superintendent of Operations
Nipak Inc.
Pryor, Okla.

THE enthalpy (heat content) of saturated air above 0° F. can be computed by the formula, $H = xt + 0.24t + x\lambda$; where H = enthalpy, BTU/lb dry air; x = saturation humidity, lb H_2O/lb air; t = temperature of air, °F.; λ = latent heat of evaporation, BTU/lb; and 0.24 = specific heat of air.

The saturation humidity is calculated by this formula:

$$x = [P_w/(p_t - p_w)]\ (M_w/M_a)$$

where p_t = total pressure, psia; p_w = saturation partial pressure of water, psia; M_w = molecular weight of water; and M_a = molecular weight of air.

This nomogram, based on those equations, will calculate the enthalpies of saturated air at various elevations.

Example. What is the saturation enthalpy of air at 75° F. where the elevation is 4,000 ft above sea level?

Solution. Connect 75° F. on Air-temperature scale with 4,000 ft on Elevation scale, and read enthalpy above 0° F. as 42.5 BTU/lb dry air where Line 1 intersects Enthalpy scale. END

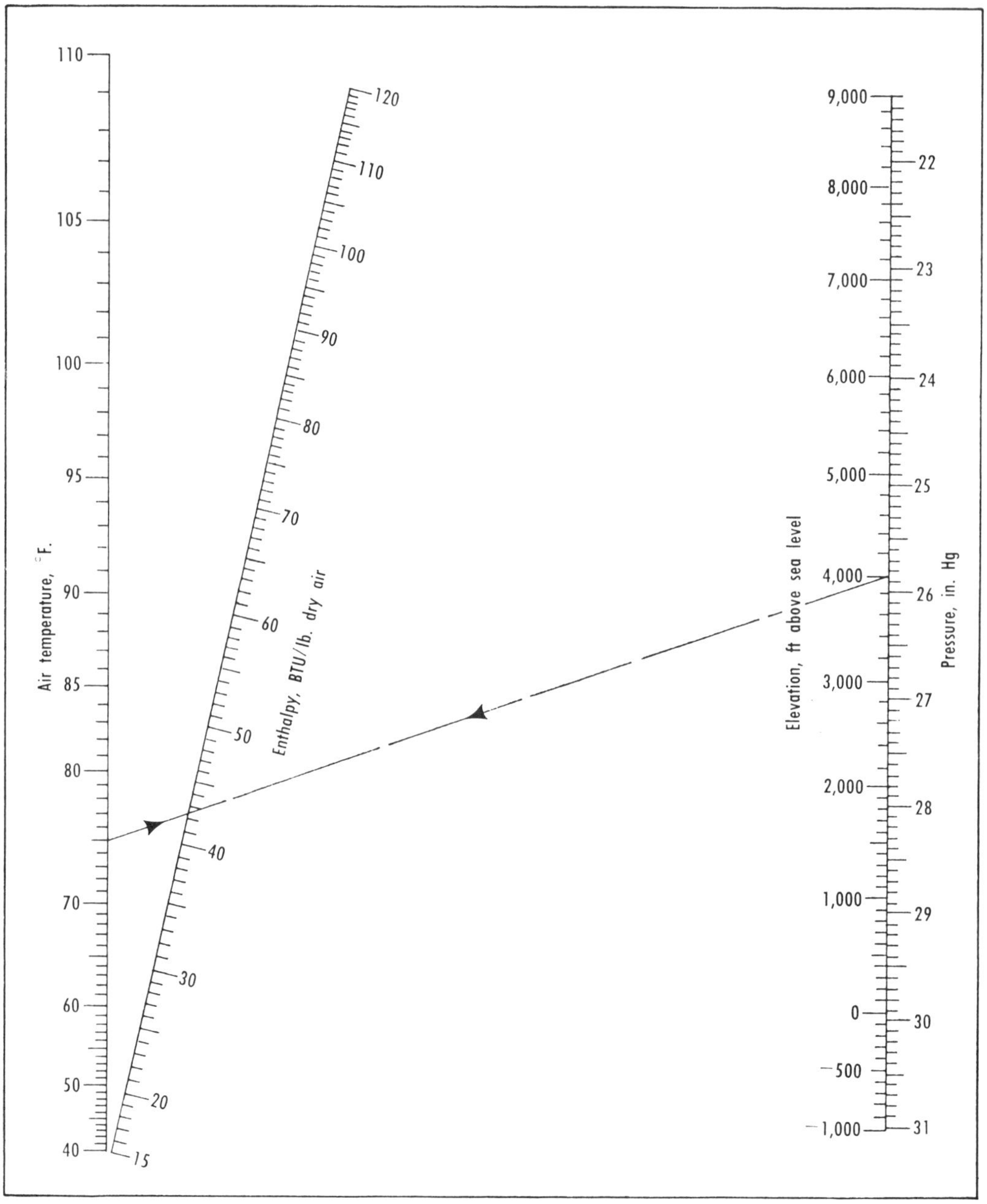

Nomogram helps select fans for heat, altitude

Bill Sisson
Superintendent of Operations
Nipak Inc.
Pryor, Okla.

A FAN with a given blade angle and speed is a constant-volume machine. The flow rate (cfm) remains constant regardless of changes in temperature and altitude.

Fan catalog ratings are for standard air at 0.075 lb/cu ft density. This is at sea level (29.92 in. barometric pressure) and is equivalent to dry air at 70° F. For changes in temperature and altitude, a correction factor must be applied.

For air of a different weight, the static pressure and horsepower change in direct proportion to the density, as shown by the following example.

Example. A catalog table lists a particular model fan for 12,000 cfm at 1 in. static pressure, 1,875 rpm, requiring 5.0 brake horsepower (bhp). What is the actual performance at 50° F. and 7,000 ft altitude?

Solution. Step 1, connect 7,000 ft on Altitude scale with 50° F. on Temperature scale and mark at intersection with Pivot line. Step 2, connect marked point on Pivot line with 1 in. on Static pressure, in H_2O, @ altitude & 70° F., scale and read actual static pressure as 0.80 in. H_2O where line crosses Static Pressure, in. H_2O, @ actual conditions scale. Step 3, connect marked point on Pivot line with 5 on bhp @ 0 altitude & 70° F. scale, and read 4 bhp required where line crosses bhp @ actual conditions scale. The cfm and rpm remain the same.

END

Fan selection

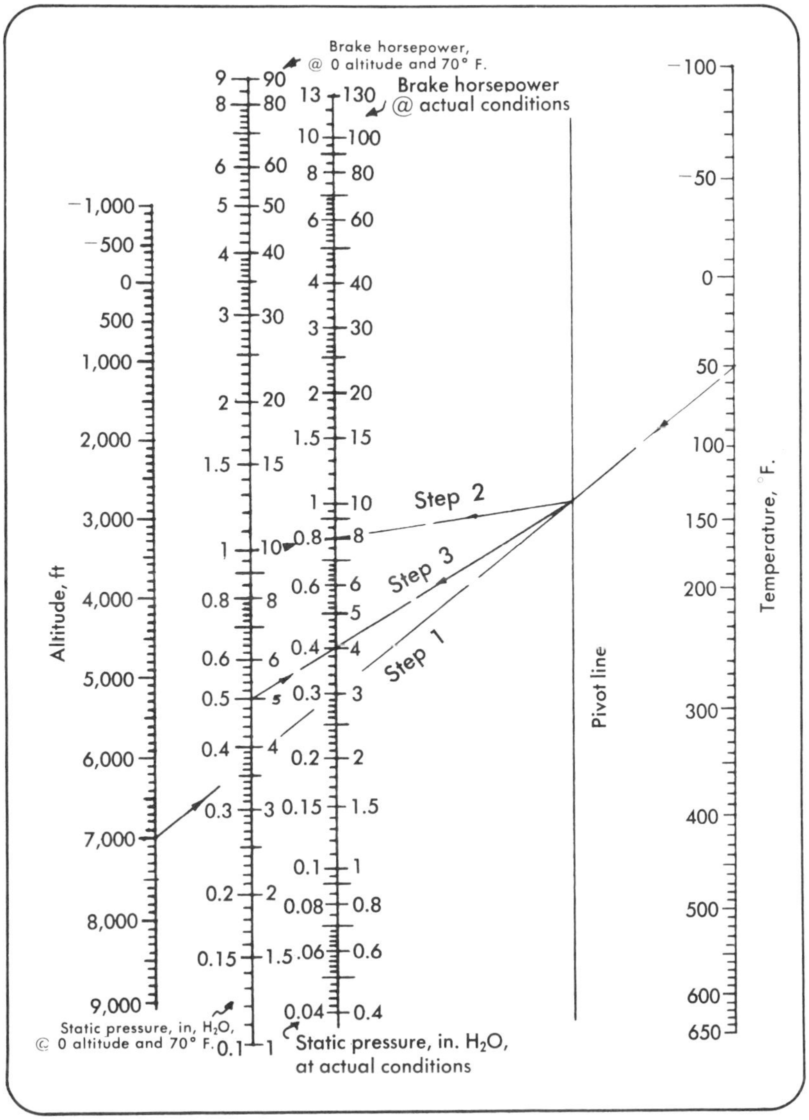

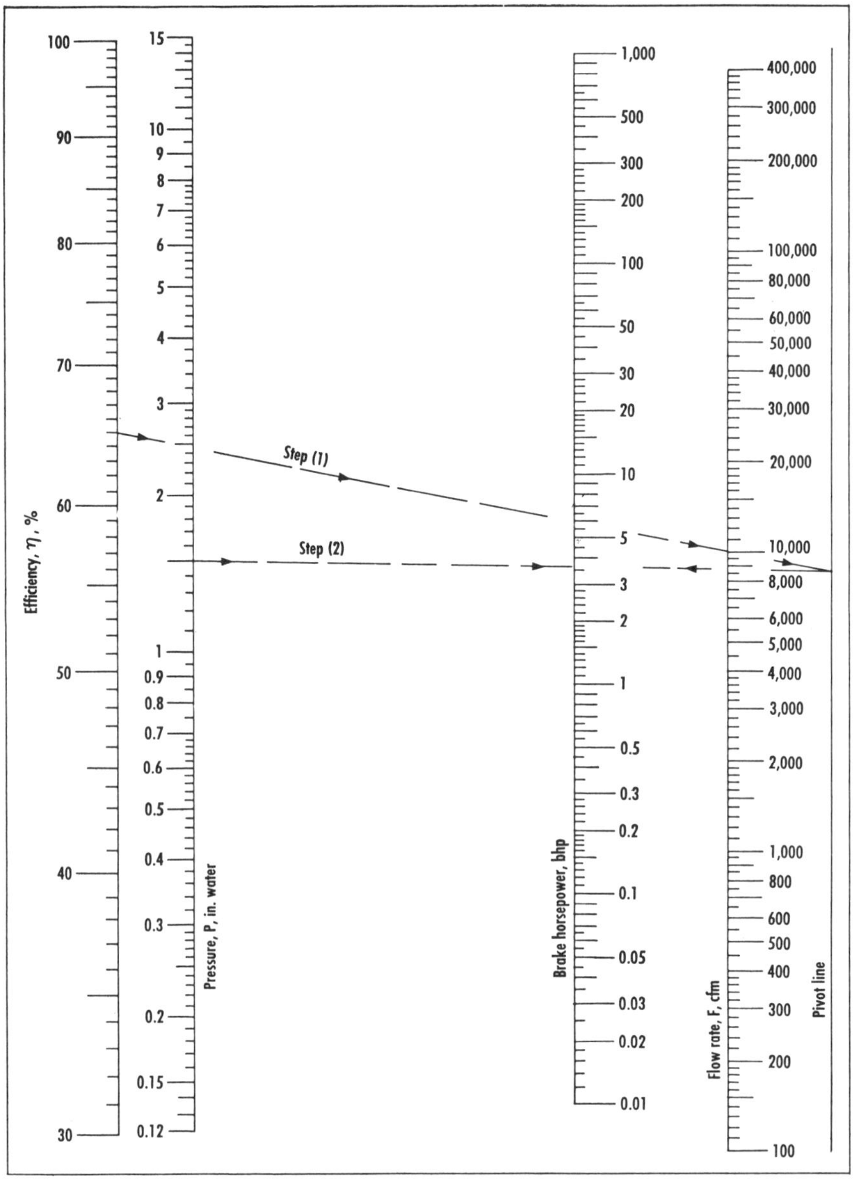

Use this nomogram to find fan-motor brake horsepower

F. Caplan
Kaiser Engineers

THIS nomogram is designed to find the correct size motor to drive a fan. It is based on the equation:

$$Bhp = 0.01573\ P\ F/\eta$$

Where:

Bhp = brake horsepower
F = Flow rate, cfm
P = Pressure, in. water
η = Fan efficiency, percent

Example: How much brake horsepower is required for a 65% efficient fan handling 10,000 cfm at a pressure of 1.5 in. water? To find the answer, align 65% on the efficiency scale with 10,000 cfm on the flow-rate scale, and mark the pivot line at the intersection with a straightedge. Next align the mark on the pivot line with 1.5 in. on the pressure scale, and read 3.6 brake horsepower on the horsepower scale.

Nomogram tells time required to drain vessels

Bill Sisson
Superintendent of Operations
Nipak Inc.
Pryor, Okla.

THE time required to drain a vessel under the head of its own liquid is easily found with the nomogram presented here. Construction of the nomogram is based on these formulas:

Vertical cylinder, $t = (D^2/d^2)\ H^{1/2}$

Horizontal cylinder, $t = 2.4\ L\ R^{3/2}\ d^2$

Spherical vessel, $t = R^{5/2}/d^2$

Where:

t = time to drain, minutes
D = cylinder diameter, ft
d = drain diameter, in.
R = radius of cylinder or sphere, ft
H = height of liquid in vertical cylinder, ft
L = length of horizontal cylinder, ft

These formulas apply only to light hydrocarbons, water, and other liquids that have relatively low viscosities.

Example 1. For a vertical cylinder with a 10-ft diameter and a 4-in. drain, how much time is required to drain vessel if height of liquid is 36 ft?

Solution. Connect 10 ft on D scale with 4-in. drain on d-v scale (Step 1-A), and extend to pivot line. Connect point found on pivot line with 36 ft on H scale (Step 1-B), and read 37.5 min drain time where line crosses T(vertical) scale.

Example 2. For a horizontal cylinder 20 ft long and 8 ft in diameter, with a 3-in. drain, what is the drain time?

Solution. Connect 8 ft on D scale with 3 in. on d-h scale (Step 2-A), and extend to pivot line. Connect point found on pivot line with 20 ft on L scale (Step 2-B), and read drain time as 43 min where line crosses T(horizontal) scale.

Example 3. For a sphere with a radius of 15 ft and an 8-in. drain, what is the drain time?

Solution. Connect 15 ft on R scale with 8 in. on d-s scale (Step 3), extend to T(spheres) scale, and read 13.7 min drain time. END

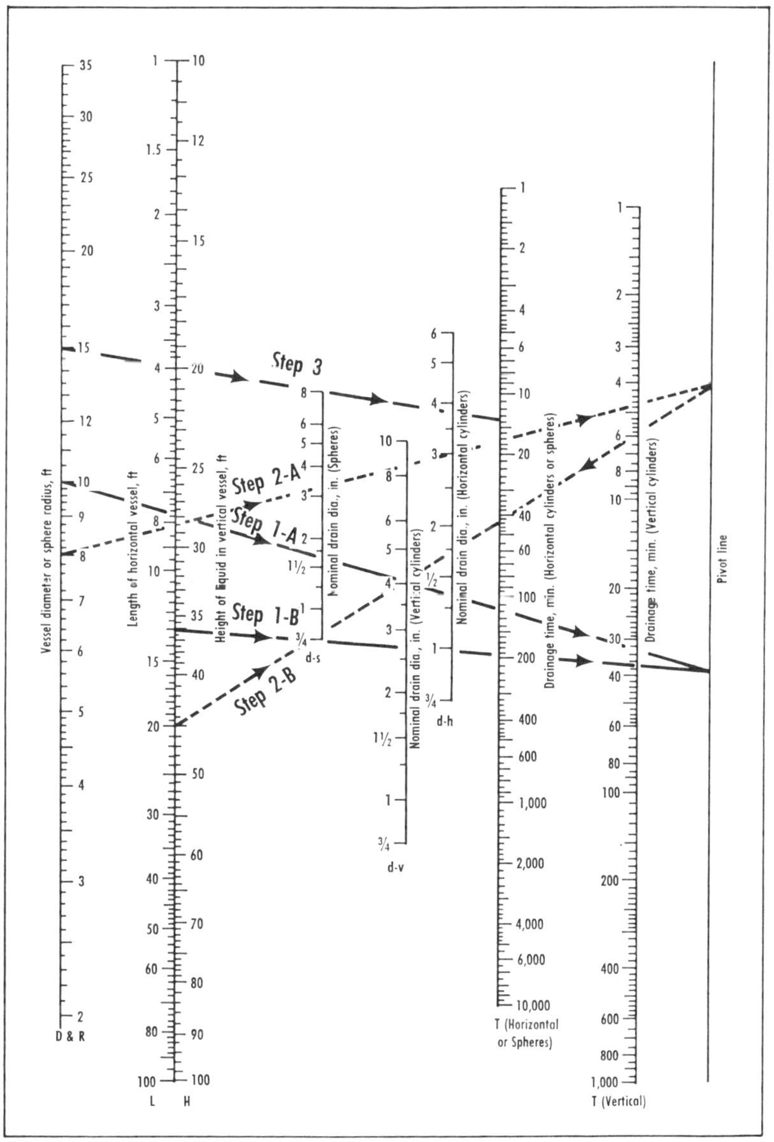

Nomogram computes volume of conical-bottomed vessels

Bill Sisson
Superintendent of Operations
Nipak Inc.
Pryor, Okla.

THE accompanying nomogram for computing the volume (V) of the contents of a conical-bottomed cylindrical vessel when the cone is full is based on the sums of the volumes in the conical and cylindrical portions of the vessel:

$$V = 5.875\ D^2\ (H - 0.667L)$$

Where:

V = volume of vessel, gal
D = diameter, ft
L = length of cone, ft
H = total depth of liquid, ft

A scale for determining volume in cu ft is also included.

Example. If the diameter is 5 ft 4 in.; length of cone is 5 ft 2 in. and depth of the liquid 12 ft 3 in., how many gallons does the vessel contain? How many cu ft?

Solution. Step 1, align 5 ft 2 in. on L scale with 12 ft 3 in. on H scale, extend to pivot line and mark. Step 2, line marked point on pivot line with 5 ft 4 in. on D scale, extend to V_g scale, and read 1,470 gal. To find cu ft, Step 3, connect 1,470 gal on V_g scale with reference point for cu ft, and read 196 cu ft where line crosses Volume, cu ft (V_{cf}) scale. END

Conical vessel volume

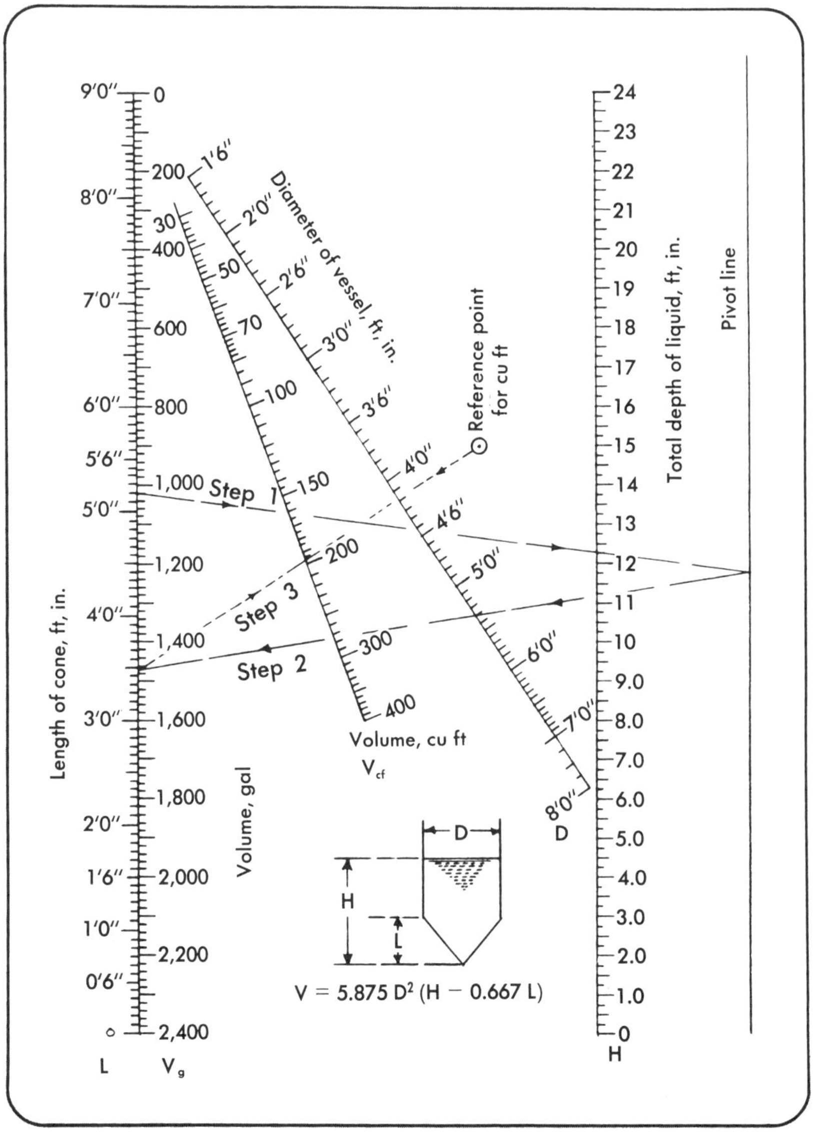

Nomogram finds tank-head volume

Bill Sisson
Superintendent of Operations
Nipak Inc.
Pryor, Okla.

THIS nomogram provides a quick method for finding the volume of part-full dished-only (bulged, bumped), elliptical, or hemispherical tank heads. Construction of the nomogram is based on the following formulas for the three types of tank heads:

$$V\ (\text{dished}) = 0.0009328\ h^2\ (1.5\ d - h)$$
$$V\ (\text{elliptical}) = 0.0034 h^2\ [d - (h/1.5)]$$
$$V\ (\text{hemispherical}) = 0.0068 h^2\ [d - (h/1.5] = 2 \times V\ (\text{elliptical})$$

Where V=volume of both heads, gal; h=height of liquid in head, in.; and d=inside diameter of tank, in.

Example. How many gallons does two, 6-ft diameter dished heads contain if they are filled to a depth of 56 in.? How many gallons would elliptical and hemispherical heads contain?

Solution. Step 1, connect 6 ft on d scale with 56 in. on h_1 (No. 1) scale, and mark at intersection with Pivot Line. Step 2, line marked point on Pivot Line with 56 in. on h_2 (No. 2) scale, extend line to V scale, and read 152 gal in dished heads and 369 gal in elliptical heads. Then 369 × 2 = 738 gal for two hemispherical heads. END

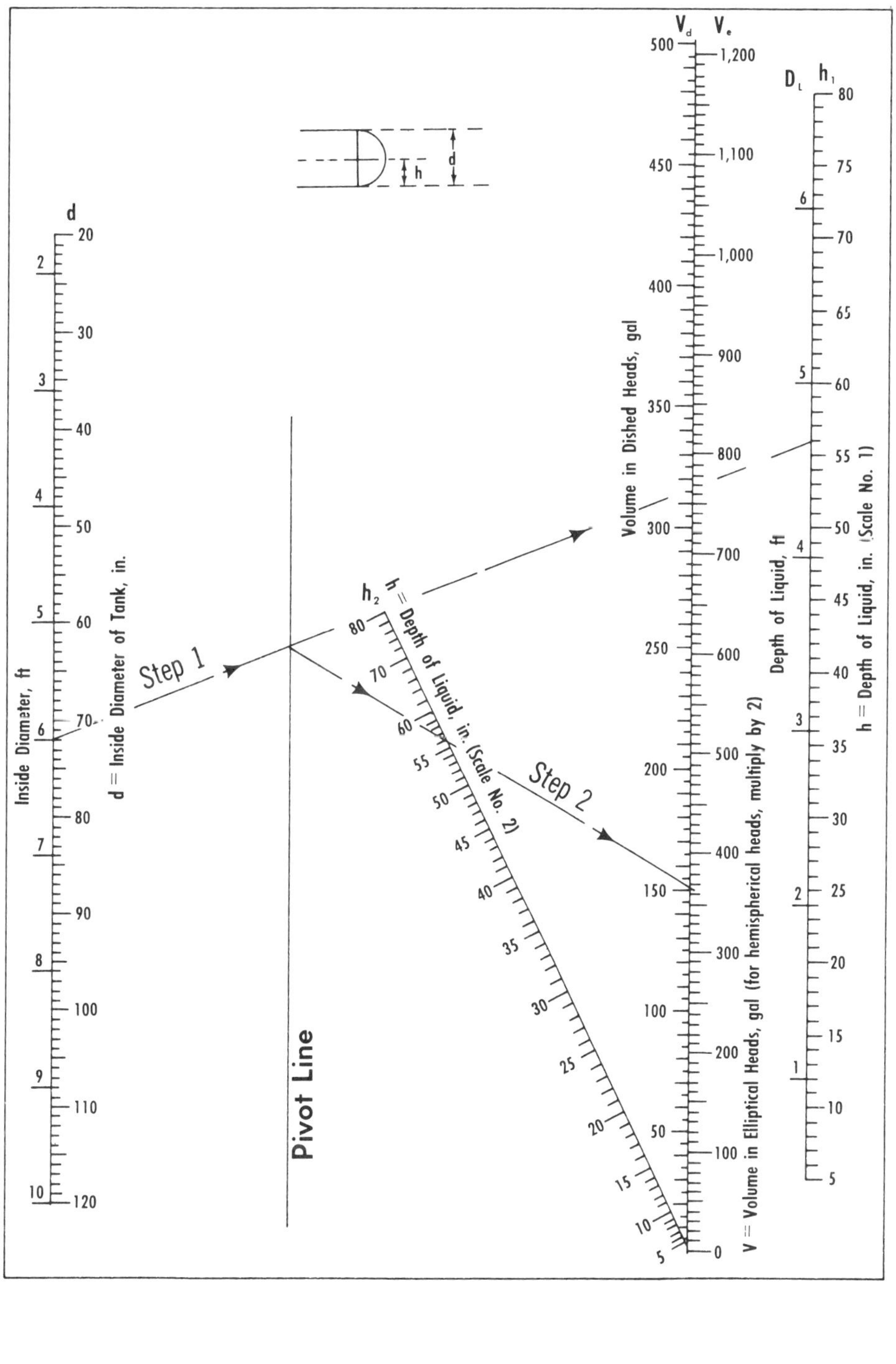

Use this nomogram to find the partial capacity of one

ASME code flanged and dished head

F. Caplan
Kaiser Engineers

THIS nomogram is designed to give the partial capacity of one ASME code flanged and dished head. It is constructed from data of Buffalo Tank Corp. The following example shows how to use it to find quickly partial capacity.

Example. What is the capacity of one 80-in. ID flanged and dished head filled to a depth of 30.5 in.?

To find the answer, connect, with a straight edge, 80 in. on the inside-diameter scale with 30.5 on the liquid-depth scale, and read 58.2 gal on the liquid-volume scale.

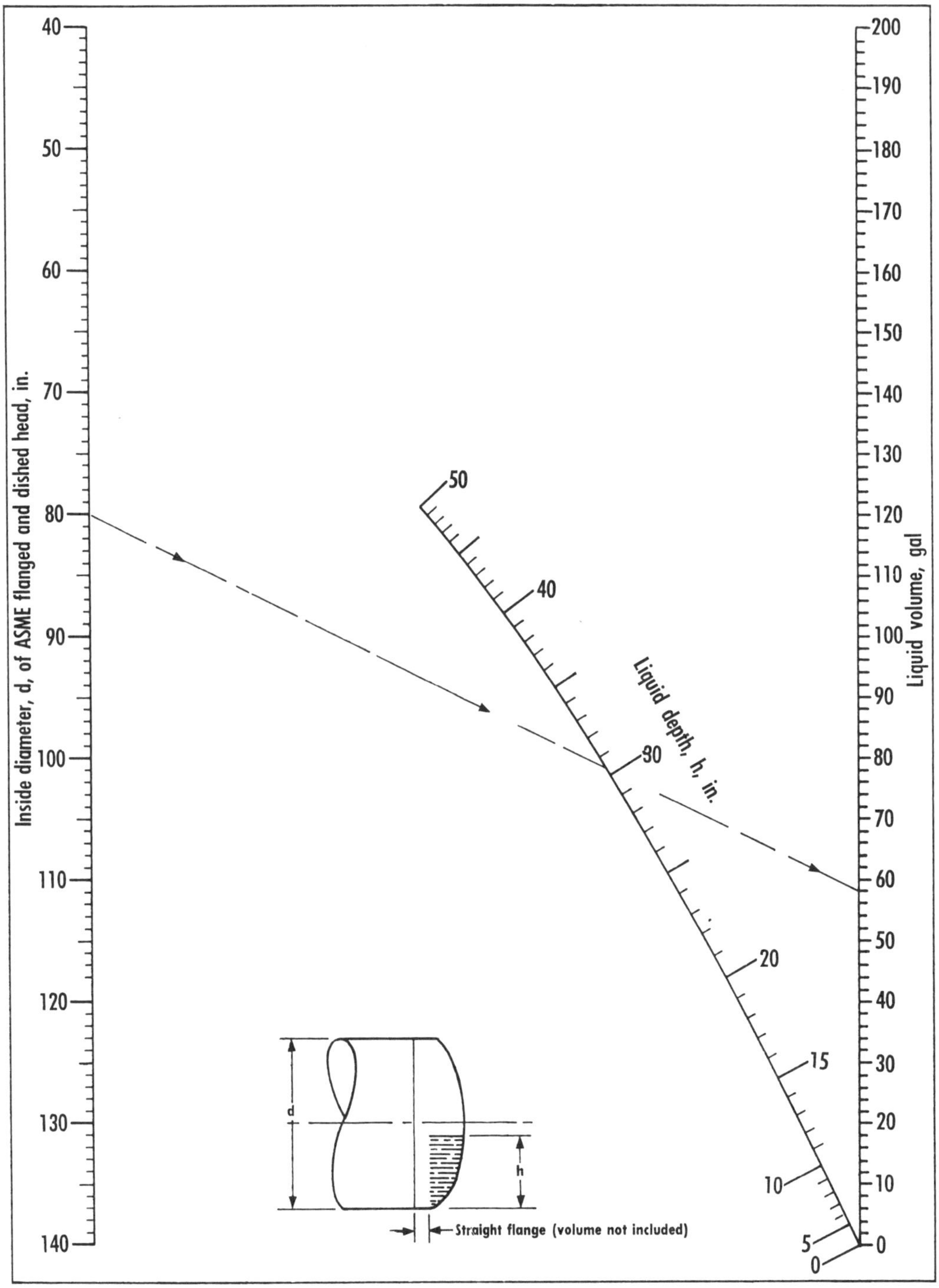

Pressure vessel head content

F. Caplan
Kaiser Engineers

This nomogram is designed to help you quickly compute the contents of two part-full hemispherical or two ASME Code elliptical (2:1) heads. The formula for the part-full contents of two hemispherical heads is this:

$$V_h = 0.0068\, h^2\, [d - (h/1.5)]$$

For two ASME Code elliptical (2:1) heads the formula is:

$$V_e = 0.0034\, h^2\, [d - (h/1.5)]$$

Where:

V_h = volume of liquid (not including straight flange) of two hemispherical heads, gal
V_e = volume of liquid of two elliptical heads, gal
h = liquid height above bottom, in.
d = tank ID, in.

Note. In using the nomogram for full head capacities, multiply the volume for $h = d/2$ by 2. If h is greater than d/2, subtract the difference from the full head capacity.

Example. What is the contents of two part-full hemispherical heads, if the tank ID is 80 in. and the liquid height above bottom is 20 in.? Align d = 80 with h = 20 and read V_h = 182 gal. Note that for two elliptical heads, the volume V_e = 92 gal. **END**

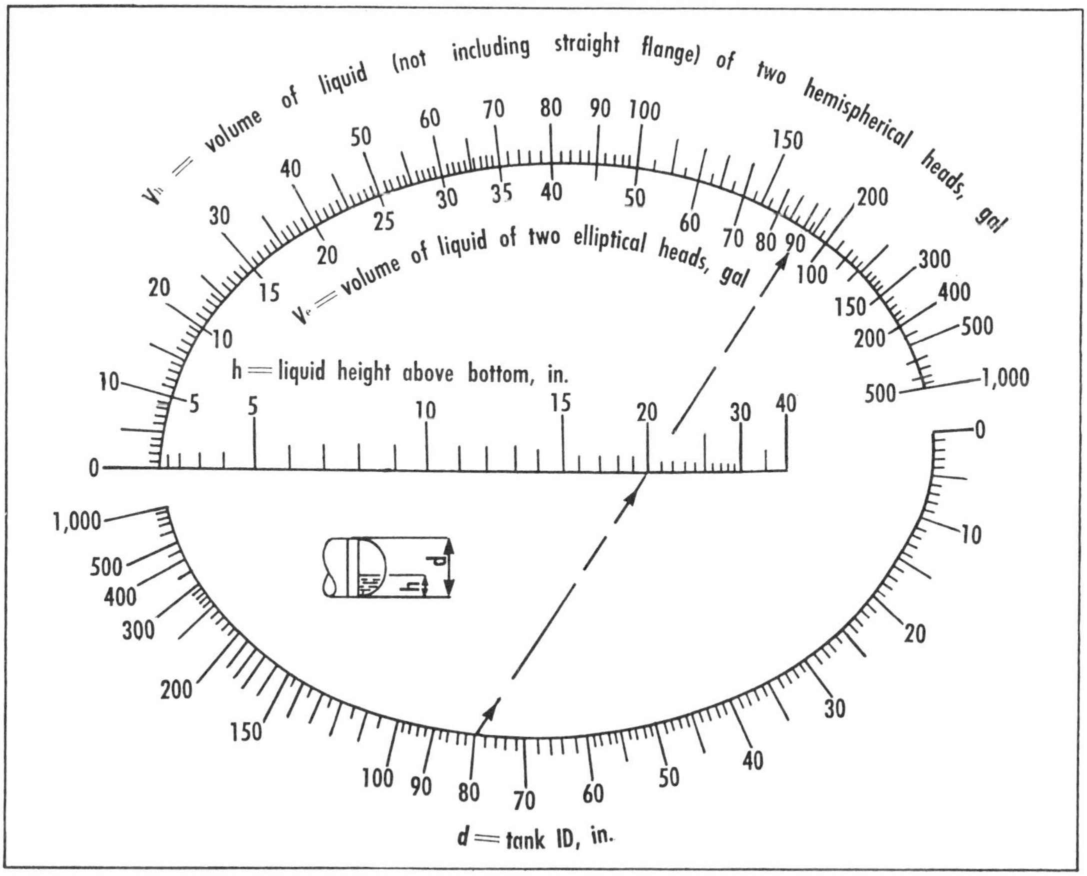

Nomogram tells vessel head thickness

F. Caplan
Kaiser Engineers

THIS nomogram is designed to help you find the minimum wall thickness of ASME Boiler & Pressure Vessel Code flanged and dished (torispherical) and semielliptical (ellipsoidal) vessel heads for unfired pressure vessels.

The Code, Section VIII, Unfired Pressure Vessels, gives the following formula for the minimum (not including corrosion allowance) required wall thickness of torispherical (flanged and dished) vessel heads:

$t = (0.885\ L\ P)\ /\ (S\ E - 0.1\ P)$

Where:

t = wall thickness, in.

L = dish radius, in. (if not known, assume $L = d$, tank diameter, in.)

P = design internal pressure, psig

S = maximum allowable stress, psi (tabulated in Code)

E = joint efficiency (tabulated in Code)

For ellipsoidal (semielliptical) formed heads, the formula is:

$t = (d/2)\ P/(S\ E - 0.1\ P)$

Example. What is the minimum wall thickness for $S = 9{,}000$ psi, $E = 1.0$ ($S\ E = 9{,}000$), $P = 100$, and $d = 60$ (or $L = 33.9$)? Step 1, align the product, SE, with P and mark intersection with Pivot Line. Step 2, align mark on Pivot Line with d and read $t = 0.333$ in. **END**

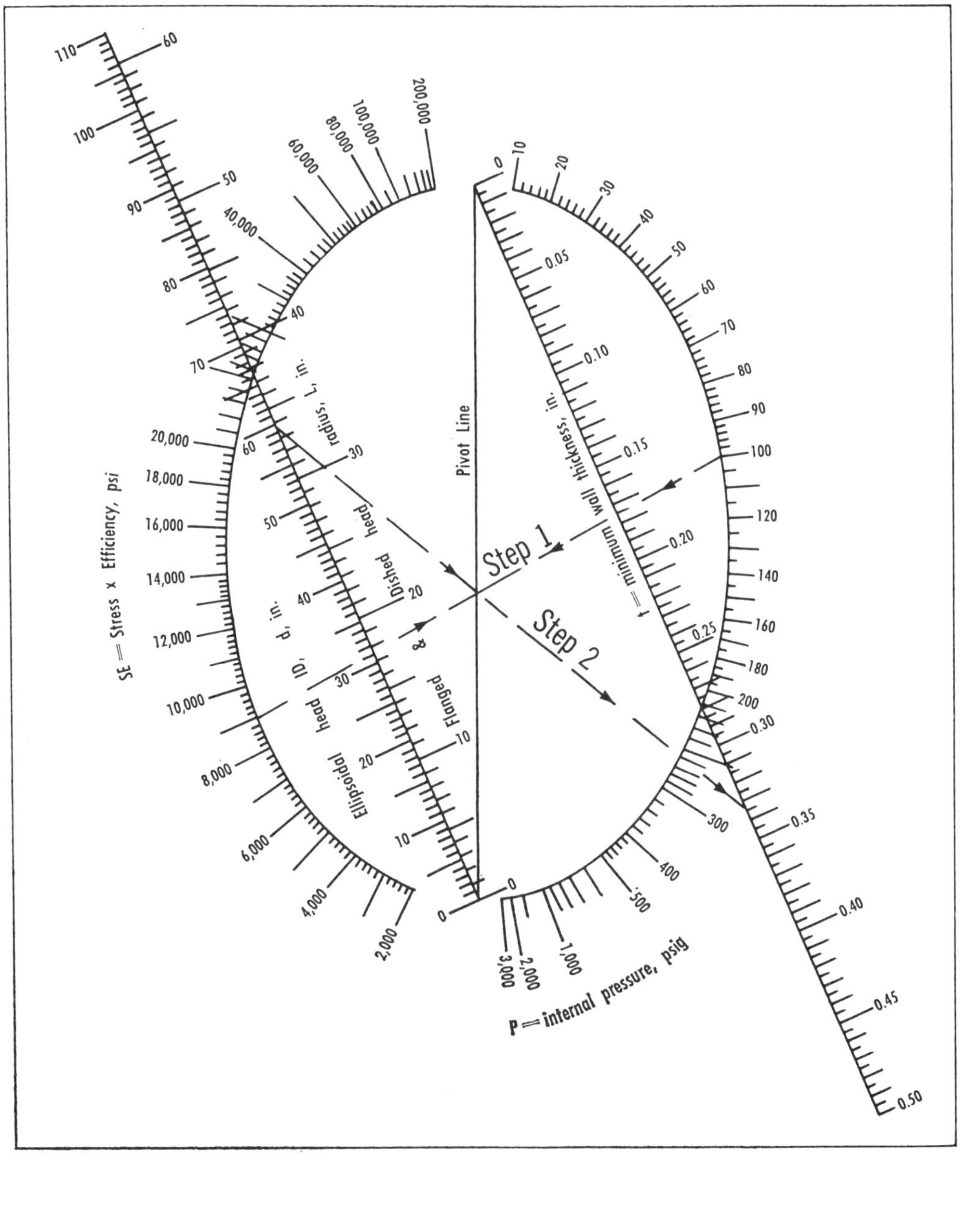

Part-full contents of horizontal tanks

F. Caplan
Kaiser Engineers

CALCULATING the contents of a part-full horizontal tank can be tedious. The following nomogram permits a rapid determination of the contents, and is accurate enough for most purposes. The nomogram is based on the equation:

$$Q=0.00433\ L\left\{\frac{\pi}{8}d^2-\left[\left(\frac{d}{2}-h\right)(d\,h-h^2)^{1/2}+\frac{d^2}{4}\ \text{arc sine}\left(\frac{d-2h}{d}\right)\right]\right\}$$

Where:

Q = contents, gal/in. of straight tank length
L = tank length, in.
d = tank ID, in.
h = depth of liquid, in.

Example. How much liquid is contained in a 120-in. diameter tank filled 40 in. above the bottom? Using a straight edge, align d = 120 in. with h = 40 in. and read Q = 14.3 gal/in. of straight tank length. Note that for full tank capacities, you multiply Q for $h = d/2$ by two. For h greater than $d/2$, subtract the void volume from the full tank. **End**

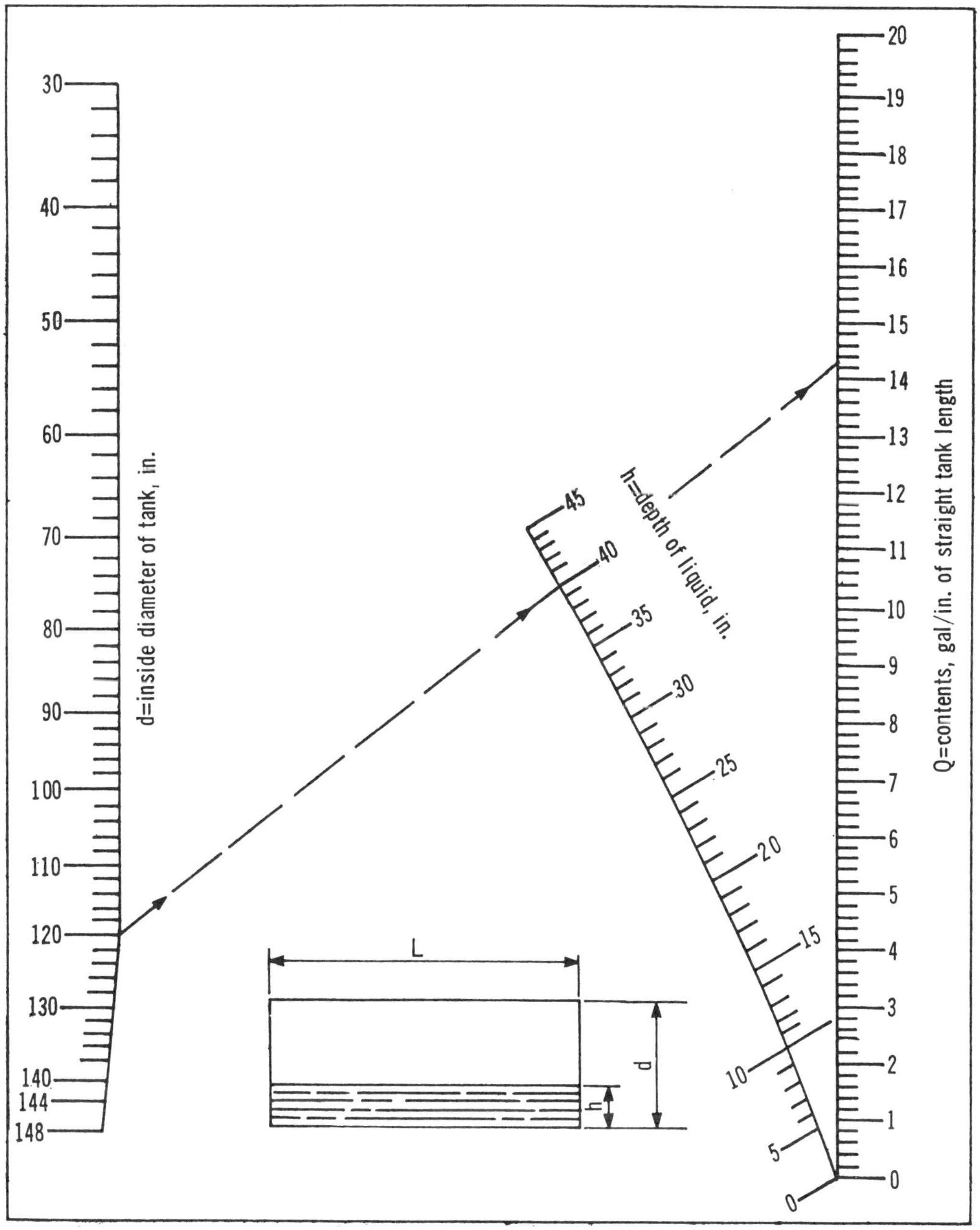

F. Caplan
Kaiser Engineers

Part-full contents of bulged heads

ONE of the more common tank heads encountered is the dished only type. It's also called a bulged or bumped head. In the usual case, the radius of the head is equal to the tank diameter. The following nomogram permits a rapid determination of the part-full contents of two such heads. It's based on the equation:

$$Q = 0.0009328\ h^2\ (1.5\ D - h)$$

Where:
Q=contents of two heads, gal
h=depth of liquid, in.
D=tank diameter, in.

Example. How much is contained in two heads filled to a depth of 23 in., if the tank diameter is 60 in.? Using a straight edge, align D=60 with h=23 and read Q=33.0 gal. as indicated on the nomogram.

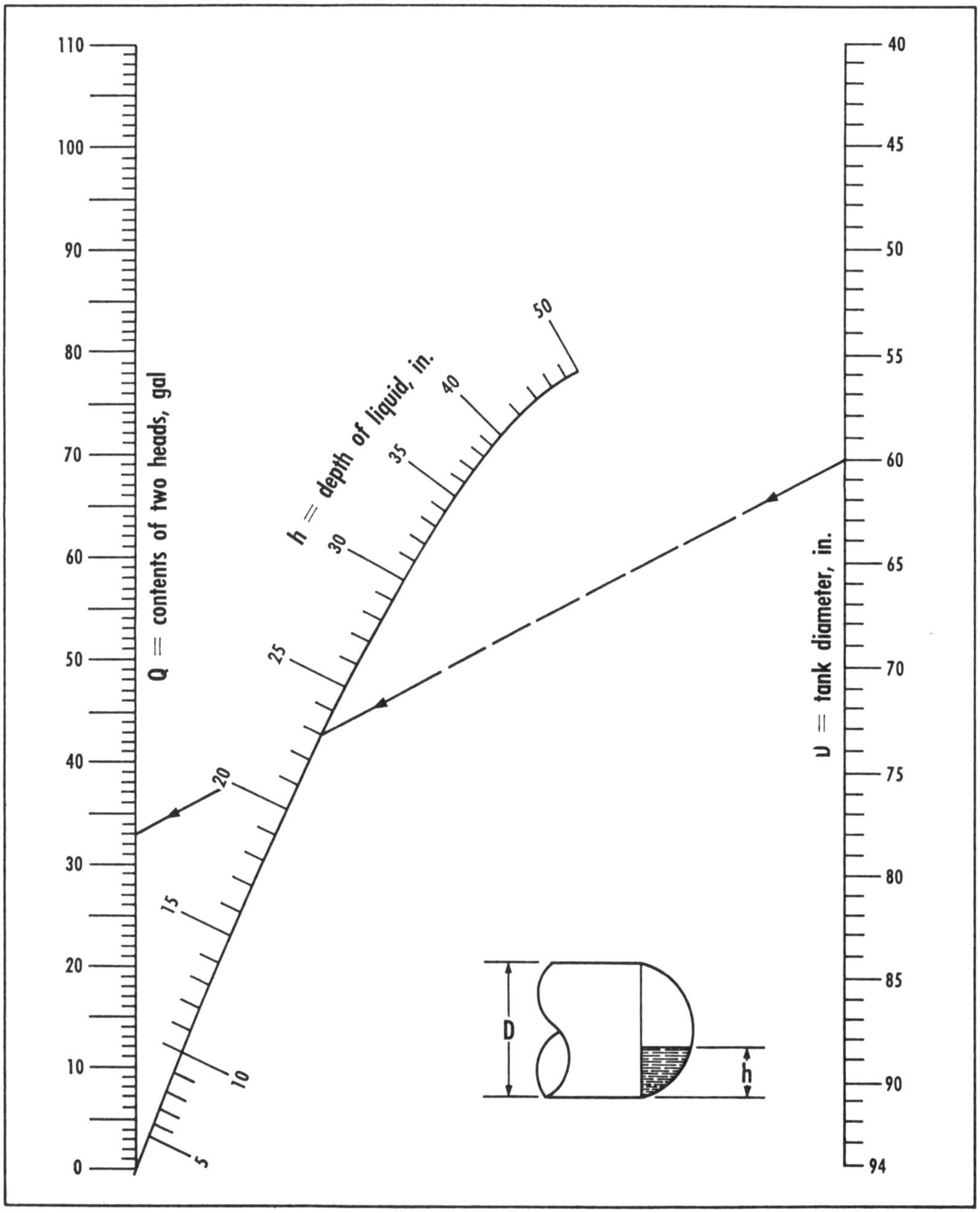

Nomogram finds heavy-wall tube thickness

F. Caplan
Kaiser Engineers

THIS nomogram solves Clavarino's formula for heavy-wall tubing. This is a commonly used equation for thick-walled tubes where the stress produced is not uniform across the wall thickness and Barlow's formula is not applicable. Clavarino's formula is this:

$$s = p\,(1.3\,D^2 + 0.4\,d^2)/(D^2 - d^2) \quad (1)$$
$$= p\,(1.3\,R^2 + 0.4\,r^2)/(R^2 - r^2) \quad (2)$$

and since

$$2\,t = D - d \quad (3)$$
$$t = (d/2)\,[([s + 0.4\,p]/[s - 1.3\,p])^{1/2} - 1] \quad (4)$$
$$= (D/2)\,(1 - [(s - 1.3\,p)/(s + 0.4\,p)]^{1/2}) \quad (5)$$

Where:

s = maximum allowable hoop stress, psi
t = wall thickness, in.
p = design internal pressure, psig.
D = outside diameter of tube, in.
d = inside diameter of tube, in.
R = outside tube radius, in.
r = inside tube radius, in.

Example. What must be the wall thickness of a 4.00-in. ID tube if the design pressure is 1,500 psig and stress shall be limited to 10,000 psi? Step 1, align s = 10,000 with p = 1,500 and mark intersection with Pivot Line. Step 2, align Pivot Line mark with d = 4.00. Read t = 0.296 in. and D = 4.59 in. Note that the t, d, and D scales can be simultaneously multiplied (or divided) by 2, 10, etc. The s and p scales can also be extended the same way. END

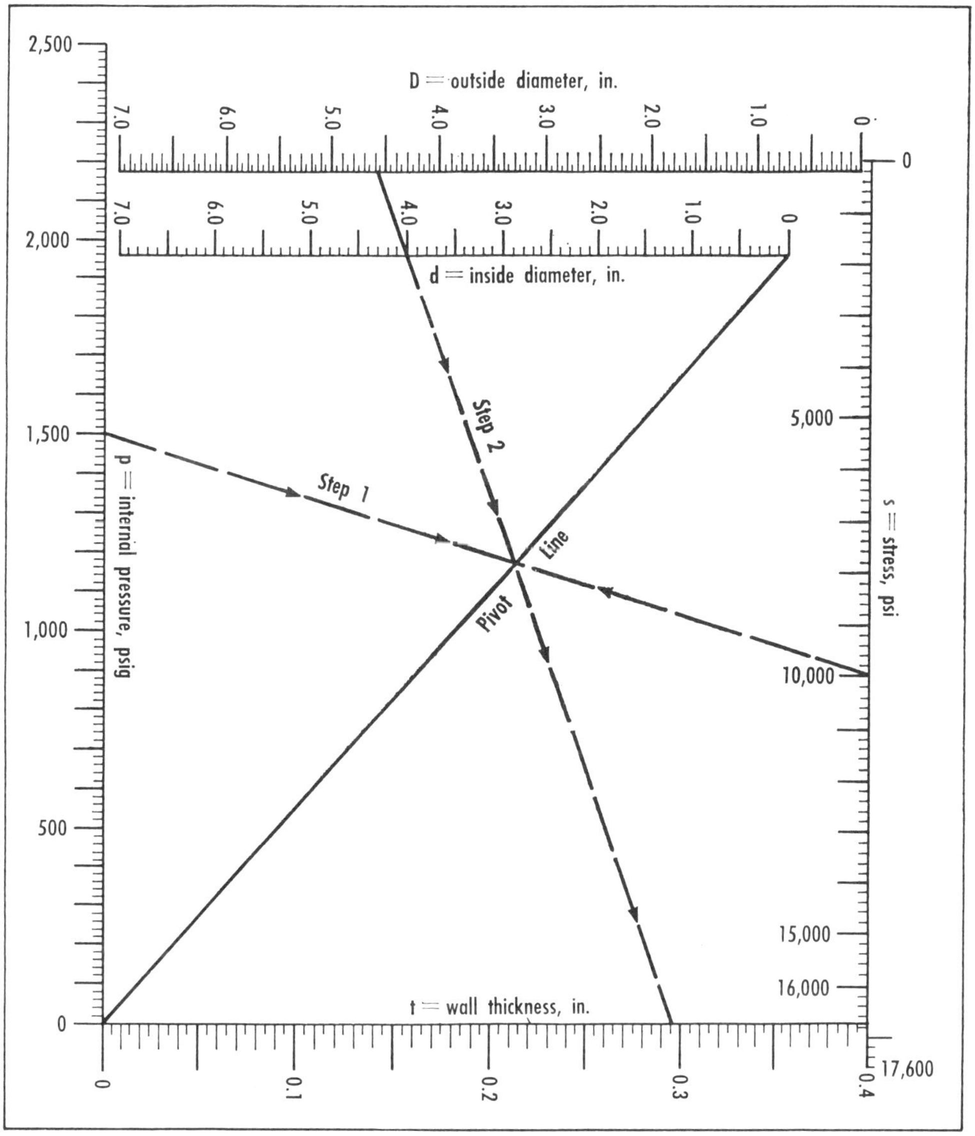

Size electric heaters for petroleum products

Bill Sisson
Superintendent of Operations
Nipak Inc.
Pryor, Okla.

THE nomogram presented here can be of considerable assistance when it is necessary to select an electric heater for heating various petroleum products. In selecting heaters for this service, you are confronted not only with temperature rise and flow rates but also with variable specific heat of the product being heated. The nomogram is constructed from the formula:

$$H = \frac{Q\,d^{1/2}\,[3.235 + 0.00185\,(t_2 + t_1)\,]\,(t_2 - t_1)}{3{,}600 \times 0.9486}$$

Where:

H = heater size, kw

Q = quantity to be heated, gph

t_1 = initial temperature, ° F.

t_2 = final temperature, ° F.

d = specific gravity (60/60 ° F.)

The formula is based on the formula for variable specific heat:

$$c = (1/d^{1/2})\,(0.388 + 0.0045t)\ \text{BTU/lb/° F.}$$

Example. Determine the kw required to heat 400 gph of oil with a specific gravity of 0.88 from an initial temperature of 60°F. to a final temperature of 190°F.

Solution. Line from 60°F. on t_1 scale with 190°F. on t_2 scale (Step 1), extend line to Transfer line No. 1 and mark. From this marked point, align with 0.88 on Sp.Gr. scale (Step 2) and mark where line intersects Transfer line No. 2. From there, line with 400 gpm on Q scale (Step 3), extend line to kw scale, and read 52.85 kw required. **END**

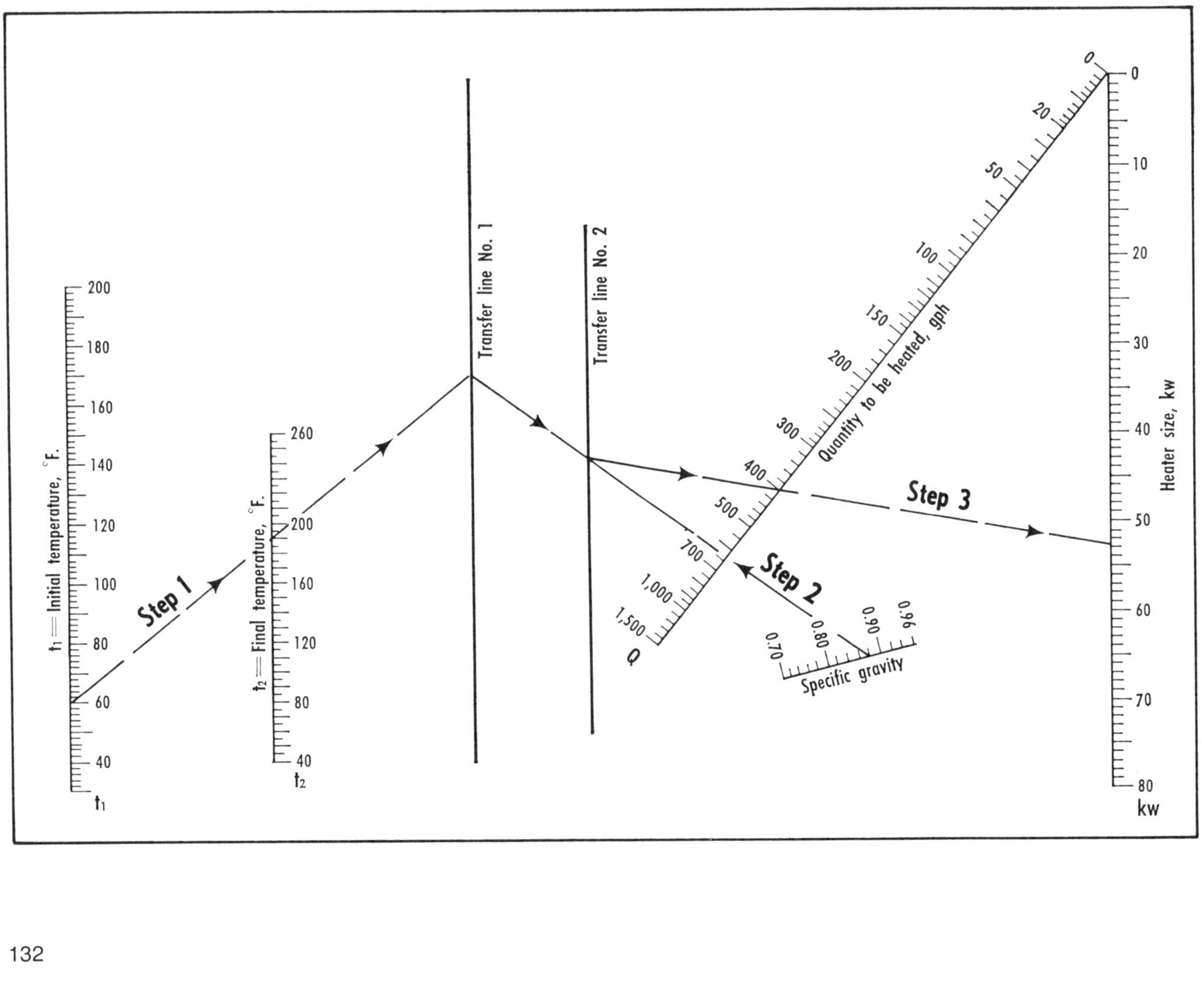

F. Caplan
Kaiser Engineers

This nomogram finds the valley angles of hoppers and chutes

UNLESS all angles in a hopper are steeper than the angle of repose of the material, flow may not occur. The valley angle, the vertical angle of the intersection of two sides, is always less than any side vertical angle but must be greater than the angle of repose of the material. It is given by the equation,

$$\cot^2\gamma = \cot^2\alpha + \cot^2\beta$$

Where:

γ = valley angle, degrees
α, β = side vertical angles, degrees

This nomogram permits a fast solution of the equation.

Example. What is the valley angle in a hopper if one vertical wall angle is 43° and the other vertical wall angle is 30°? Using a straightedge, align 43° on the larger hopper-wall-angle scale with 30° on the smaller hopper-wall-angle scale. At the intersection with the valley-angle scale, read $\gamma = 26.1°$.

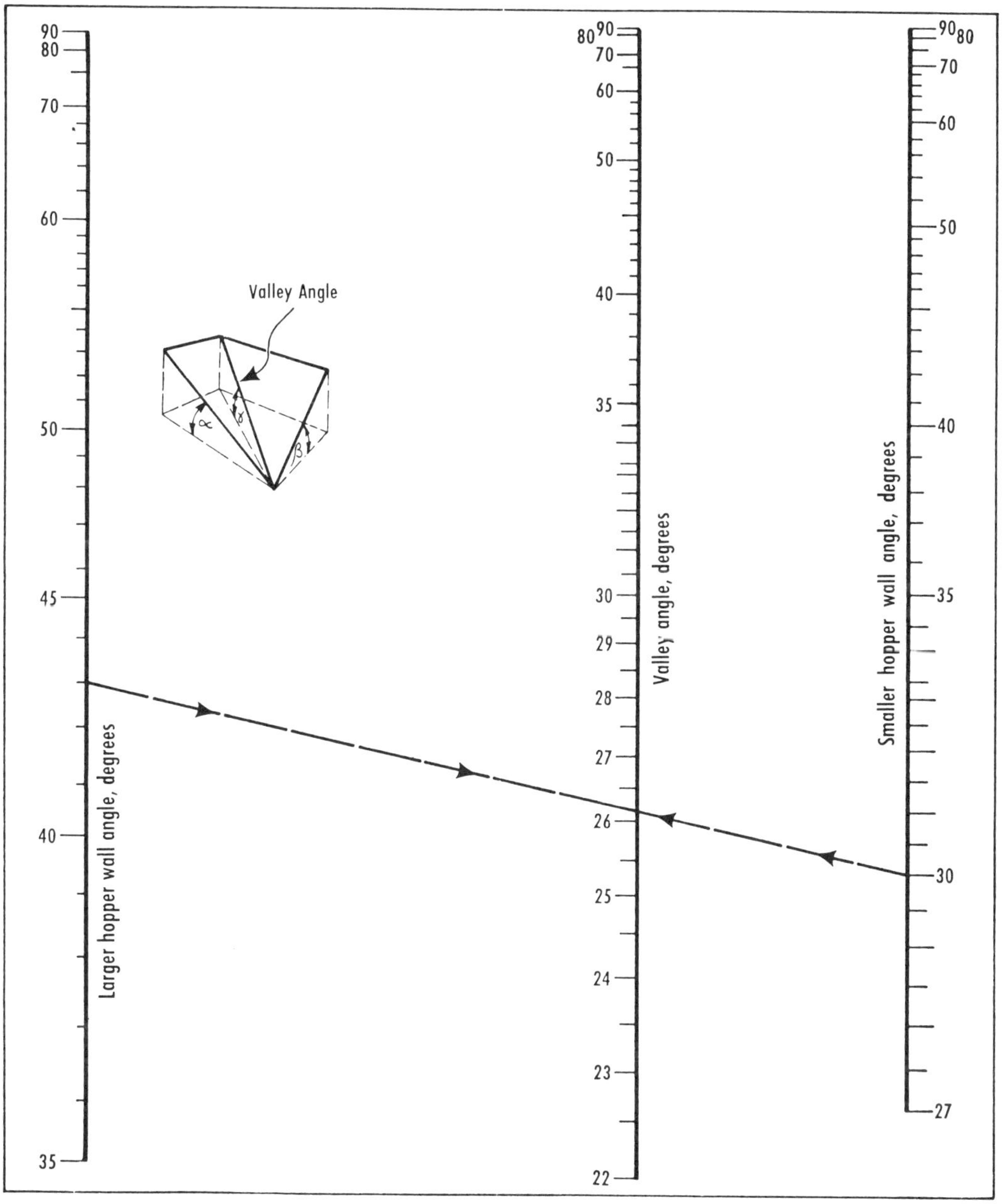

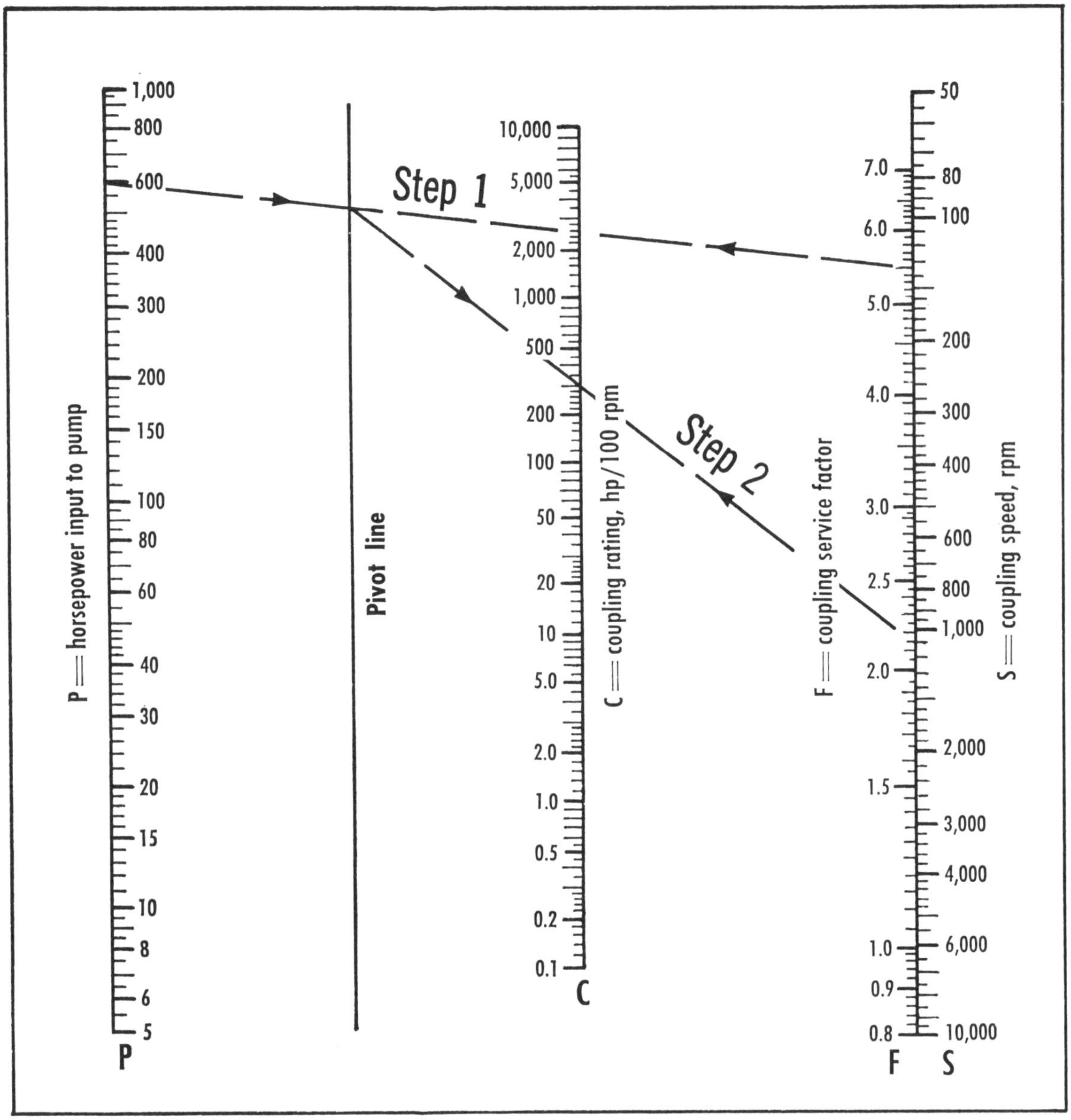

Nomogram selects flexible couplings for pumps

Bill Sisson
Superintendent of Operations
Nipak, Inc.
Pryor, Okla.

WHEN it is necessary to choose a flexible coupling for a pump, the horsepower transmitted, speed of rotation, and applicable service factor must be known. The service factor compensates for shock loads and other variations in power output. Couplings are usually rated in hp/100 revolutions/min. The rating may be determined from the formula:

$$C = 100(PF)/S$$

Where:

C = coupling rating, hp/100 rpm
P = horsepower input to pump
F = coupling service factor
S = coupling speed, rpm

Service factors can be found in manufacturers' literature. One manufacturer uses the following: turbine-driven centrifugal pump, 1.25; motor-driven centrifugal pump, 1.5; motor-driven duplex and triplex pumps, 3.5; engine-driven centrifugal pumps, 3.0; and engine-driven duplex and triplex pumps, 5.5. When choosing a coupling be sure to use the correct service factor. Values vary from one manufacturer to another. The nomograph presented here permits rapid determination of the proper coupling rating.

Example. What rating coupling should be chosen for an engine-driven triplex pump rated at 600 hp at 1,100 rpm?

Solution: Using the service factor (5.5) given above for an engine-driven triplex pump, line 600 hp on P scale with 5.5 on F scale and mark where line crosses pivot line (Step 1). From this marked point, align with 1,100 rpm on S scale and read 300 hp/100 rpm where line crosses C scale (Step 2). END

Nomogram gives total surface area, metal volume, weight of tanks

Bill Sisson
Superintendent of Operations
Nipak Inc.
Pryor, Okla.

THIS nomogram will help you quickly determine the total surface area and volume of metal in cylindrical tanks. Scales are also provided for finding the weight of the tank for various materials of construction. The formulas used for construction of the nomogram are:

$$A = 2\pi\ (D^2/4 + \pi\ DH$$

$$V = A\ (t_w/12)$$

$$W = Vd$$

Where: A = total surface area of tank, sq ft; D = diameter of tank, ft; H = height or length of tank, ft; V = volume of metal in tank wall, cu ft; t_w = thickness of tank wall, in.; d = density of vessel wall, lb/cu ft; and $\pi = 3.1416$.

Example. If a tank has a diameter of 20 ft and height of 40 ft, what is the total surface area? If the thickness of the tank wall is ⅜ in., what is the volume of metal in the vessel wall? If the tank is constructed of mild steel, how much does it weigh?

Solution. Line 40 ft on H scale (Step 1) with 20 ft on D scale, extend to A scale, and read total surface area as 3,142 sq ft. Line 3,142 sq ft on A scale (Step 2) with ⅜ in. on T_w scale, extend to V scale, and read metal volume as 98.2 cu ft. To find tank weight, align 98.2 on V scale (Step 3) with Mild Steel reference point on T_w scale, extend to W scale, and read tank weight as 48,000 lb. END

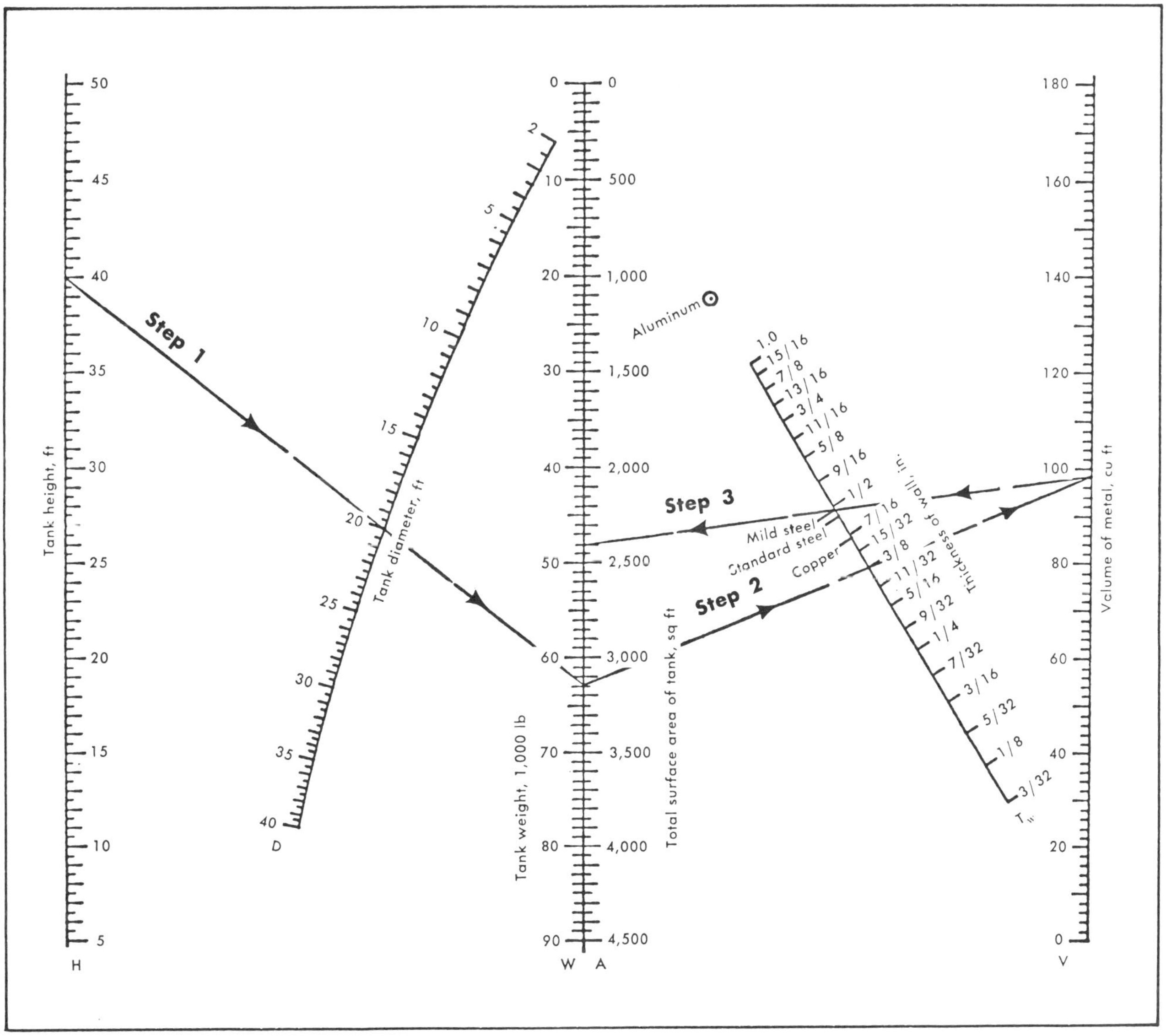

Nomogram gives heat loss through 85% magnesia insulation

Bill Sisson
Superintendent of Operations
Nipak Inc.
Pryor, Okla.

THE function of a heat insulation is to lower the temperature of exposed surfaces and so minimize heat losses. This is done by covering the heated surface with a material that retards heat flow. The most widely used insulation for hot surfaces up to 600° F. is 85% magnesia.

This nomogram, based on data from Magnesia Insulation Manufacturers Association, will determine the heat loss for standard and double-standard thicknesses of 85% magnesia insulation applied to pipelines. It can also be used approximately for such insulations as mineral wool and sponge-felted asbestos.

The standard thicknesses are: 7/8-in. for 1/2, 3/4, 1, 1 1/4, and 1 1/2-in. pipes; 1 1/32-in. for 2, 2 1/2, 3, and 3 1/2-in. pipes; 1 1/8-in. for 4, 5, and 6-in. pipes; and 1 1/4-in. for 8 and 10-in. pipes. The double-standard thicknesses are: 1-29/32-in. for 1/2, 3/4, 1, 1 1/4, and 1 1/2-in. pipes; 2 5/32-in. for 2, 2 1/2, 3, and 3 1/2-in. pipe; 2 1/4-in. for 4-in. pipe; 2 5/16-in. for 5 and 6-in. pipe; and 2 1/2-in. for 8 and 10-in. pipe.

Example. A 150-ft run of 6-in. pipe insulated with double-standard insulation carries 600-psig saturated steam (489° F.) through a room where the temperature is 80° F. What is the hourly heat loss? What is the heat loss for 1,500 ft of pipeline?

Solution. Connect (Step 1) 80° F. on T_1 scale with 600 psig (489° F.) on T_2 scale, extend to Pivot line 1 and mark. From marked point (Step 2) line with 6-in. on Double-Standard pipe-size scale and extend to Pivot line 2. From this point (Step 3) connect with 150 ft on L scale, extend to H scale, and read heat loss at 26,000 BTU/hr. The heat loss for 1,500 ft of pipe (150×10=1,500) is 26,000×10=260,000 BTU/hr.

END

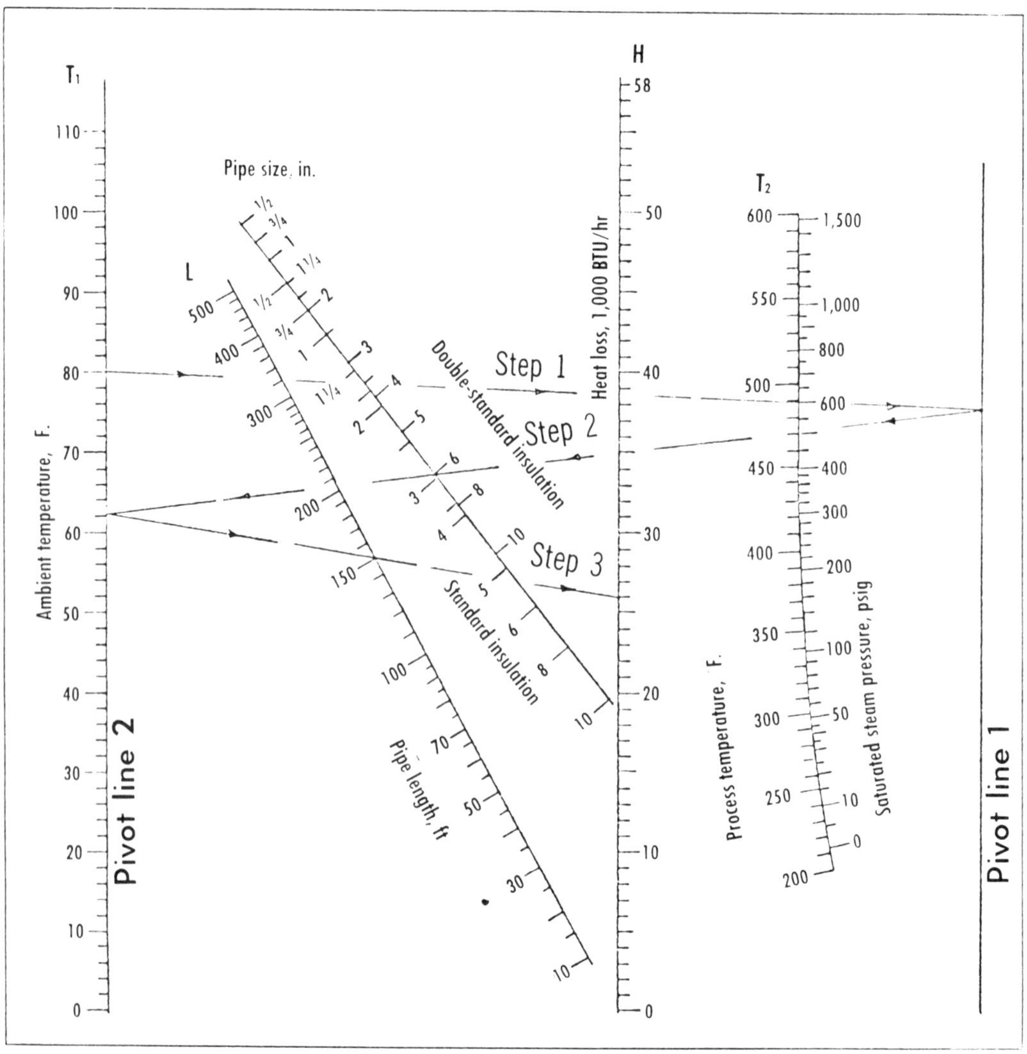

Nomogram gives properties of air

Bill Sisson
Superintendent of Operations
Nipak, Inc.
Pryor, Okla.

MANY engineering calculations require a knowledge of the properties of air at different temperatures and elevations. The nomogram presented here provides a convenient way to determine such properties as actual vapor pressure, dew point, specific humidity, and relative humidity. It is based on the following equations:

$$p_v = p_w - B\,(t_d - t_w)/2{,}700$$
$$R = p_v/p_d$$
$$w = p_v/1.61\,(B - p_v)$$
$$\Delta t = t_d - t_w$$

Where:

p_v = actual vapor pressure, in. Hg (corresponds with dew point temperature)
t_w = wet bulb temperature, °F.
t_d = dry bulb temperature, °F.
p_w = vapor pressure, in. Hg, corresponding with t_w
p_d = vapor pressure, in. Hg, corresponding with t_d
B = total vapor pressure (barometric pressure), in. Hg
w = specific humidity, lb water vapor/lb dry air
R = relative humidity (calculated as a decimal)
Δt = wet-bulb depression, °F.

Example. Determine the actual vapor pressure, dew point, specific humidity, and relative humidity of air at an altitude of 750 ft above sea level (barometric pressure = 29.1 in. Hg), when dry-bulb temperature is 95° F. and wet-bulb temperature is 80° F.

Solution. Align 750 ft (29.1 in. Hg) on Scale A or B with wet-bulb depression = 15 on Scale F (Δt = 95-80), and mark where line crosses Scale E (Pivot line) (Step 1). From the marked point on Scale E align with 80° F. on Scale L (Step 2) and read 0.87 in. Hg vapor pressure on Scale H and a dew point of 74.5° F. on Scale K. Next, from the 0.87 point on Scale H, align with 750 ft (29.1 in. Hg) on Scale A or B (Step 3) and read 0.019 specific humidity where the line crosses Scale G. Again, from 0.87 on Scale H align with 95° F. on Scale D, extend the line to Scale C (Step 4) and read 52.4% relative humidity. END

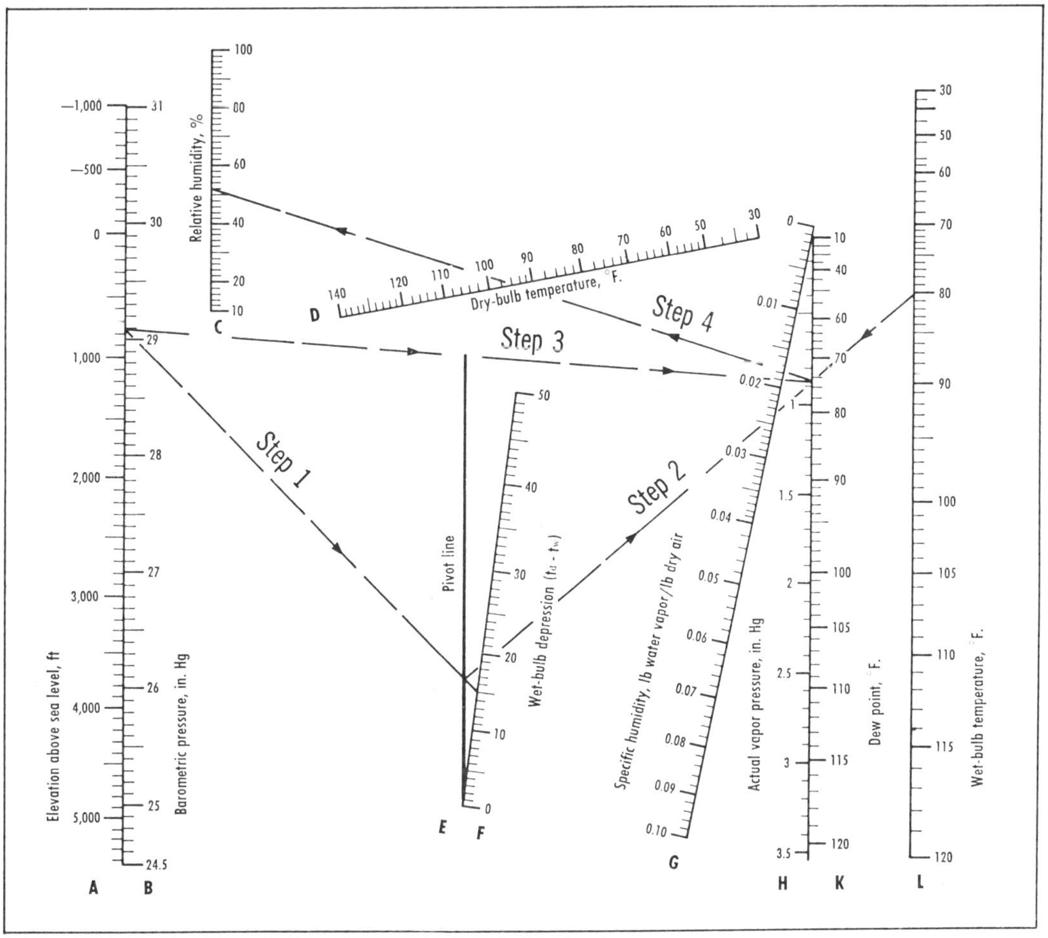

Nomogram gives allowable short-circuit currents for insulated copper and aluminum conductors

Bill Sisson
Superintendent of Operations
Nipak Inc.
Pryor, Okla.

WITH the ever-increasing kva capacity of power systems, the possible short-circuit currents are becoming so high that it is often necessary to consider the effect of these short circuits on the heating of cables. The conductor size must be large enough to carry the short-circuit current for a long enough time to permit the circuit breakers to open before the conductor is heated to the point where it damages the insulation. The nomogram shown here gives the maximum currents to which various size copper and aluminum conductors can be subjected for various amounts of time without injuring the insulation. The formulas used for construction of the nomogram are:

Copper-conductor current = 0.0549 cm $(1/T)^{1/2}$

Aluminum conductor current = 0.0366 cm $(1/T)^{1/2}$

Where cm = circular mil area of conductor; T = time, sec; allowable temperature rise = 80° C.; and no heat transfer from conductors.

Example 1. What is the maximum allowable short-circuit current for AWG 4/0 copper cable carrying 60-cycle current and lasting 8 cycles?

Solution. Connect AWG 4/0 on Copper-conductor-size scale (Step 1) with 8 cycles on Time scale, and read 32,000 amp as maximum current allowable where line crosses current scale.

Example 2. What is the maximum allowable short-circuit current for AWG 4 aluminum cable carrying 60-cycle current and lasting 2 sec?

Solution. Connect AWG 4 on Aluminum-conductor-size scale (Step 2) with 2 sec on Time scale, and read 1,080 amp as maximum current allowable where line crosses Current scale. END

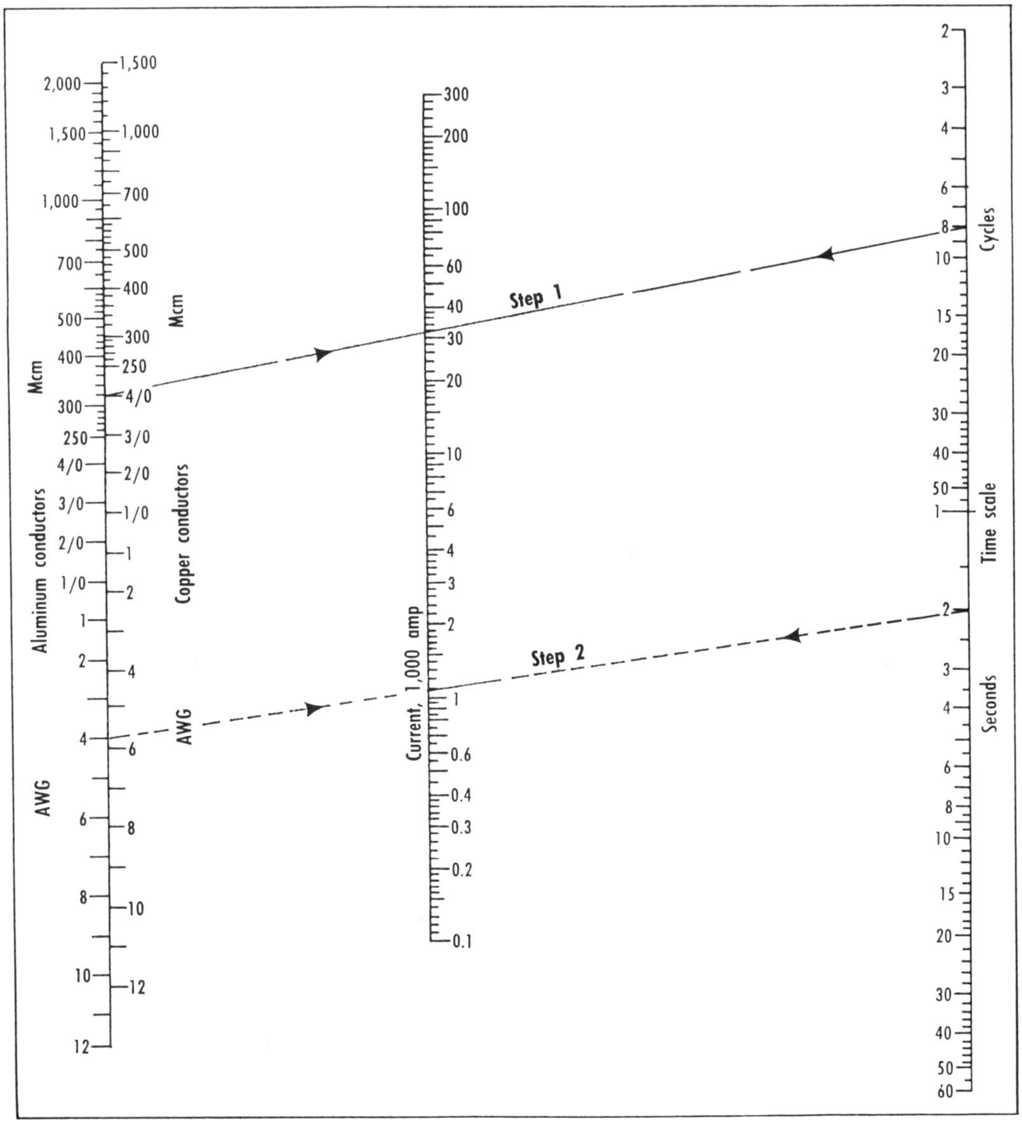

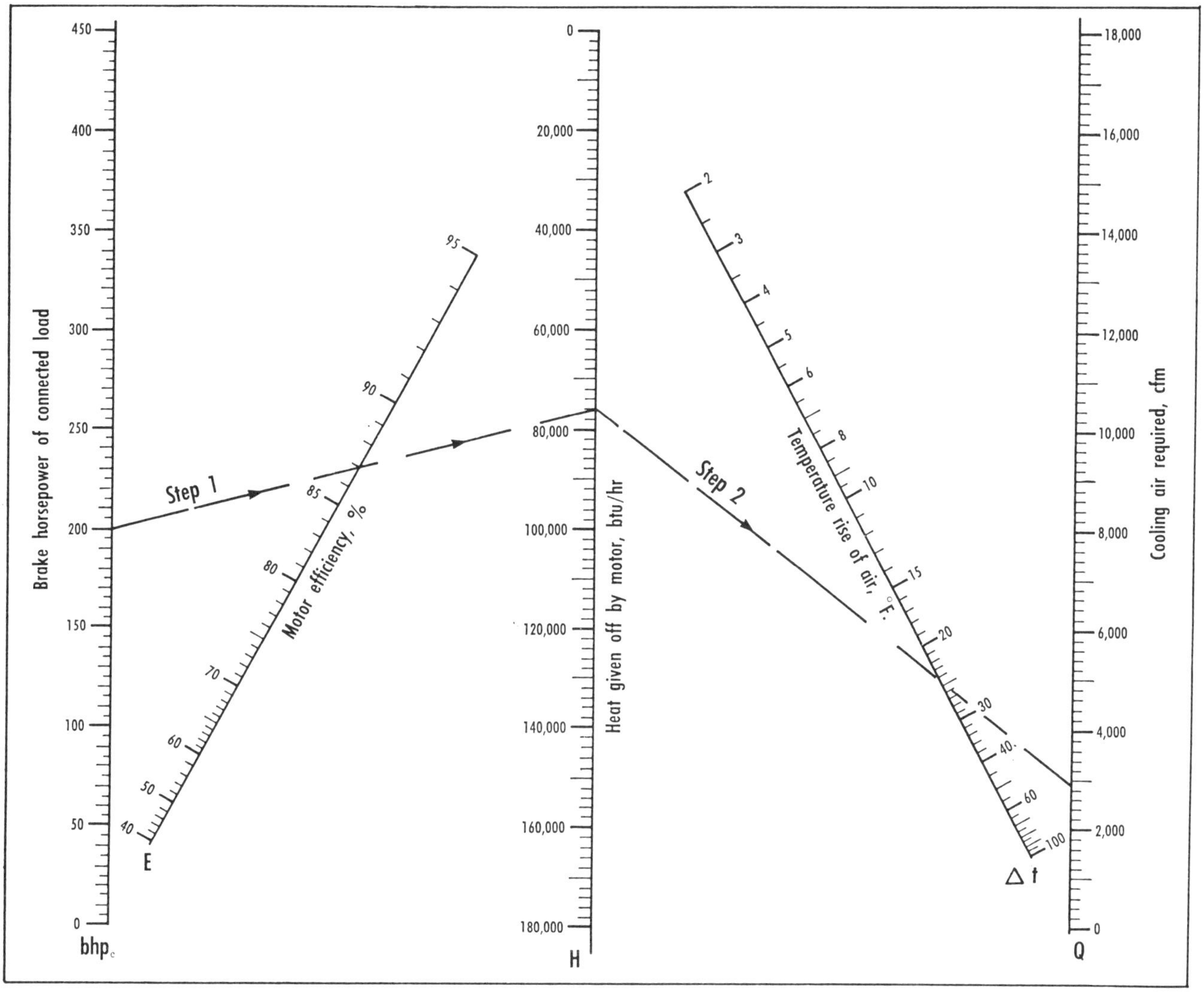

Nomogram gives motor heat, cooling air needs

Bill Sisson
Superintendent of Operations
Nipak, Inc.
Pryor, Okla.

ELECTRIC motors are rated on their operating temperature. High temperature can reduce their life span. Motor temperature is the result of ambient temperature and temperature rise due to losses within motor in converting electrical energy into mechanical energy. Many electric motors are rated at a temperature rise of 40° C. (72° F.) in an ambient temperature not to exceed 40° C. (104° F.).

Heat generated by motor inefficiency and drive transmission must be removed to prevent reduced motor life. Motor heat that must be removed is the difference between the input of electrical energy and the output of mechanical energy. It can be expressed as:

$$H = 2{,}545\ bhp_c\ [(100-E)/E]$$

Where:

H = heat to be removed, BTU/hr
E = motor efficiency, %
bhp_c = brake horsepower of connected load

If the heat generated by motor and connected load is not removed, ambient temperature and motor operating temperature will both increase. The amount of air required to cool electric motors can be found with the formula:

$$Q = [(2{,}358)\ (bhp_c)\ (100\text{-}E)]/(\Delta t)\ (E)$$

Where:

Q = air required for cooling, cfm
Δt = temperature rise of air, °F.

The nomogram shown here provides a quick method for finding heat load and air required for cooling.

Example. What is the heat load and amount of air required to keep room temperature from exceeding 104° F., if brake horsepower of connected load is 200, motor and drive losses result in an efficiency of 87%, and entering air temperature is 80° F.? What is heat load and air required if bhp_c is 2,000?

Solution. Line 200 on bhp_c scale with 87% on E scale, extend line to H scale, and read 76,000 BTU/hr heat load where line intersects H scale (Step 1). Now, connect point found on H scale with 24° on Δt scale (Δt = 104° F. − 80° F. = 24° F.), extend line to Q scale, and read 2,950 cfm cooling air required (Step 2).

If bhp_c is 2,000 then heat load = 76,000 × 10 = 760,000 BTU/hr, and cooling air required = 2,950 × 10 = 29,500 cfm.

END

Nomogram finds power factor

F. Caplan
Kaiser Engineers

This nomogram is designed for use with a common method of measuring the power factor of three-phase electrical circuits in which two single-phase wattmeters are used. Each wattmeter has its current coil connected into one of the phase conductors and its potential coil connected between its own phase conductor and the third conductor, as shown in diagram in nomogram.

The equation for the power factor by this method is:

$$\text{P.F.} = \cos \arctan [3^{1/2}(W_2 - W_1)/(W_2 + W_1)]$$
$$= (W_2 + W_1)/[(W_2 + W_1)^2 + 3(W_2 - W_1)^2]$$

Where:

W_1 = Smaller wattmeter reading
W_2 = Larger wattmeter reading
P.F. = Power factor

Example. What is the power factor of a three-phase system using two single-phase wattmeters for power-factor measurement in which the larger wattmeter reading is 100 and the smaller wattmeter reading is 20?

Align $W_2 = 10$ with $W_1 = 2$ and read the power factor as 0.655. Note that the W_1 and W_2 scales can be simultaneously multiplied by 0.1, 2, 10, 100, etc. END

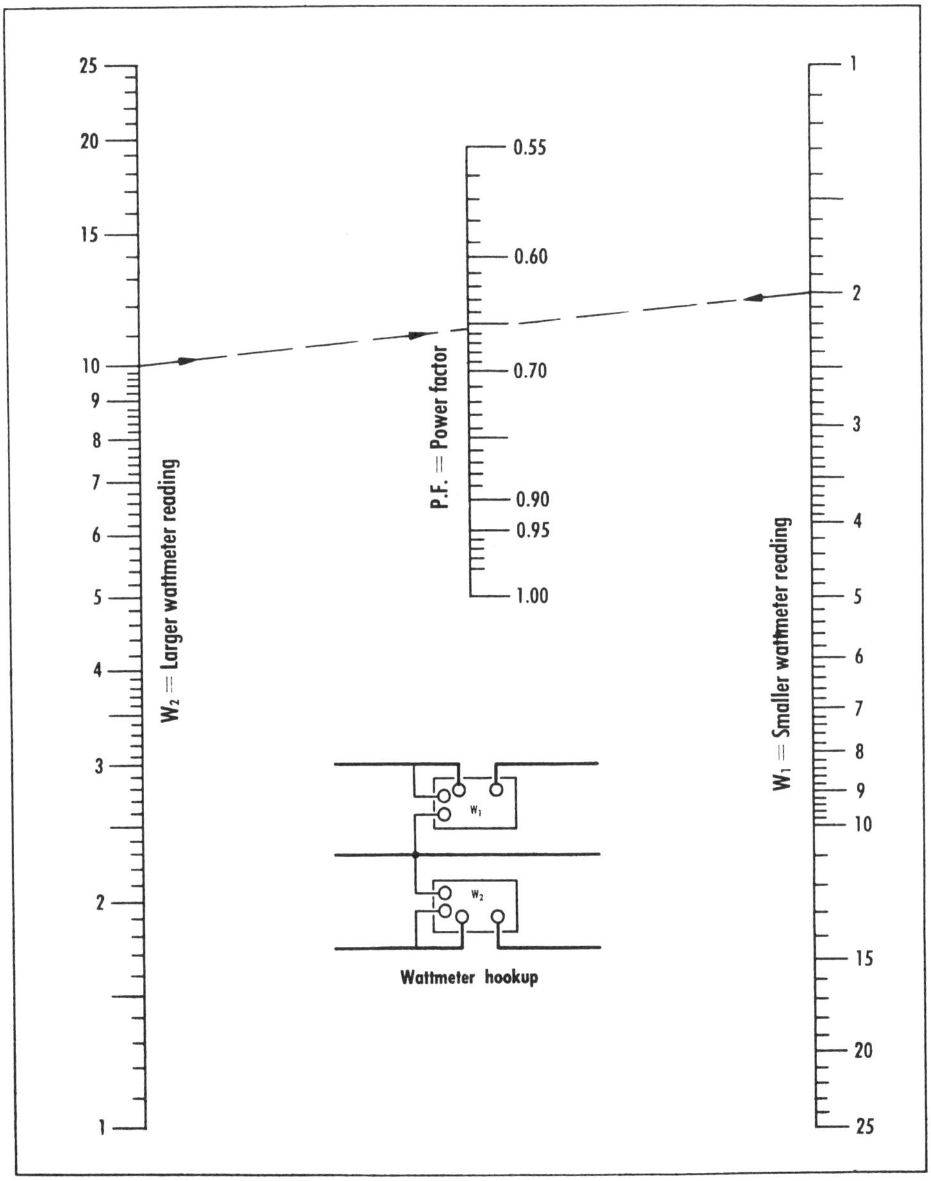

Pump-motor brake horsepower

F. Caplan
Kaiser Engineers

THIS nomogram is designed to find the correct size motor to drive a pump. It is based on the equation:

$$\text{Bhp} = \text{F H s}/3{,}960\,\eta$$

Where:

Bhp = brake horsepower
F = flow rate, gpm
H = head, feet of fluid
s = specific gravity
η = pump efficiency (decimal)

Example. How much pump-brake horsepower is required to move 60 gpm of liquid at a 400-ft head, if the liquid specific gravity is 0.9 and the pump efficiency is 70%?

To find the answer, align 0.9 on the specific-gravity scale with 400 ft on the fluid-head scale and mark the intersection of pivot line (1) with a straightedge.

Next align 0.70 on the pump-efficiency scale with the mark on pivot line (1) and mark the intersection of the straightedge with pivot line (2). Next align the mark on pivot line (2) with 60 gpm on the flow-rate scale, and read 7.8 bhp on the brake-horsepower scale.

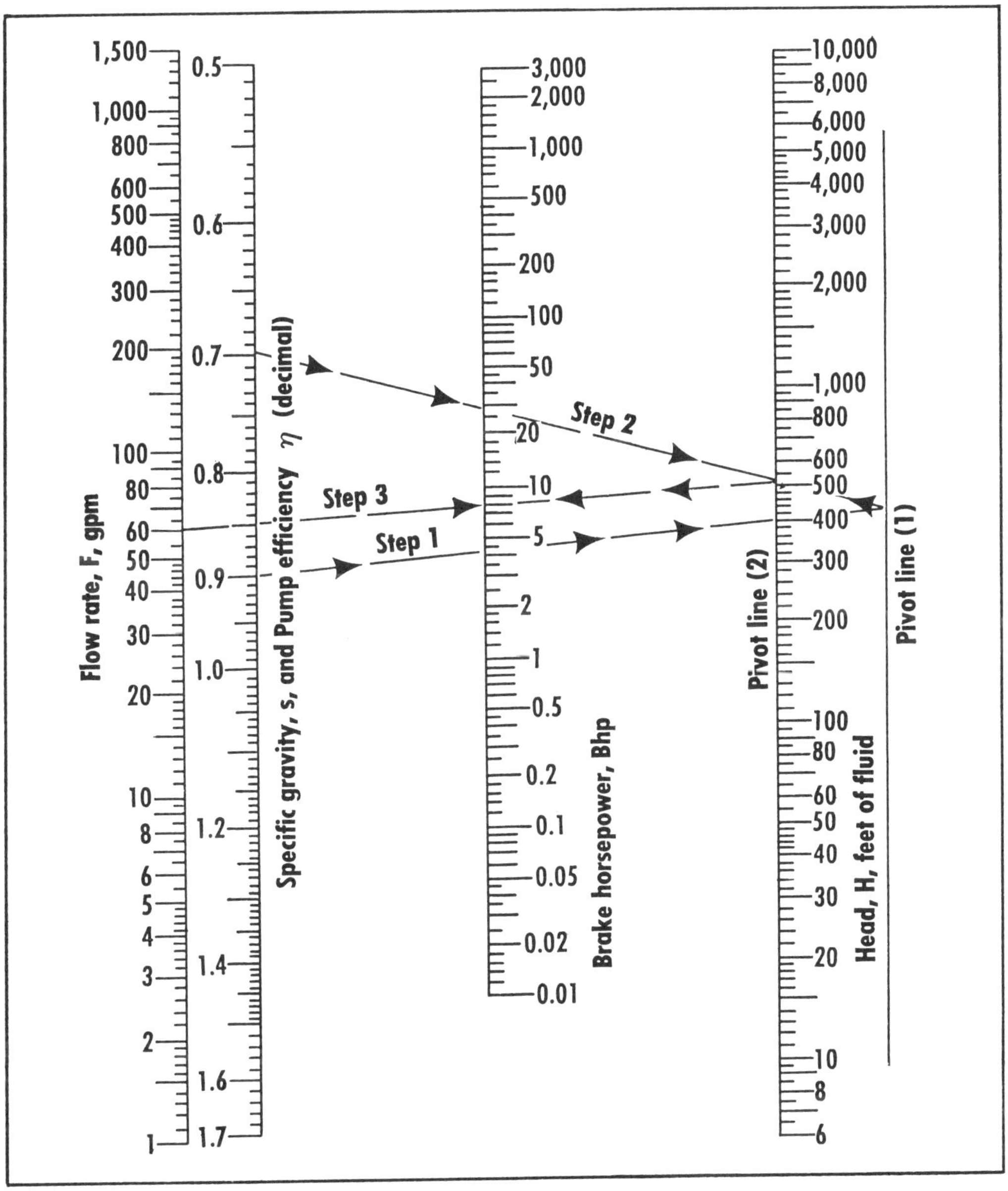

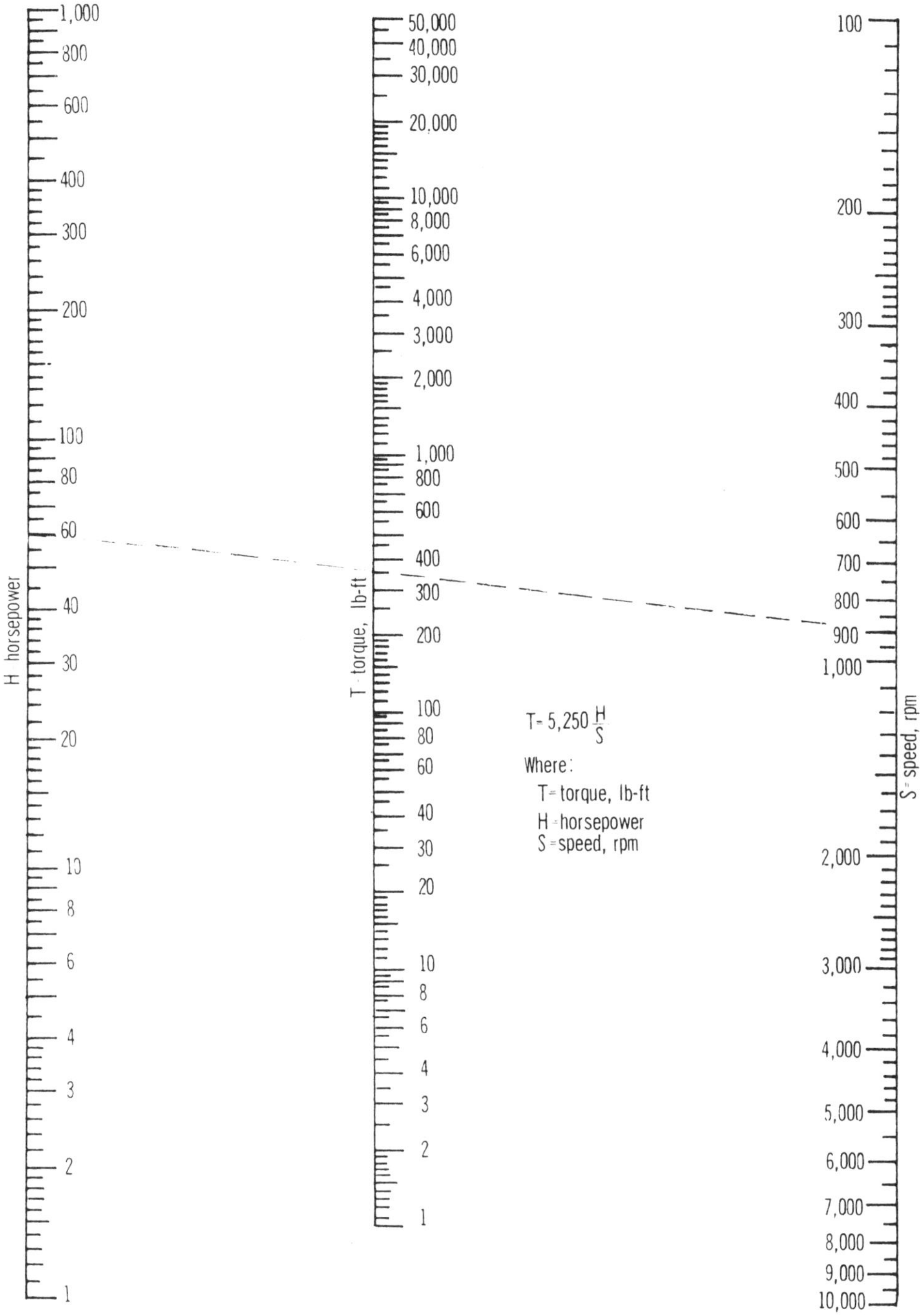

Use this nomogram to find torque

TORQUE of a rotating machine is directly proportional to horsepower and inversely proportional to speed. Or

$$T = 5{,}250\ H/S$$

Where:

T=torque, lb-ft
H=horsepower
S=speed, rpm

The nomogram solves this equation.

Example: Find the torque when the horsepower is 60 and the speed is 900 rpm.

Align a straightedge at 60 on the H scale with 900 on the S scale, find the intersection of the straight edge on the T scale, and read the answer of 3,500 lb-ft on the T scale, as indicated by the dashed line.

Nomogram determines approximate length of V-belts

Bill Sisson
Superintendent of Operations
Nipak Inc.
Pryor, Okla.

WHEN you are required to select a V-belt drive, the known quantities are usually the rpm of the driver and of the driven machine and the approximate center distance between the sheaves.

After you have selected the pitch diameter of the small sheave, the approximate length of the belt can be found with the nomogram as shown by the following.

Example. A 1,400-rpm engine is to drive a centrifugal pump at 380 rpm. If the shaft centers are 52 in. apart and the pitch diameter of the small sheave is 13½ in., what is the speed ratio and belt length?

Solution. Align 1,400 on Faster rpm (F) scale (Step 1) with 380 on Slower (S) rpm scale, extend to R scale, and read speed ratio as 3.68. Line 3.68 on R scale (Step 2) with 13.5 in. on d scale, extend to Pivot line, and mark. Connect marked point (Step 3) with 52 in. on C scale, and read belt length as approximately 208 in. where line intersects L scale. END

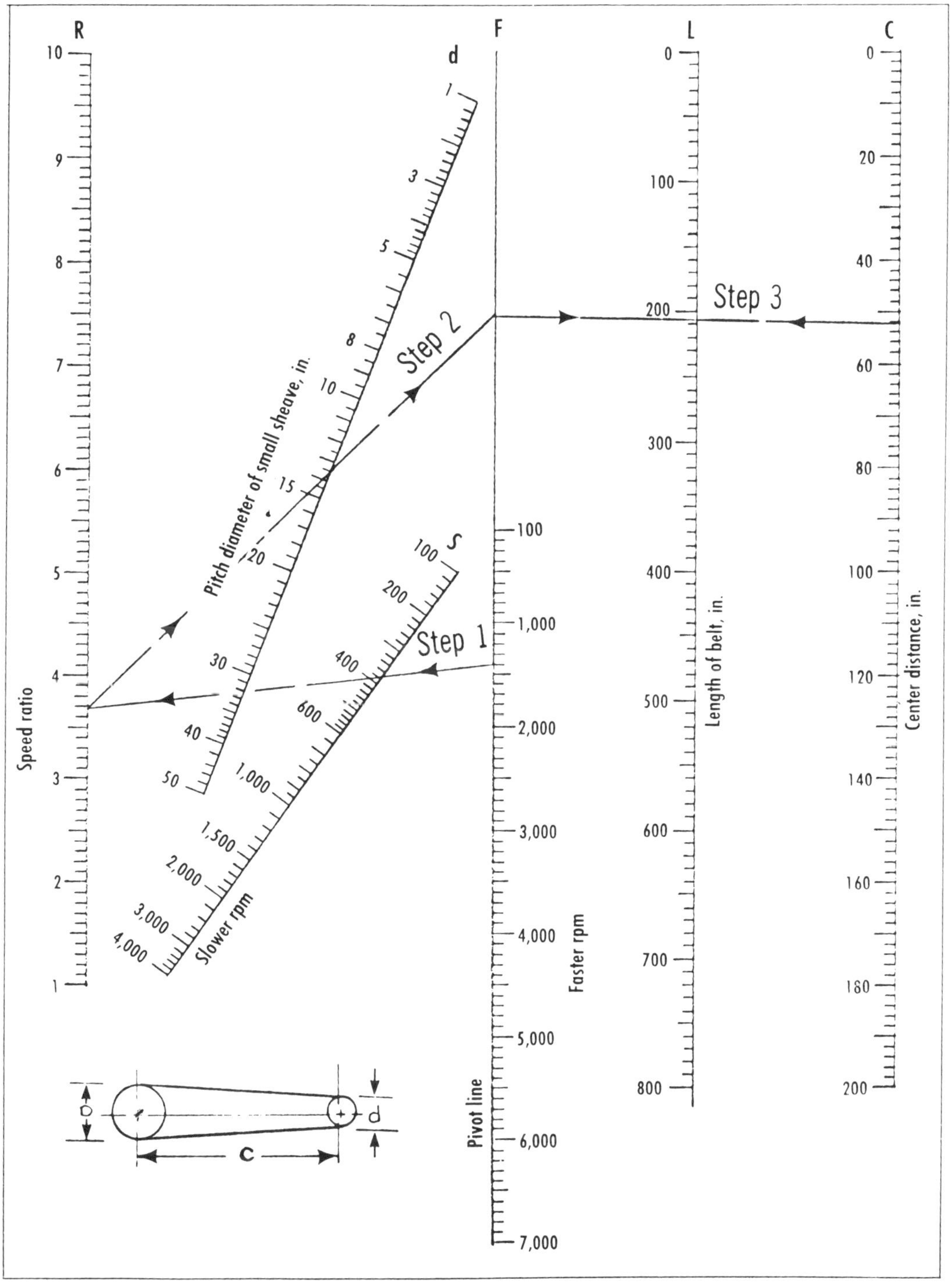

This nomogram determines length of crossed belts

Bill Sisson
Superintendent of Operations
Nipak, Inc.
Pryor, Okla.

THE NOMOGRAM shown here provides a quick, easy method for determining required length for crossed-belt drives. It is based on the formula:

$$L = 2C + 1.57\,(D + d) + (D + d)^2/4C$$

Where:

L = belt length, in.
C = center distance, in.
D = large-sheave pitch diameter, in.
d = small-sheave pitch diameter, in.

Example: Assume the pitch diameter of the larger sheave in a crossed-belt drive is 50 in., pitch diameter of smaller sheave is 14 in., and center distance is 72 in. what length belt is required?

Solution. Connect 72 in. on C scale with 64 in. (50 + 14 = 64) on D + d scale, and read required length as 259 in. where line crosses L scale. END

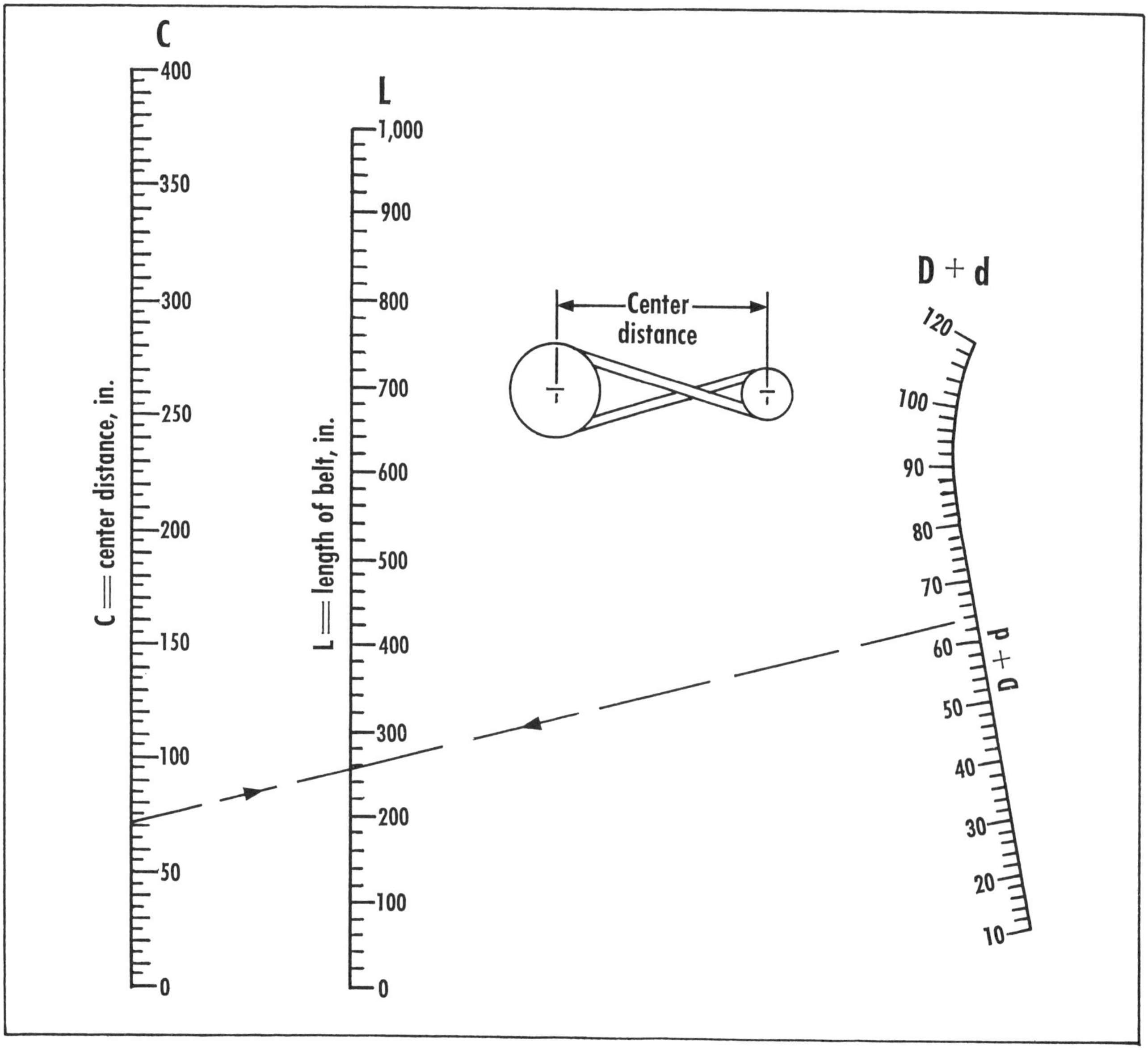

Nomogram gives V-belt drive tension

Bill Sisson
Superintendent of Operations
Nipak, Inc.
Pryor, Okla.

V-BELT tension is required to provide a wedging action between a V-belt and sheave in order to transmit power from a driver to a driven sheave. The force on the shaft, when the drive is running is $T_1 + T_2$, where T_1 = tight-side tension, and T_2 = slack-side tension. When the drive operates under load, T_1 is greater than T_2. The difference between the two, called effective tension T_e, is the tension required to turn the driven sheave.

Tight-side tension T_1 must always be greater than effective, T_e, but the question is how much greater. If the tension ratio, $R = T_1/T_2$, is too much, the belt will slip. Through experimentation, ideal tension ratios have been established, based on the arc of contact on the sheave. Ideal tension ratios, R and R/(R − 1), can be obtained from Durkee-Atwood's "Maintenance Manual for Belt-Drive Systems."

The nomogram presented here will calculate the desired tight-side and slack-side tensions for a drive when the horsepower, speed (rpm), and arc of contact are known. It is constructed for one belt. To obtain total tight and slack-side tensions for multiple-belt drives, multiply by the number of belts. The nomogram is based on these formulas:

Belt speed, S, fpm = $0.262 \times PD \times rpm$

$T_e = hp \times 33{,}000/\text{belt speed}$

$T_1 = T_e\,[R/(R - 1)]$

Where:

hp = total design hp/belt

PD = pitch diameter of sheave.

You may also find it convenient to use the rpm, PD, and S scales on the nomogram to choose sheave diameters. To this, on the PD scale connect the diameter of the sheave you start with and its rpm with a straight edge. Where the straight edge crosses the S scale at the belt speed, mark that point. Now align this marked point with the rpm of the other machine, and read the sheave diameter for that machine where the straight edge crosses the PD scale.

Example. Determine belt speed, effective tension T_e, tight-side tension T_1, and slack-side tension T_2 for a one-belt drive with a design hp of 5, a 1,750-rpm driver motor, and 6.0-in. sheaves on both driver and driven shafts.

Since the diameter of both the driver and driven sheaves are equal, the arc of contact must be 180°. Align (Step 1) 1,750 rpm with 6-in. on the PD scale, extend the line to the S scale, and read speed as 2,770 fpm. From this point (Step 2) align with 5 on the hp scale, extend the line to the T_e scale, and read 60 lb. From $T_e = 60$, align with 180° on A scale (Step 3) and read $1_1 = 75$ lb where the line crosses the T_1 scale. Since $T_2 = T_1 - T_e$, then $T_2 = 75 - 60 = 15$ lb.

END

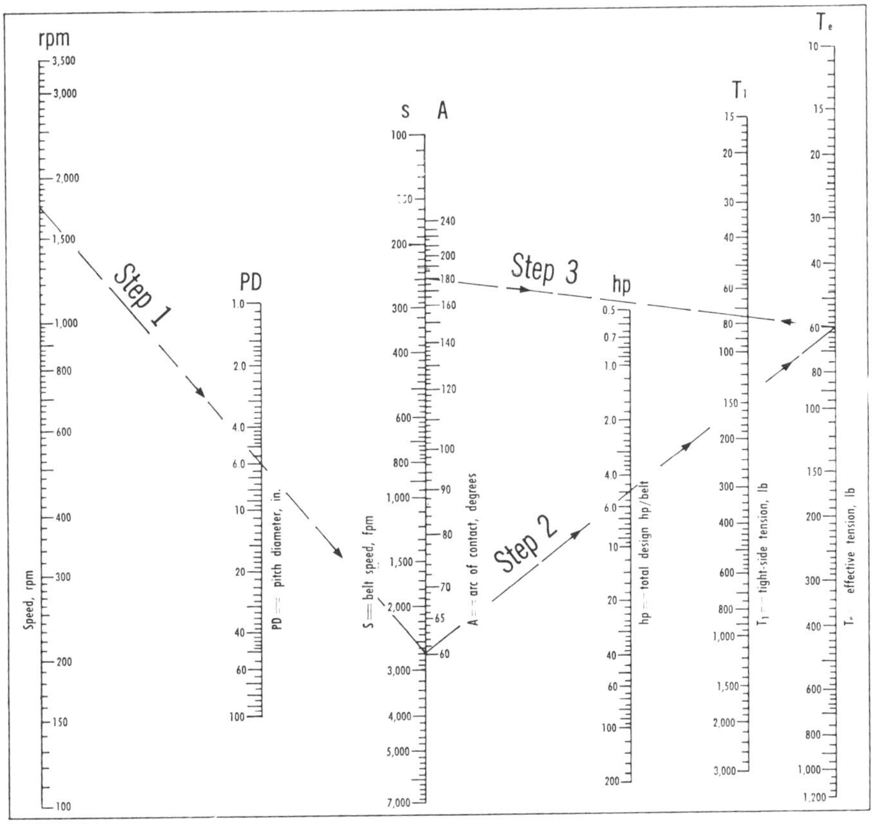

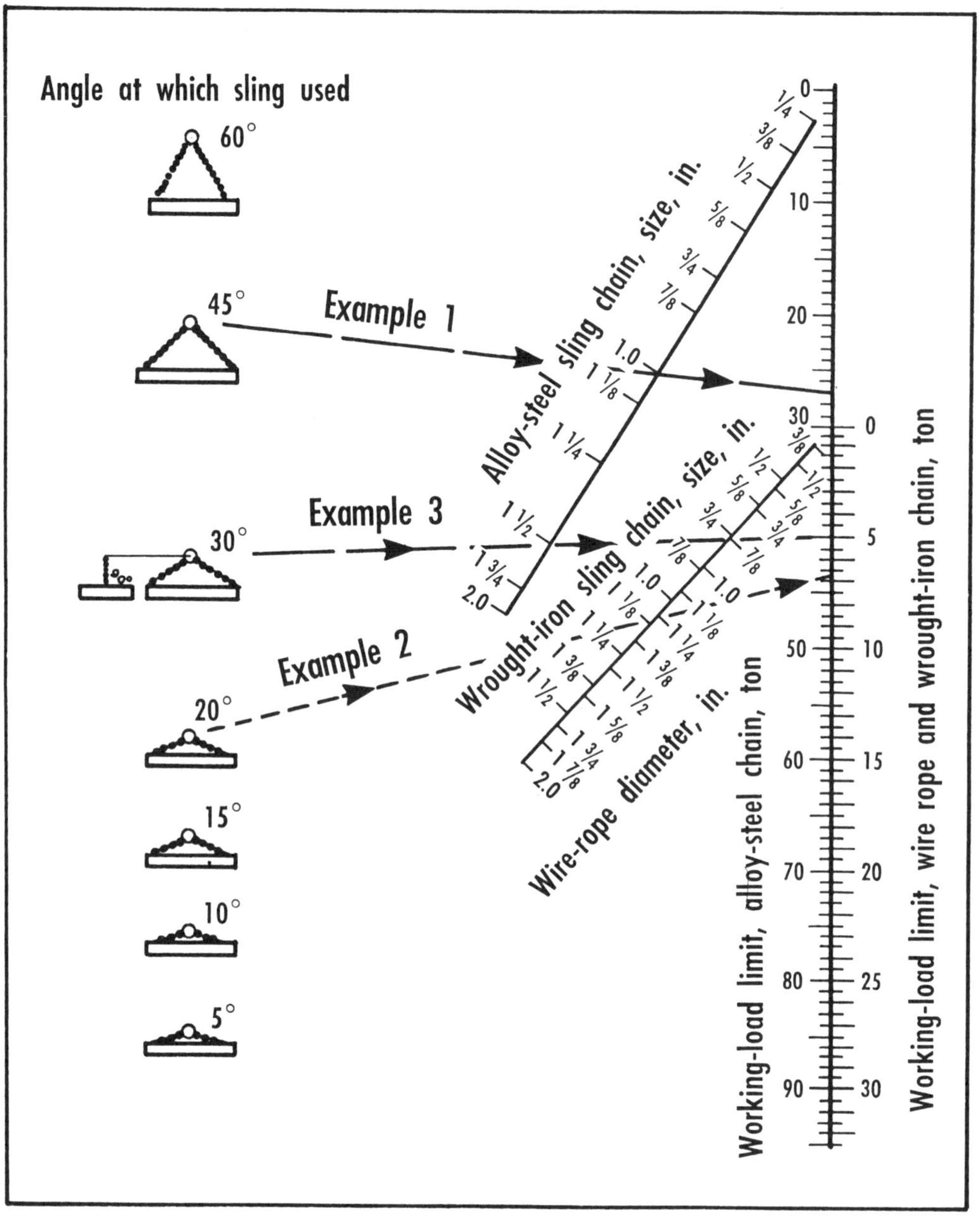

Nomogram gives working limits for slings

Bill Sisson
Superintendent of Operations
Nipak, Inc.
Pryor, Okla.

OFTEN it can be helpful to know the working limit of wire rope or sling chain. A simple method for finding working limits is with the nomogram shown here. It will determine working-load limits for alloy-steel and wrought-iron sling chains and for 6x37 improved-plow-steel wire rope. Note that the number "6" equals the number of strands in the rope, and the number "37" equals the number of wires in a strand.

Example 1. What is the working limit of a 1-in. alloy-steel double-sling chain when used at a 45° angle?

Solution. Connect the point for a 45° angle with 1 on the alloy-steel sling-chain size scale, extend the line to alloy-steel working-load-limit scale, and read 27.3 ton (54,600 lb).

Example 2. What is the working limit for $1\frac{1}{8}$-in. wrought-iron double-sling chain when used at a 20° angle?

Solution. Connect the point for a 20° angle with $1\frac{1}{8}$-in. on wrought-iron sling-chain size scale, extend to wrought-iron-chain working-load-limit scale, and read **6.85** ton (13,700 lb).

Example 3. What is the working limit for a $\frac{7}{8}$-in. single-wire rope used in a vertical position?

Solution. Connect 30° (which is also the 90° point) with $\frac{7}{8}$ in. on wire-rope-diameter scale, extend to wire-rope working-load-limit scale, and read 5 ton (10,000 lb).

Note that the 30° point on the angle scale also applies for single rope and single alloy or wrought-iron sling in a vertical lift. END

Nomogram solves ideal gas laws

F. Caplan
Kaiser Engineers

THIS nomogram permits a fast solution to the equations stating the combination of Boyle's and Charles' laws for ideal gases. It lets you quickly solve for changes in pressure and/or temperature and/or volume of a constant weight of a confined gas. The nomogram is based on these equations:

$$P_2 V_2/(t_2 + 460) = P_1 V_1/(t_1 + 460)$$

Where:

P = Absolute pressure of gas
V = Volume of gas
t = Temperature of gas, °F.

Example. What is the final volume when nitrogen in a constant pressure accumulator is heated from 250° F. to 800° F.? Pressure is 1,150 psia and initial volume in accumulator is 30 cu in.

For Step 1, align P_1 = 115 with V_1 = 30 and mark intersection with Pivot Line 1. Align this mark with t_1 = 250 and mark intersection with Pivot Line 2, Step 2. For Step 3, align that mark with t_2 = 800 and mark the intersection with Pivot Line 1. For the fourth and final step, align the last mark on Pivot Line 1 with P_2 = 115 (constant pressure) and read V_2 = 53 cu in. Note that the Pressure Scale values can be multiplied by 2, 10, 100, etc. if the Volume Scale values are divided by the same number. END

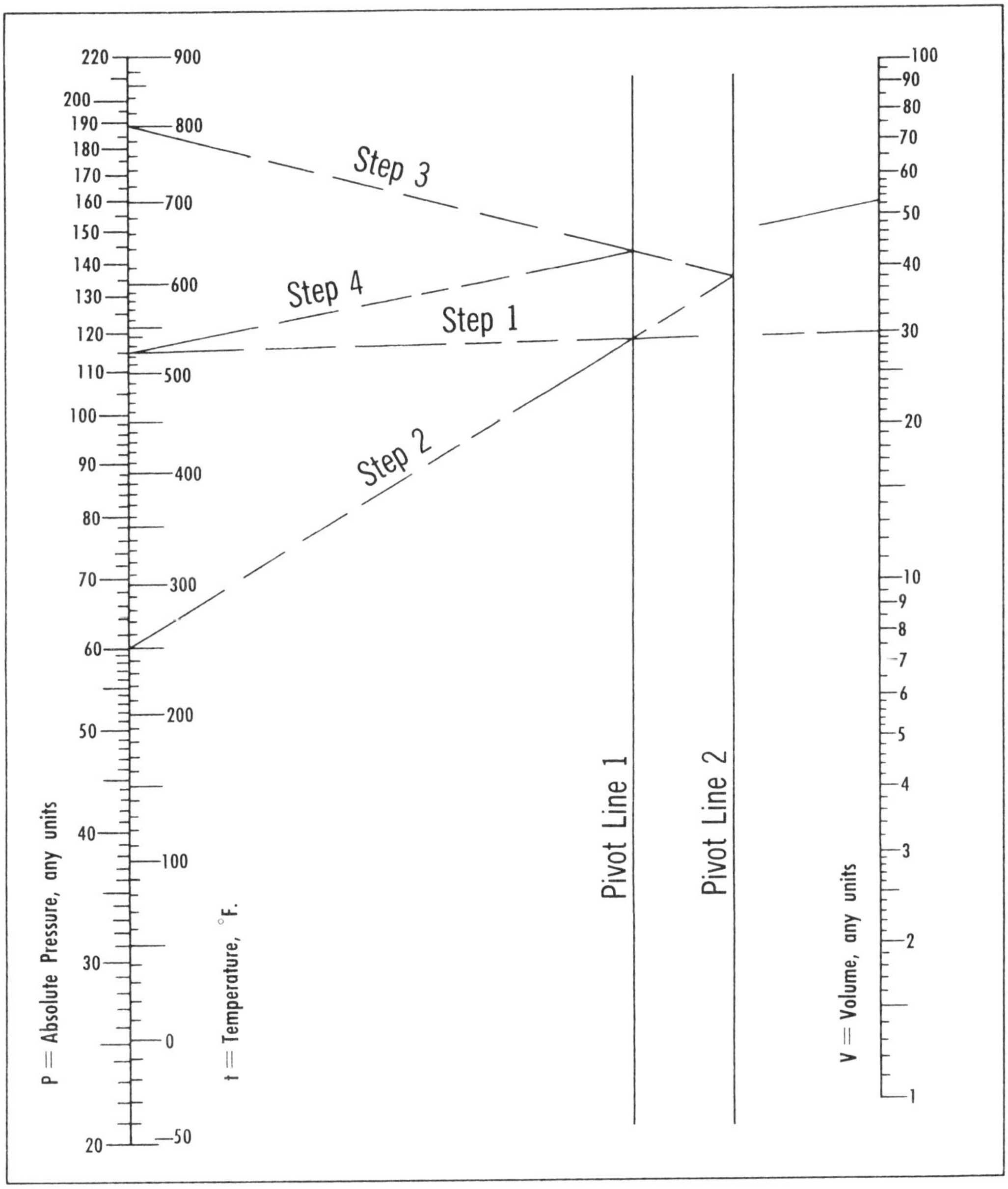

Nomogram solves gas relationships

F. Caplan
Kaiser Engineers

THIS NOMOGRAM solves the adiabatic or polytropic relationship for gases:

$$T_2/T_1 = (P_2/P_1)^{(k-1)/k} = (V_1/V_2)^{k-1}$$
$$P_2/P_1 = (T_2/T_1)^{k(k-1)} = (V_1/V_2)^{k}$$
$$V_2/V_1 = (T_1/T_2)^{1/(k-1)} = (P_1/P_2)^{1/k}$$

Where:
- T = Absolute temperature
- P = Absolute pressure
- k = specific heat ratio, c_p/c_v

EXAMPLE. When 7.86 cu in. of air (k = 1.40) at 100° F. is compressed from 14.7 psia to 85.3 psig (100 psia), what is its final temperature and volume?

Align P_1 = 14.7 with P_2 = 100 and mark intersection with pivot line, Step 1. Next align pivot-line mark with k = 1.40 and read $(V_1/V_2)_P$ = 3.93, Step 2. Final volume = 7.86/3.93 = 2.0 cu in.

Next align k = 1.40 with $(V_1/V_2)_T$ = 3.93 and mark intersection with pivot line, Step 3. Then align pivot-line mark with t_1 = 100° F. and read t_2 = 510° F. Step 4.

END

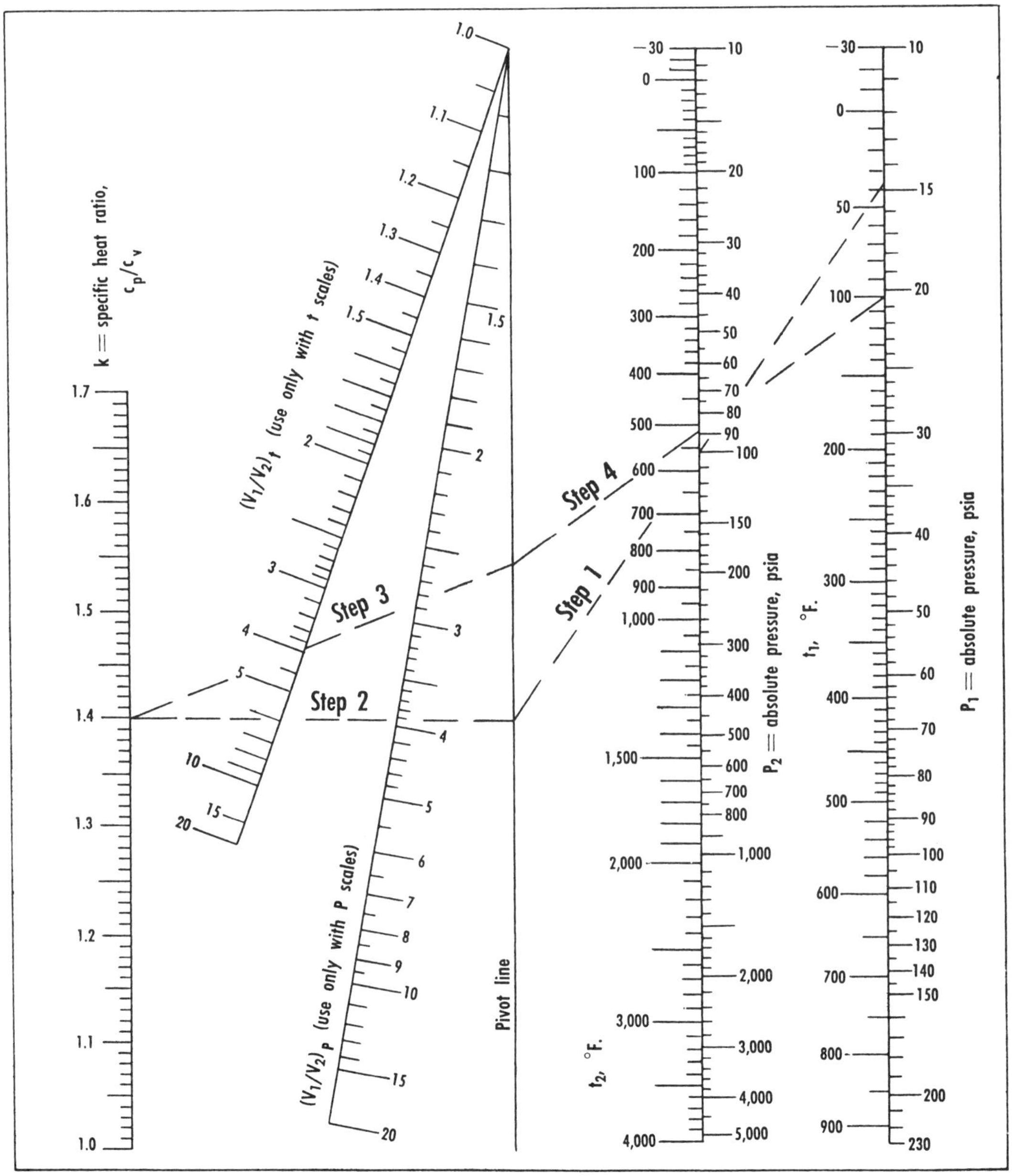

Use this nomogram to convert weight fractions to mol fractions

F. Caplan
Kaiser Engineers

THIS nomogram can be useful in dealing with mixtures of two compounds to find the mol fraction when the two weight fractions and molecular weights are known or to find the weight fraction when the mol fractions and molecular weights are known. It provides solutions to either of these two equations:

$$x_a = \frac{(W_a/M_a)}{(W_a/M_a) + [(1 - W_a)/M_b]}$$

$$W_a = (x_a M_a)/[x_a M_a + (1 - x_a) M_b$$

Where:

x_a = mol fraction of compound a
W_a = weight fraction of compound a
M_a = molecular weight of compound a
M_b = molecular weight of compound b

Example. What is the mol fraction of compound in a mixture of two compounds if its molecular weight is 18, its weight fraction is 0.30, and the molecular weight of compound b is 28? To find the answer to this question, align molecular weight, $M_a = 18$, with weight fraction $W_a = 0.30$, Step (1), and mark the intersection of a straight edge with the pivot line.

Then align the mark on the pivot line with molecular weight, Step (2), $M_b = 28$, and read mol fraction of compound a, $x_a = 0.40$. Note that the molecular weight values, M_a and M_b, can be simultaneously multiplied by such numbers as 0.1, 0.5, 2, 10, etc. to extend the range of the nomogram.

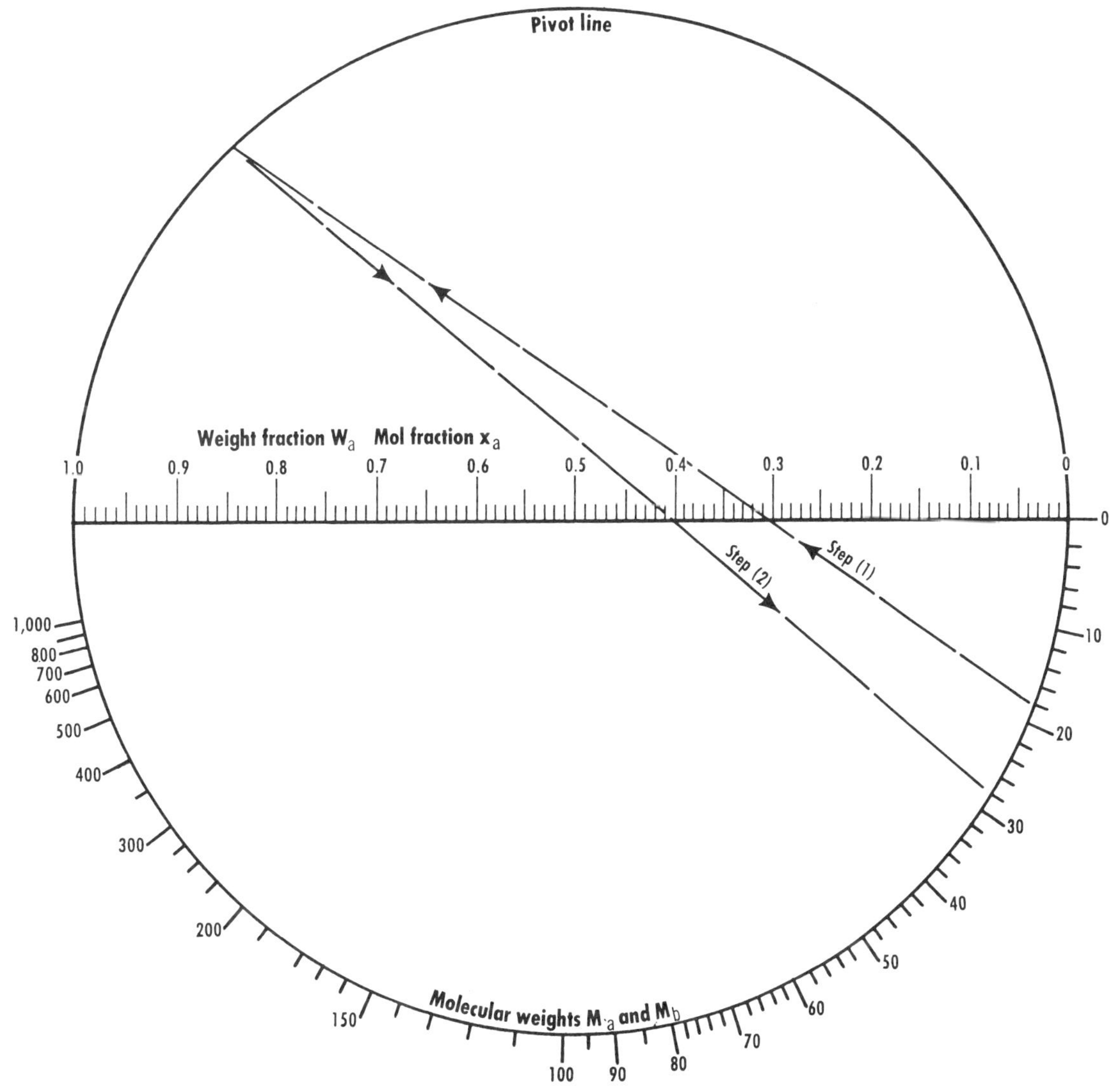

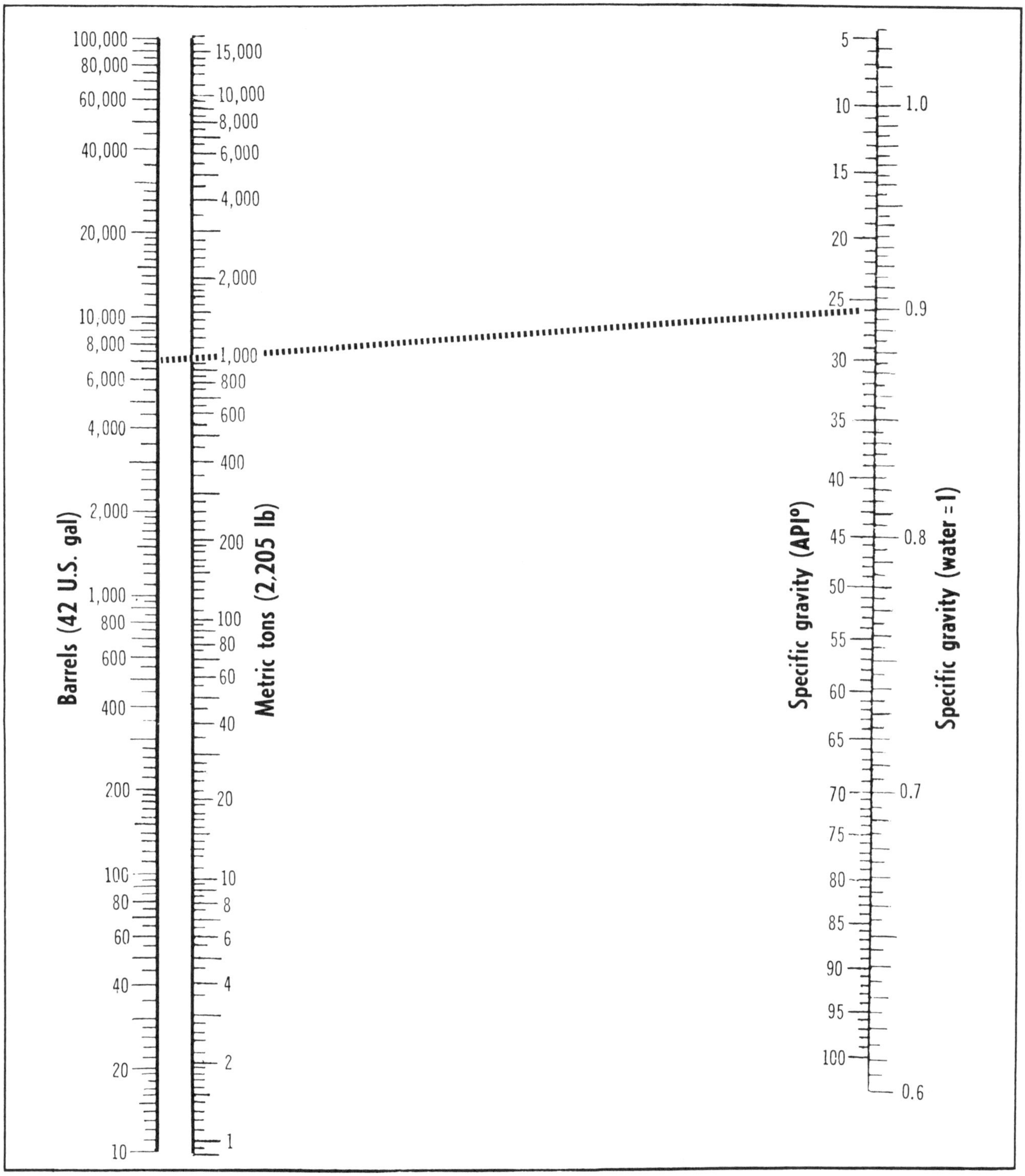

convert barrels of oil to metric tons

THIS nomogram solves the equation, T = 0.1587 GB, where T = metric tons of oil, G = specific gravity of oil (water = 1), and B = barrels (42 U.S. gal). For ease of conversion, the gravity, °API, has been placed alongside the specific-gravity scale.

To convert barrels to tons, connect the barrels on the left scale with specific gravity on the right scale using a straightedge. The intersection of the straightedge with the center scale gives the metric tons of oil (2,205 lb advp./ton).

Example: Convert 7,000 bbl to metric tons. Specific gravity = 0.9 (° API = 26). Align a straightedge at 0.9 on the right (specific gravity) scale with 7,000 bbl on the left scale, find the intersection of the straightedge on the tons scale and read 1,000 tons.

Use this dual nomogram to...

• Reduce API gravity to 60° F.
• Find cost of burning fuel oils

Bill Sisson
Superintendent of Operations
Nipak Inc.
Pryor, Okla.

PETROLEUM products are sold in the United States by barrels of 42 gal corrected to 60° F. Their specific gravity is expressed on an arbitrary scale termed degrees API.

Since it is generally impractical to determine an API gravity at exactly 60° F., it is necessary to convert the gravity at the observed temperature to the corresponding gravity at 60° F. In this conversion, two corrections are necessary. The first arises from the change in volume of the glass hydrometer with temperature and the second from the change in volume of the oil. Both have been applied in the nomogram shown here.

The price of fuel oil is of prime concern. Since the heating value of fuel oil varies with API gravity, it is important to have a quick method to check the effect on cost per million BTU.

This nomogram will convert the gravity found at the observed temperature to gravity at 60° F. and determine the cost per million BTU as shown by the following example.

Example. If the API gravity reading observed on a hydrometer in an oil at 78° F. is 25°, what is its API gravity at 60° F.? If the cost of this oil is $9.30/bbl, what is the cost per million BTU at the higher heating value (HHV)?

Solution. Step 1, line 78° F. on Observed-temperature scale, with 25° on API-gravity-at-observed-temperature scale, extend line to API-gravity-at-60° F. scale, and read 23.9°. Step 2, connect this 23.9° with $9.30 on Cost-per-barrel scale, and read $1.52 where line intersects Cost per million BTU scale.

Note that commercial standard fuel oils are shown on the opposite side of the API-gravity-at-60° F. scale. Values for reduction of observed API gravity to API gravity at 60° F. are based on Table 5, ASTM-1P, Petroleum Measurement Tables. END

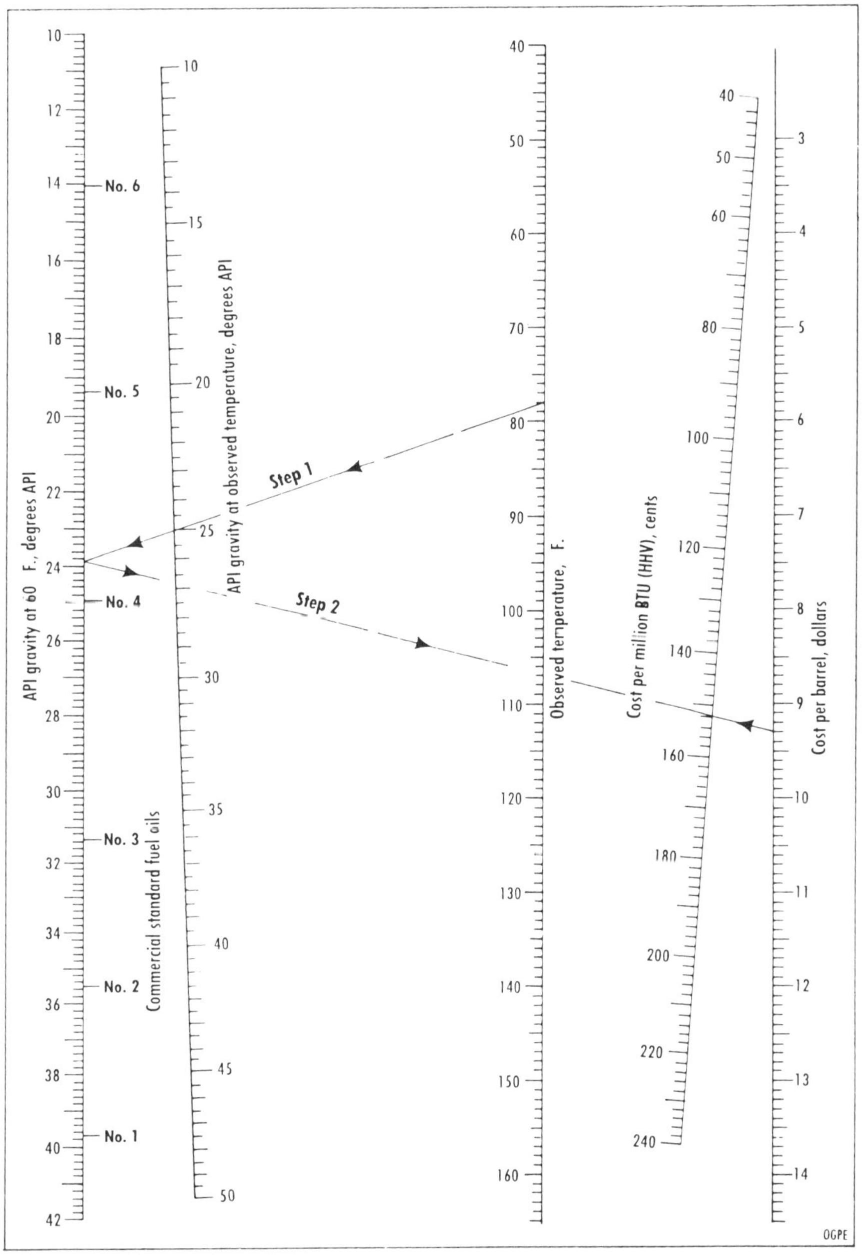

Nomogram determines fuel savings

Bill Sisson
Superintendent of Operations
Nipak Inc.
Pryor, Okla.

THE rising cost of fuel is causing plant engineers to increase their efforts to find methods to decrease fuel consumption in their heating units. Reduction in the amount of excess combustion air is one approach to reduce fuel cost, since excess air is that air which is over and above that necessary for complete combustion of the fuel. Excess air in heating operations represents a fuel waste because the excess air must be heated to the temperature of the operation, and, as it leaves the furnace, the air carries with it the heat necessary to bring it to this temperature. The amount of heat loss attributable to excess air depends on the quantity and temperature of the waste gases.

This nomogram determines the amount of fuel that can be saved by reduction of excess air through the heating unit. The percent figures on the nomogram are the percent fuel savings that would exist if zero percent excess air were used in place of the actual excess air used, based on 60° F. outside air temperature and sea-level site elevation.

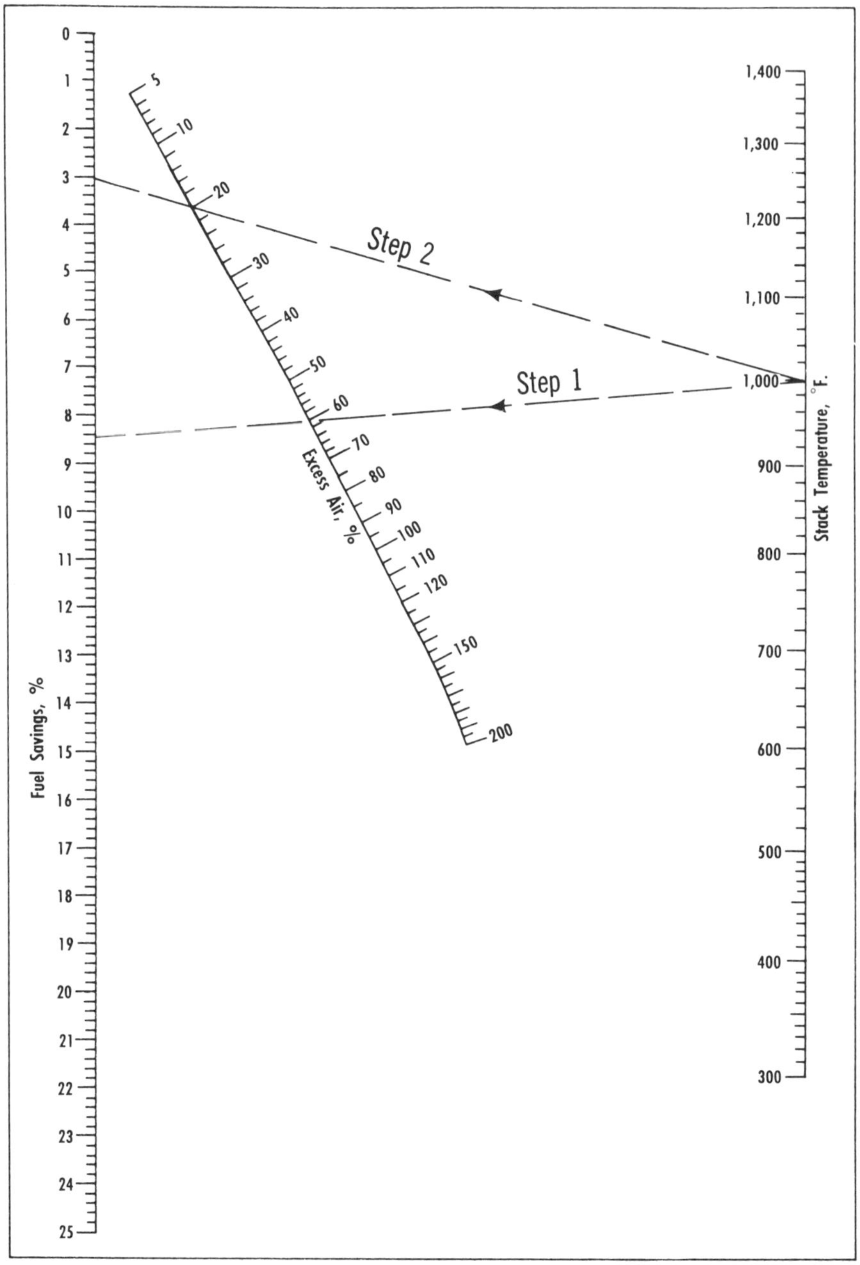

Example. Find the percent fuel savings effected by reducing the excess air to 20% from 60%, with 1,000° F. stack temperature.

Solution. Step 1, align 1,000° F. on Stack Temperature scale with 60% on Excess-Air scale, extend to Fuel Savings scale, and read 8.43%. Step 2, align 1,000° F. on Stack Temperature scale with 20% on Excess Air scale, extend to Fuel Savings scale, and read 3%. Then 8.43% −3% = 5.43 fuel savings affected by the reduction from 60 to 20% excess air.

Nomogram gives viscosity correction

F. CAPLAN
Kaiser Engineers

THIS nomogram is designed to help you in heat-transfer and fluid-flow calculations when the fluid viscosity at the wall, μ_W, is different from the viscosity of the bulk fluid, μ_b. In such a situation, the equations for the heat-transfer coefficient, h, the Colburn j factor, and the Nusselt Number, ν, have a correction factor of the form:

$$h_c = h\,(\mu_b/\mu_W)^{0.14}$$

Note that the h, j, ν, or μ scales can be multiplied by 0.1, 2, 10, etc.

Example. What is the correct value of h if $\mu_b = 65$, $\mu_W = 10$, and h = 100? Step 1, align μ_b with h and mark intersection with Pivot Line. Next align this point on Pivot Line with $\mu_W = 10$ and read corrected heat-transfer coefficient, $h_c = 130$. END

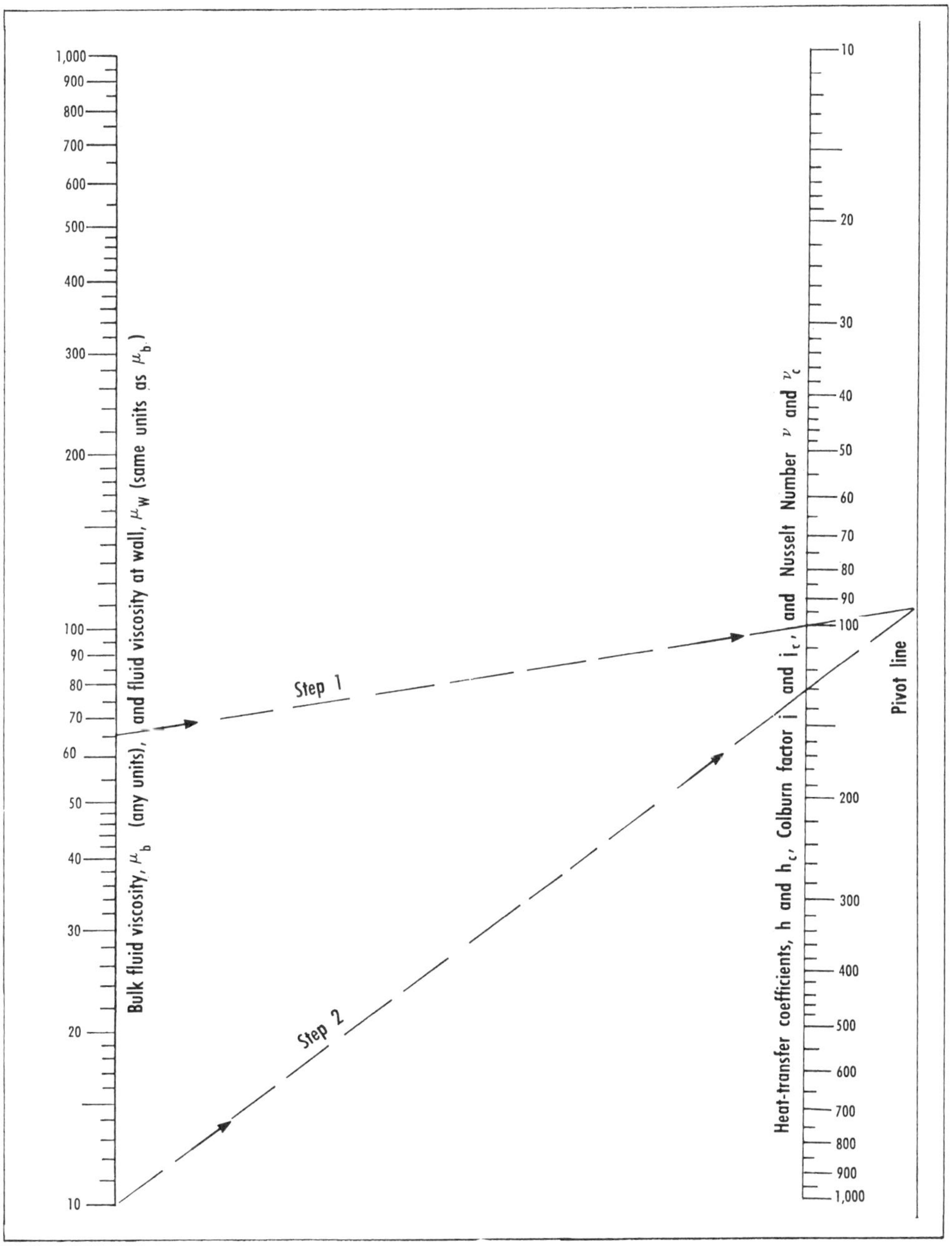

Estimate hydrogen content of fuel

Bill Sisson
Superintendent of Operations
Nipak Inc.
Pryor, Okla.

HERE'S a quick method for estimating the hydrogen content of fuel from an Orsat analysis of the flue gas, applicable for complete combustion of hydrocarbons. Construction of the nomogram is based on the formula:

$$\text{Weight \% H} = 100 - (1{,}126)/[(N_2 - 3.78\,O_2)/CO_2] + 7.48$$

Where: N_2, CO_2, and O_2 are the volumetric percentages of these constituents in the flue gas.

Example. Estimate the hydrogen content of the fuel when an Orsat analysis of the flue gas shows 16% CO_2, 4% O_2 and 80% N_2.

Solution. Connect 4% (Step 1) on O_2 scale with 80% on N_2 scale, and mark at intersection with Reference line R. From marked point on R (Step 2), line with 16% on CO_2 scale, extend to H_2 scale, and read hydrogen content of the fuel as approximately 2.4%.

END

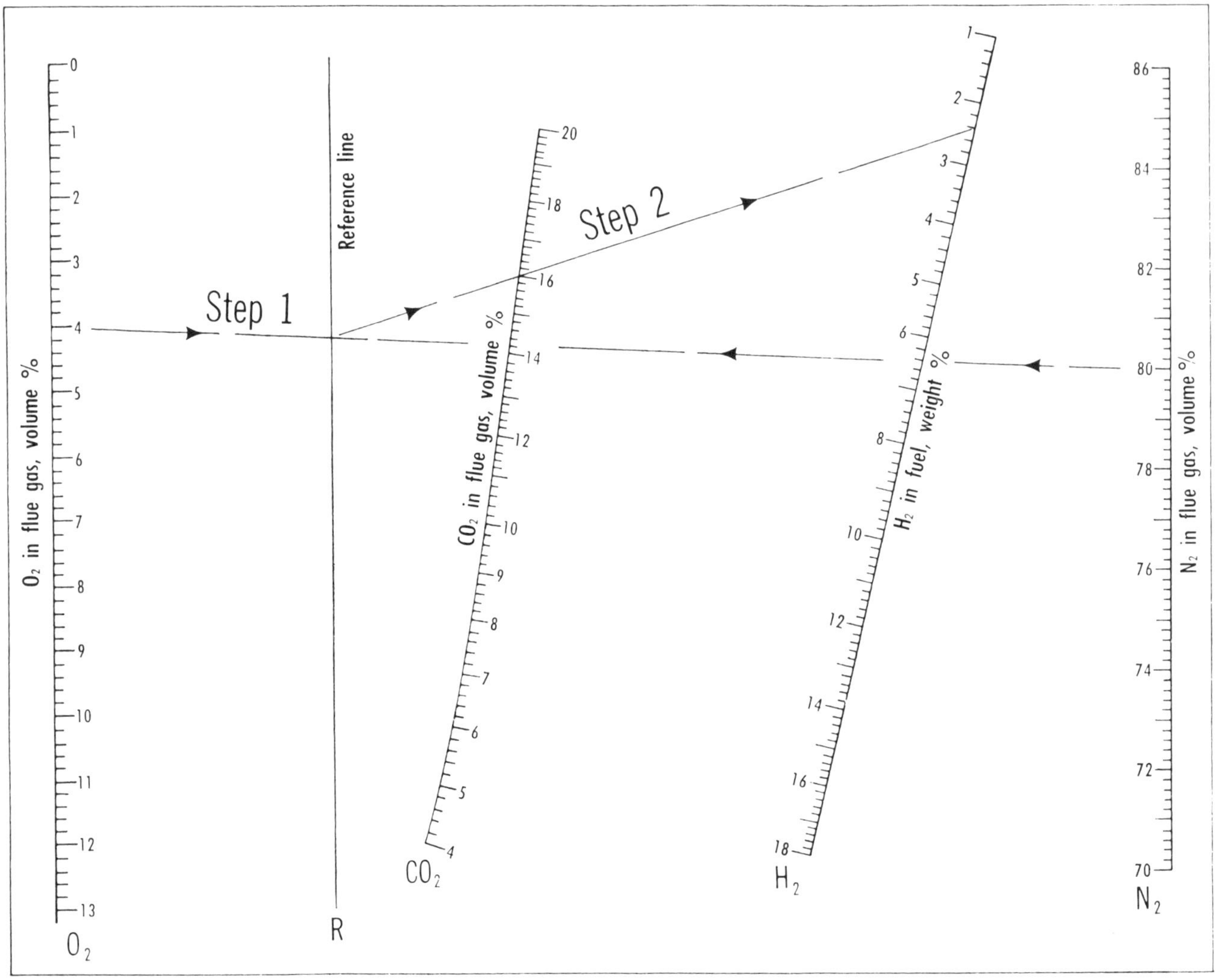

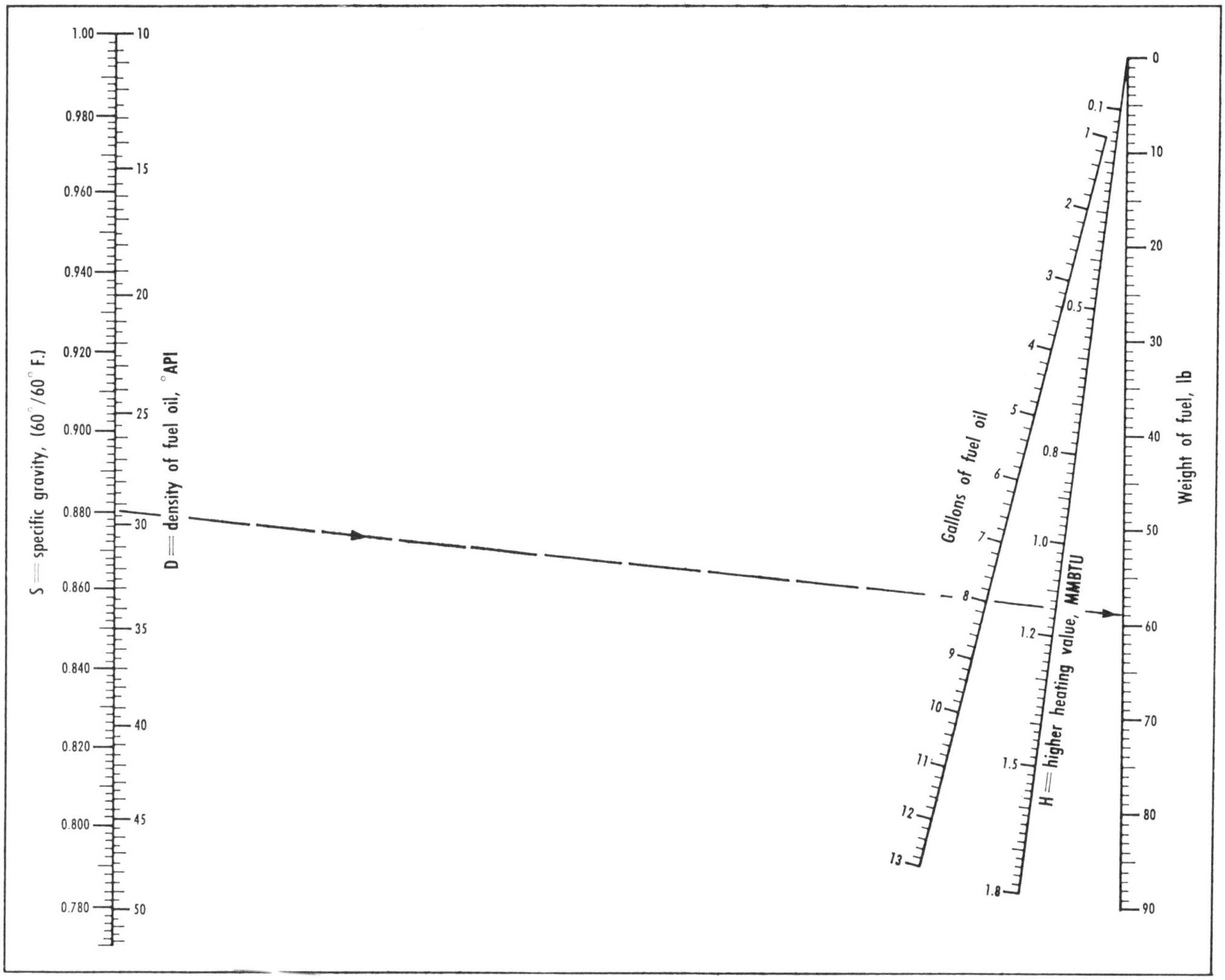

Nomogram tells density, heat value of fuel oil

Bill Sisson
Superintendent of Operations
Nipak, Inc.
Pryor, Okla.

THE NOMOGRAM presented here provides a convenient aid in determining densities and heating values of fuel oils. Densities and heating values are required for design and operation of oil-combustion equipment and for sales and purchasing. The nomogram is based on the degrees API formula and a formula which expresses gross or higher heating value. The heating value expressed in this manner is adequate as the hydrogen constituent of fuel oils is nearly constant. The degrees API formula is:

$$D = (141.5/Sp) - 131.5$$

Where:
D = density, °API
Sp = specific gravity (60°/60° F.)

The higher heating value of fuel oils is expressed closely by the equation:

$$H = 17{,}780 + 54\ D$$

Where:
D = density, °API
H = higher heating value, BTU/lb

The nomogram can be extended to higher ranges by multiplying gal, weight, and BTU scales by 10,100, etc.

Example. Assume a fuel oil has a specific gravity of 0.880. What is the °API density, weight, and heating value of 8 gal and 80 gal of this oil?

Solution. Line from 0.880 on specific-gravity scale (29.30 °API) with 8 on gal-of-fuel-oil scale, extend line to weight-of-fuel scale, and read weight as 59 lb. Where line crossed higher-heating-value scale, read 1,142,000 BTU. For 80 gal of oil: 8 gal × 10 = 80; 59 lb × 10 = 590; 1,142,000 BTU × 10 = 11,142,000 BTU. END

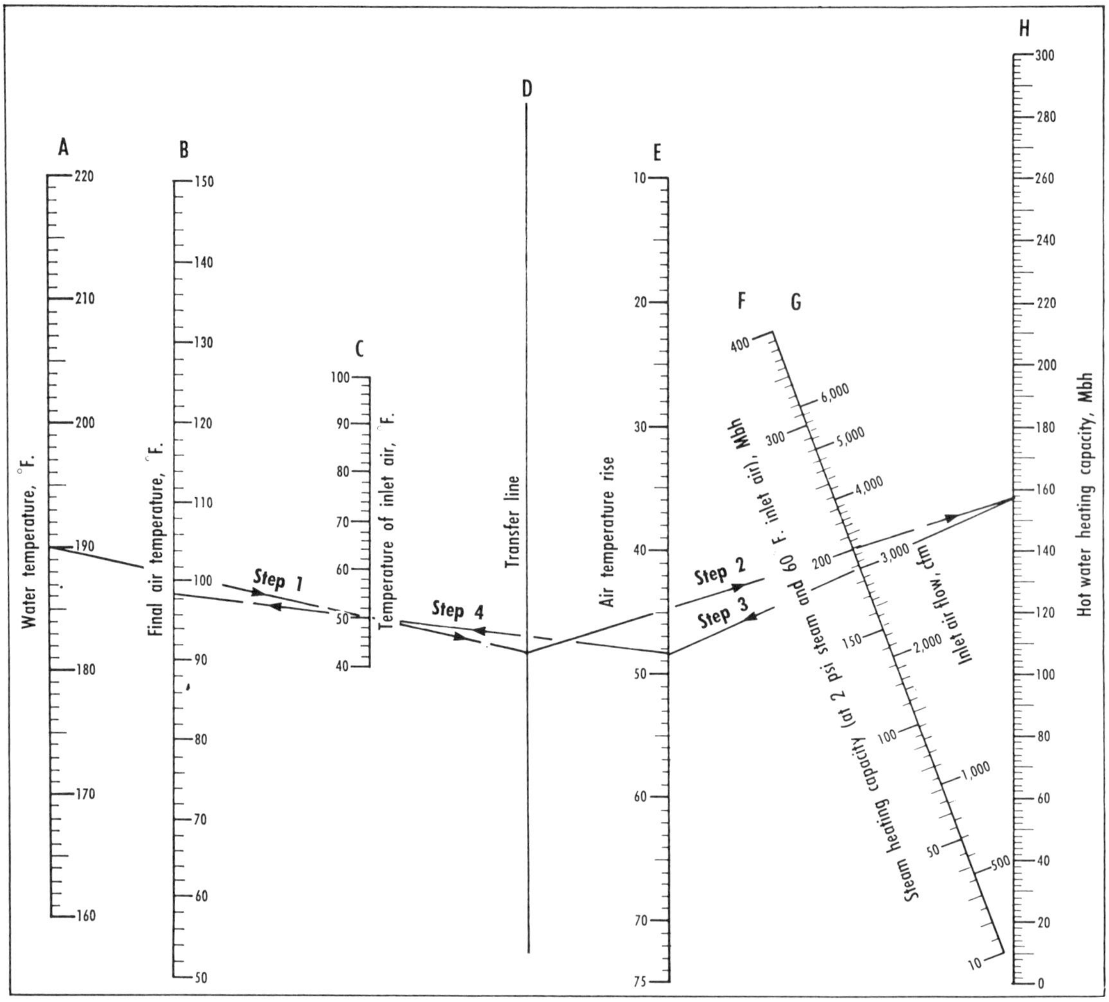

Hot-water heating capacity of unit heaters

Bill Sisson
Superintendent of Operations
Nipak, Inc.
Pryor, Okla.

MANUFACTURERS' literature and tables usually show the capacities of unit heaters at 60° F. and 2 psi steam pressure. When a unit heater is to use hot water to supply heat, the nomogram presented here determines capacity, air-temperature rise, and final temperature when capacity of unit at 60° F. entering-air temperature and 2 psi steam pressure are known. It is based on the formula:

Hot-water capacity, Mbh = capacity (at 60° F. inlet air and 2 psi steam pressure) × F

Temperature rise, °F. = BTU/cfm × 1.08

Final temperature, °F. = entering-air temperature, °F. + temperature rise

Where:
Mbh = 1,000 BTU/hr
F = conversion factor
BTU = capacity at 60° F. inlet air and 2 psi
cfm = Inlet air flow, cu ft/min

Note that water temperature is the average temperature of water in the coil. For example, 200° F. entering water with a 20° F. temperature drop equals 190° F. average water temperature for calculating final air temperature.

Example. Assume the capacity of a unit heater is 200,000 BTU/hr (at 60° F. entering-air temperature and 2 psi steam pressure), with 3,000 cfm of inlet air at 50° F., and average water temperature of 190° F. in coil. What is the hot-water heating capacity, temperature rise, and final air temperature?

Solution. Line from 190° F. on A scale with 50° F. on C scale, extend line to D scale, and mark (Step 1). From this marked point, line with 200,000 BTU/hr (200 Mbh) on F scale, extend line to H scale, and read 157,000 (157 Mbh) capacity (Step 2). From this capacity, align with 3,000 cfm on G scale, extend line to E scale, and read air temperature rise as 48.4° F. (Step 3). To find final air temperature, align 48.4 on E scale with 50° F. on C scale, extend line to B scale, and read final air temperature as 98.4° F. (Step 4). END

Nomogram gives properties of air

Bill Sisson
Superintendent of Operations
Nipak, Inc.
Pryor, Okla.

MANY engineering calculations require a knowledge of the properties of air at different temperatures and elevations. The nomogram presented here provides a convenient way to determine such properties as actual vapor pressure, dew point, specific humidity, and relative humidity. It is based on the following equations:

$$p_v = p_w - B\,(t_d - t_w)/2{,}700$$

$$R = p_v/p_d$$

$$w = p_v/1.61\,(B - p_v)$$

$$\Delta t = t_d - t_w$$

Where:

p_v = actual vapor pressure, in. Hg (corresponds with dew point temperature)

t_w = wet bulb temperature, °F.

t_d = dry bulb temperature, °F.

p_w = vapor pressure, in. Hg, corresponding with t_w

p_d = vapor pressure, in. Hg, corresponding with t_d

B = total vapor pressure (barometric pressure), in. Hg

w = specific humidity, lb water vapor/lb dry air

R = relative humidity (calculated as a decimal)

Δt = wet-bulb depression, °F.

Example. Determine the actual vapor pressure, dew point, specific humidity, and relative humidity of air at an altitude of 750 ft above sea level (barometric pressure = 29.1 in. Hg), when dry-bulb temperature is 95° F. and wet-bulb temperature is 80° F.

Solution. Align 750 ft (29.1 in. Hg) on Scale A or B with wet-bulb depression = 15 on Scale F (Δt = 95-80), and mark where line crosses Scale E (Pivot line) (Step 1). From the marked point on Scale E align with 80° F. on Scale L (Step 2) and read 0.87 in. Hg vapor pressure on Scale H and a dew point of 74.5° F. on Scale K. Next, from the 0.87 point on Scale H, align with 750 ft (29.1 in. Hg) on Scale A or B (Step 3) and read 0.019 specific humidity where the line crosses Scale G. Again, from 0.87 on Scale H align with 95° F. on Scale D, extend the line to Scale C (Step 4) and read 52.4% relative humidity. END

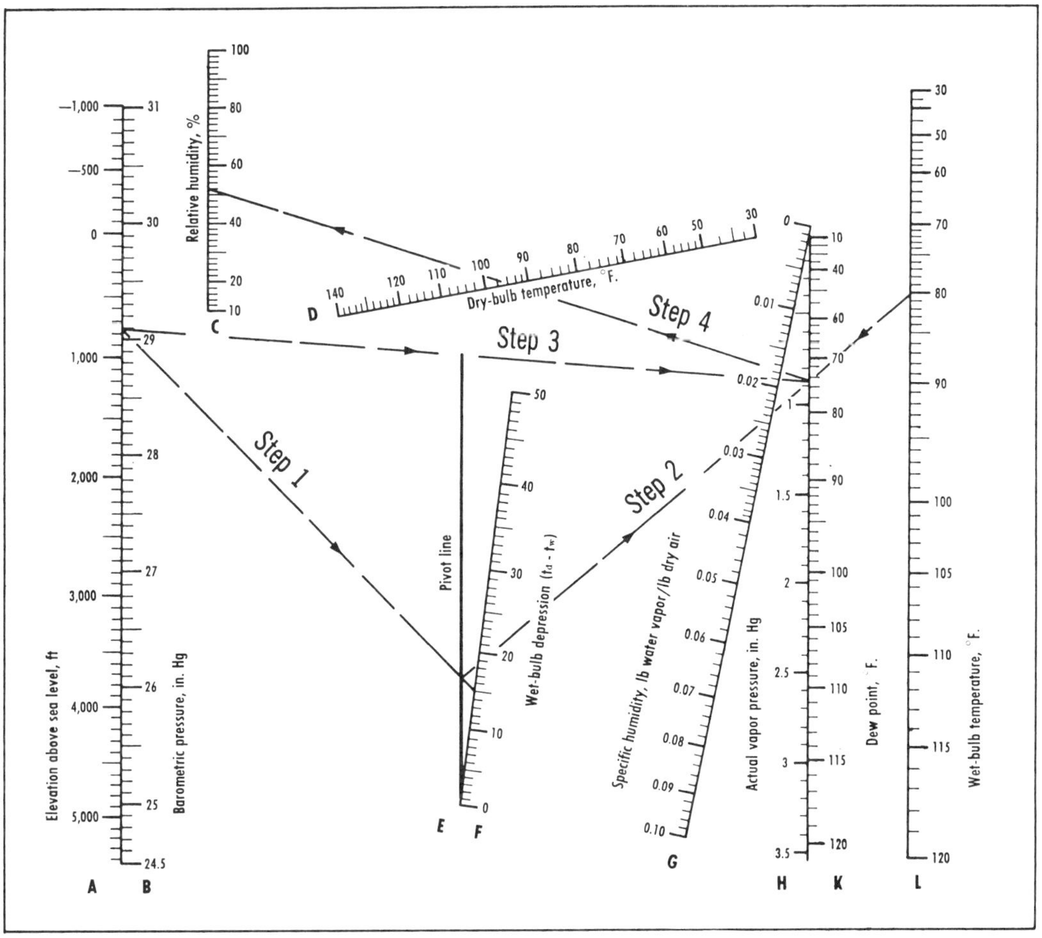

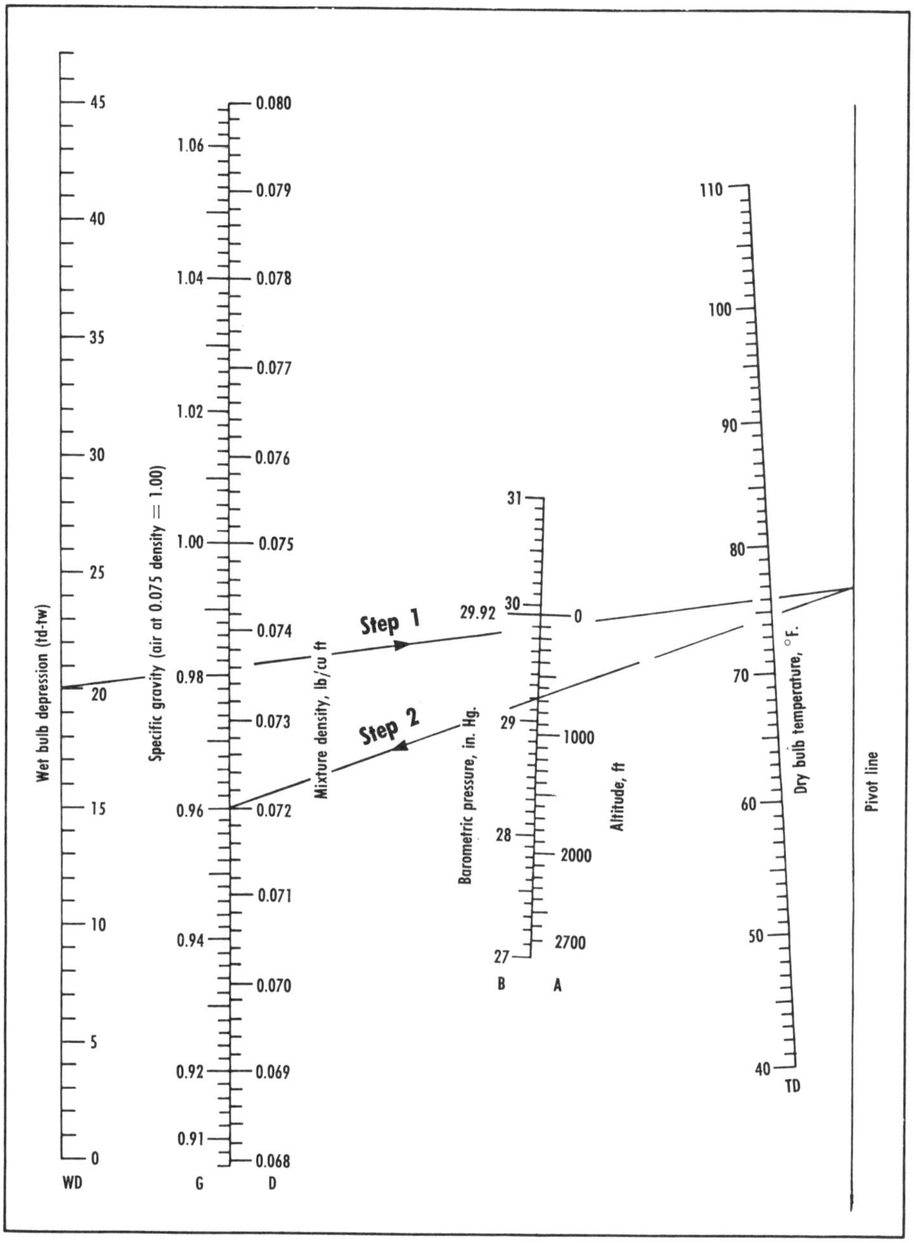

Nomogram determines air densities

Bill Sisson
Superintendent of Operations
Nipak Inc.
Pryor, Okla.

CALCULATIONS relating to air systems, fans, air preheaters, air compressors, etc., require a knowledge of actual air densities at a given set of conditions.

The nomograph presented here based on the following formula provides a fast, accurate method for solving problems where air densities are required:

$$d_m = \frac{B - 0.38\,pv}{0.7541\,(td + 460)}$$

Where:

d_m = density of air mixture, lb/cu ft
pv = vapor pressure of water, in. Hg
B = barometric pressure, in. Hg
td = dry bulb temperature, °F.

Example: What is the atmospheric density and corresponding specific gravity if dry-bulb temperature is 76° F., wet bulb temperature is 56° F., and barometric pressure is 29.2 in. Hg?

Solution: Line 20 (td − tw = 76 − 56 = 20) on WD scale with 76° F. on TD scale, extend to Pivot Line and mark (Step 1). From marked point, line with 29.2 in. Hg on B scale, extend to D and G scale, and read density as 0.072 lb/cu ft and specific gravity as 0.96 (Step 2). END

Nomogram gives air density, moisture content

Bill Sisson
Superintendent of Operations
Nipak Inc.
Pryor, Okla.

THE nomogram presented here provides a convenient method for finding the density, lb/cu ft, of compressed dry air over a wide range of temperatures and pressures and the grains of water/cu ft of saturated air (14.7 psia) at various temperatures and relative humidities.

The density-of-compressed-dry-air portion of the nomogram (identified by scales 1A, 1B, and 1C, and based on air at an atmospheric pressure of 14.7 psia and molecular weight of 28.97) can be useful in compressed-air flow calculations.

The grains of water/cu ft of saturated air scales (2A, 2B, and 2C) can be used in solving problems relating to air-conditioning, humidification, etc.

Example 1. What is the density of dry air at 600 psig and 400° F.?

Solution. Line 400° F. on 1-A scale with 600 psig on 1-C scale (Step 1-solid line), and read density as 1.95 lb/cu ft where line crosses 1-B scale.

Example 2. How many grains of water/cu ft of saturated air if temperature is 64° F. and relative humidity is 40%?

Solution. Line 64° F. on 2-A scale with 40% relative humidity on 2-C scale (Step 2-dashed line), and read 2.65 grains/cu ft of air where line crosses 2-B scale. END

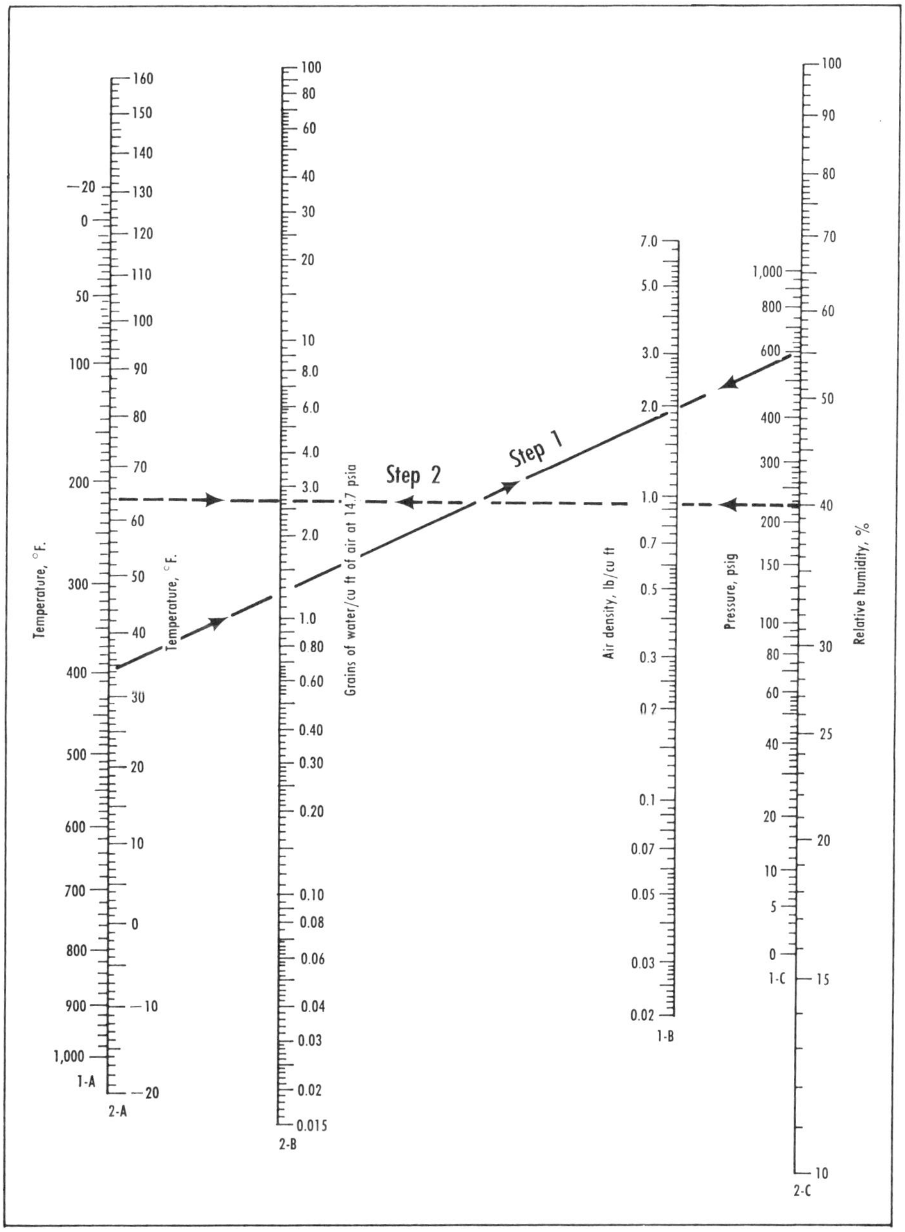

Nomogram gives dew point, relative humidity at various elevations

Bill Sisson
Superintendent of Operations
Nipak Inc.
Pryor, Okla.

IN the analysis of humidification and dehumidification problems, and when selecting or specifying drying or ventilating equipment, it is frequently necessary to know the water-vapor content of air. One simple approach is to determine the dry and wet-bulb temperatures and to use a psychrometric chart or tabulated data. Since most published charts and tables are based on a barometric pressure of 29.92 in. Hg (sea level), they would be inaccurate if your plant is at an elevation where the barometric pressure is not 29.92 in. Hg because barometric pressure of atmospheric air varies widely with weather conditions and geographical locations, notably altitude.

For the solution of relative humidity problems relating to different elevations, the nomogram shown here will provide fast, accurate solutions. Construction of the nomogram is based on the following formulas:

$$p_v = p_w - 0.00037B\ (t_d - t_w)$$
$$RH = p_v/p_d$$

Where B = barometric pressure, in. Hg; t_d = dry-bulb temperature, °F.; t_w = wet-bulb temperature, °F.; t_c = dew point temperature, °F.; p_w = vapor pressure, in. Hg, corresponding with t_w; p_d=vapor pressure, in. Hg, corresponding with t_d; p_v = actual vapor pressure, in. Hg, corresponds to t_c; and RH = relative humidity, % (calculated as decimal).

Example. If the dry-bulb temperature is 90° F., and the wet-bulb temperature is 65° F. at an elevation of 8,000 ft (barometric pressure equals 22.22 in. Hg), what is the actual vapor pressure, dew-point temperature, and relative humidity?

Solution. On the nomogram, Step 1, connect 8,000 ft on Altitude scale with 25° F. (90° − 65° = 25°) on Wet-Bulb Depression scale, and mark at intersection with Pivot Line. Step 2, from the marked point on Pivot Line, connect with 65° F. on Wet-Bulb Temperature scale, and read actual vapor pressure as 0.42 in. Hg and the corresponding dew-point temperature as 53.7° F. where line crosses Dew-Point scale. Step 3, align this point with 90° F. on Dry-Bulb Temperature scale, and read 29.3% where line intersects Relative Humidity scale. END

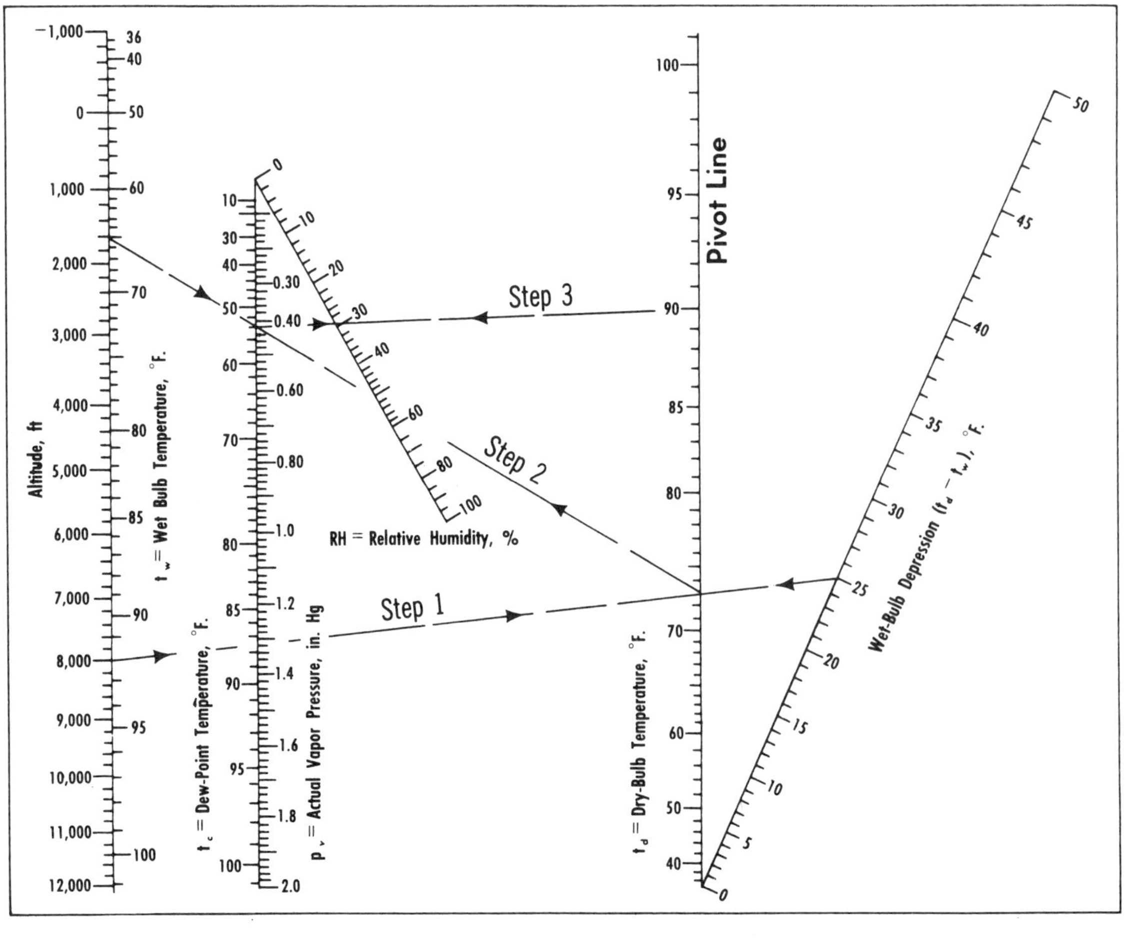

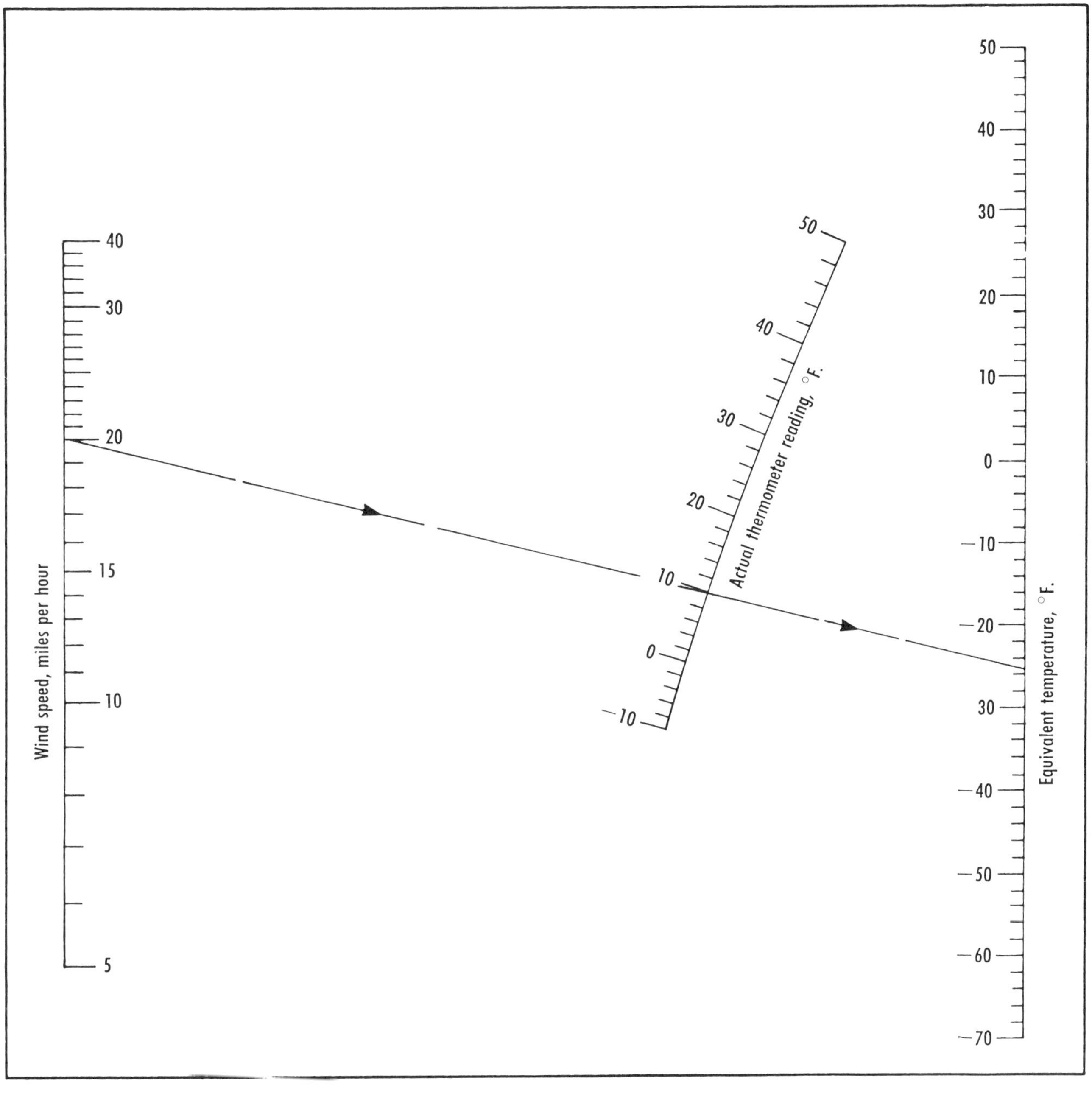

Nomogram gives wind-chill index

BILL SISSON
Superintendent of Operations
Nipak Inc.
Pryor, Okla.

DURING the winter months you often hear the radio or TV announcer use the term "wind chill index" and give the actual thermometer-reading temperature and the equivalent temperature for the prevailing wind speed. If you know or can estimate the wind speed and atmospheric temperature, the nomogram presented here will provide the equivalent temperature.

Example. What is the equivalent temperature when the actual thermometer reading is 10° F. and the wind speed is 20 miles per hour?

Solution. Connect 20 mph on the wind-speed scale with 10° F. on the actual thermometer-reading scale, extend to the equivalent-temperature scale, and read − 25° F. (25 below zero). END

Nomogram estimates room heat loss

Bill Sisson
Superintendent of Operations
Nipak Inc.
Pryor, Okla.

A NUMBER of published reference books give methods for finding the heat loss from a building or room. For a quick estimate, the following Carpenter formula is frequently used:

$$Q=[0.02\ n\ V+G\ +\ 0.25W]\ \Delta t$$

Where Q=heat loss, BTU/hr; n=number of air changes/hr; V=room volume, cu ft; G=exposed glass surface, sq ft; W=exposed wall surface, sq ft; and Δt=temperature difference between room and outdoors, °F.

The nomograph shown here, based on this formula, provides a fast method for estimating heat loss as illustrated by the following example:

Example. What is the heat loss in a room where the air changes 2 times/hr, if the room has a volume of 12,000 cu ft, an exposed wall surface of 2,000 sq ft, an exposed glass surface of 320 sq ft, the outside temperature is 20° F., and the room temperature is 64° F?

Solution. Align (Step 1) 12,000 cu ft on V scale with 2 on N scale, extend to Pivot line 1 and mark intersection. From marked point (Step 2) connect with 320 sq ft on G scale, and mark at intersection with Pivot line 2. Connect this point (Step 3) with 2,000 sq ft on W scale and mark where line crosses Pivot line 1. From this last marked point (Step 4), line with 44° F. (64 −20=44) on Δt scale, extend to Q scale, and read heat loss as 57,200 BTU/hr. END

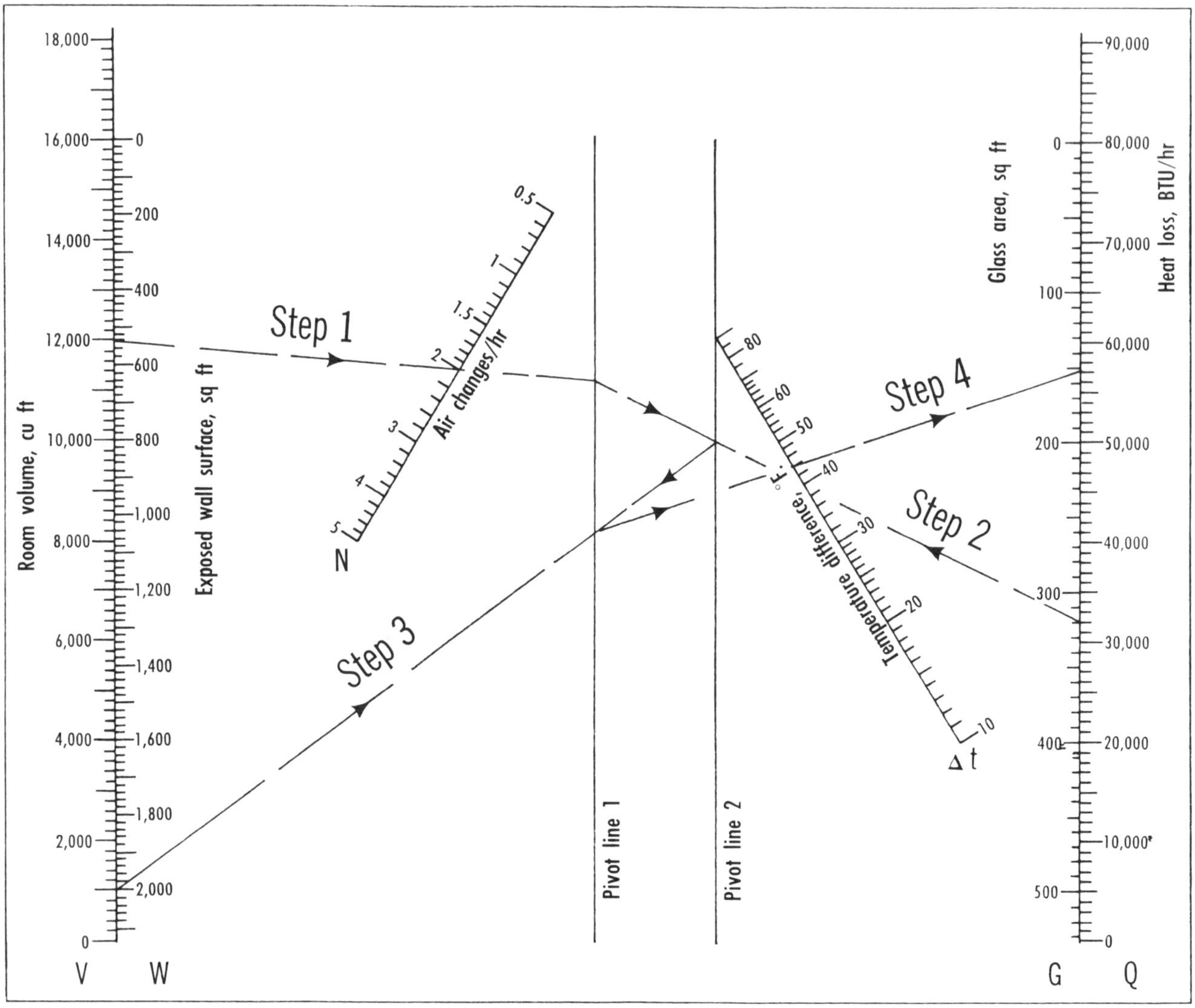

Use this nomogram. . .

to determine relative humidity the easy way

BILL SISSON
Superintendent of Operations
Nipak Inc.
Pryor, Okla.

RELATIVE humidity is usually shown in increments of 10% on a psychrometric chart, and this requires considerable interpolation when determining the relative humidity from wet and dry-bulb readings. Also, the number of other values found on a well-constructed psychrometric chart results in a great many lines that leads to confusion when determining relative humidity.

A faster, less confusing method for finding relative humidity at standard barometric pressure of 29.92 in. Hg (14.7 psig) from dry and wet-bulb temperature readings is provided with the nomogram shown here.

Example. What is the percent relative humidity, if the dry-bulb temperature is 72° F. and the wet-bulb temperature is 60° F.?

Solution. On the nomogram, connect 60° F. wet-bulb temperature on the WB scale with 72° F. dry-bulb temperature on the DB scale, extend to the R scale, and read 50% relative humidity. END

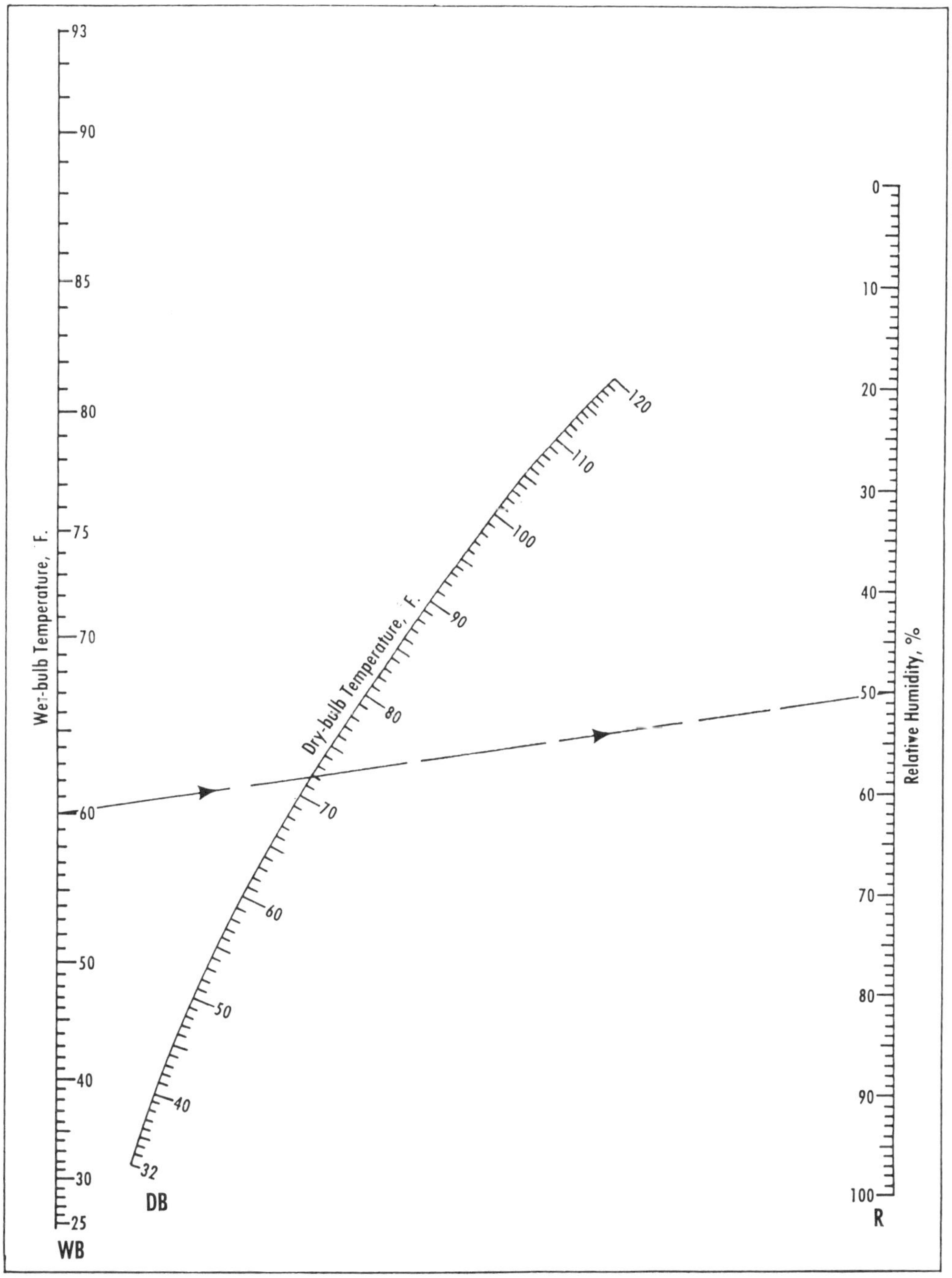

Nomogram gives maximum room relative humidity

BILL SISSON
Superintendent of Operations
Nipak Inc.
Pryor, Okla.

WHENEVER there is a difference of temperature and pressure of water vapor across a structure, conditions may develop that lead to a condensation of moisture. This condensation occurs at the point of saturation temperature.

This condensation, or sweating, results from the contact of humid air with a surface whose temperature is below the dewpoint of the air. The maximum room relative humidity that can be maintained and yet avoid condensation on wall, roof, or glass can be calculated with this nomogram based on the equations,

$$t_{dp} = t_i - [U(t_i - t_o)] / 1.46$$

$$R = 100(p_v / p_d)$$

Where t_{dp} = dew point temperature of room air = dry-bulb temperature of inside surface, °F.; t_i = room temperature, °F.; t_o = outdoor temperature, °F.; U = transmission coefficient, BTU/hr – sq ft/°F.; 1.46 = inside air film or surface conductance, BTU/hr – sq ft/°F.; R = relative humidity, percent; p_v = vapor pressure at t_{dp}, in. Hg; and p_d = vapor pressure at t_i, in. Hg.

Example: If a room that has a 12-in. stone wall with ⅝-in. sand aggregate plaster is at 70° F., and the outdoor temperature is –20° F., what is the dewpoint temperature of the room, and what is the maximum room relative humidity that can be maintained without condensation?

Solution. The 1965-66 ASHRAE Guide and Data Book gives the transmission coefficient (U) for 12-in. (150 psf) stone wall as 0.52 BTU/hr-sq ft/°F. On the nomogram, Step 1, connect 70° F. on t_i scale with –20° F. on t_o scale, and mark at intersection with Pivot line 1. Step 2, line this marked point with 0.52 on U scale, and extend to Pivot line 2. Step 3, from the point found on Pivot line 2, connect with 70° F. on t_i scale, and read the maximum room relative humidity as 30.5% where line crosses R scale, and dew point of room air as 38° F. where line crosses t_{dp} scale. END

Maximum room relative humidity

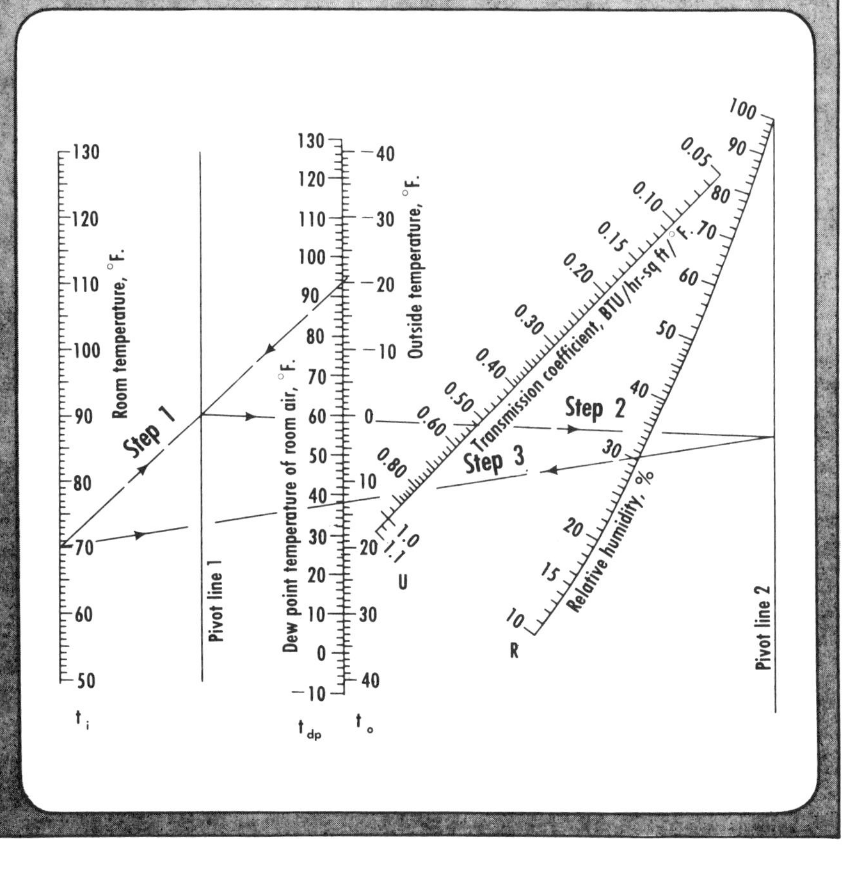

Nomogram tells unit-heater output

Bill Sisson
Superintendent of Operations
Nipak, Inc.
Pryor, Okla.

ONE of the problems when you install new unit heaters in your plant is to determine the size of the condensate piping and associated equipment. The nomogram presented here was developed as an aid in doing just that.

Where only the cu ft/min capacity of a unit heater and the air temperature are known, you can find the actual BTU output by using the following formula:

BTU output = cfm × 1.08 × temperature rise

When you know the BTU output, it is only necessary to divide the BTU output by the latent heat of the steam at the steam pressure used to determine the actual weight of steam condensed:

Condensate, lb = BTU output/Latent heat of steam at given pressure

The nomogram presented here was designed on the basis of the preceding formulas.

Example. What is the actual BTU output and how much condensate is condensated in a unit heater with a fan capacity of 7,000 cfm? Air temperature rise is 90° F., and steam pressure is 30 psig.

Step 1, align 7,000 cfm on the A scale with 90 on the E scale, and read 680,400 BTU where the line crosses the D scale. Step 2, from that point, align with 30 psig on the B scale, and read 733 lb of condensate where the line crosses the C scale.

END

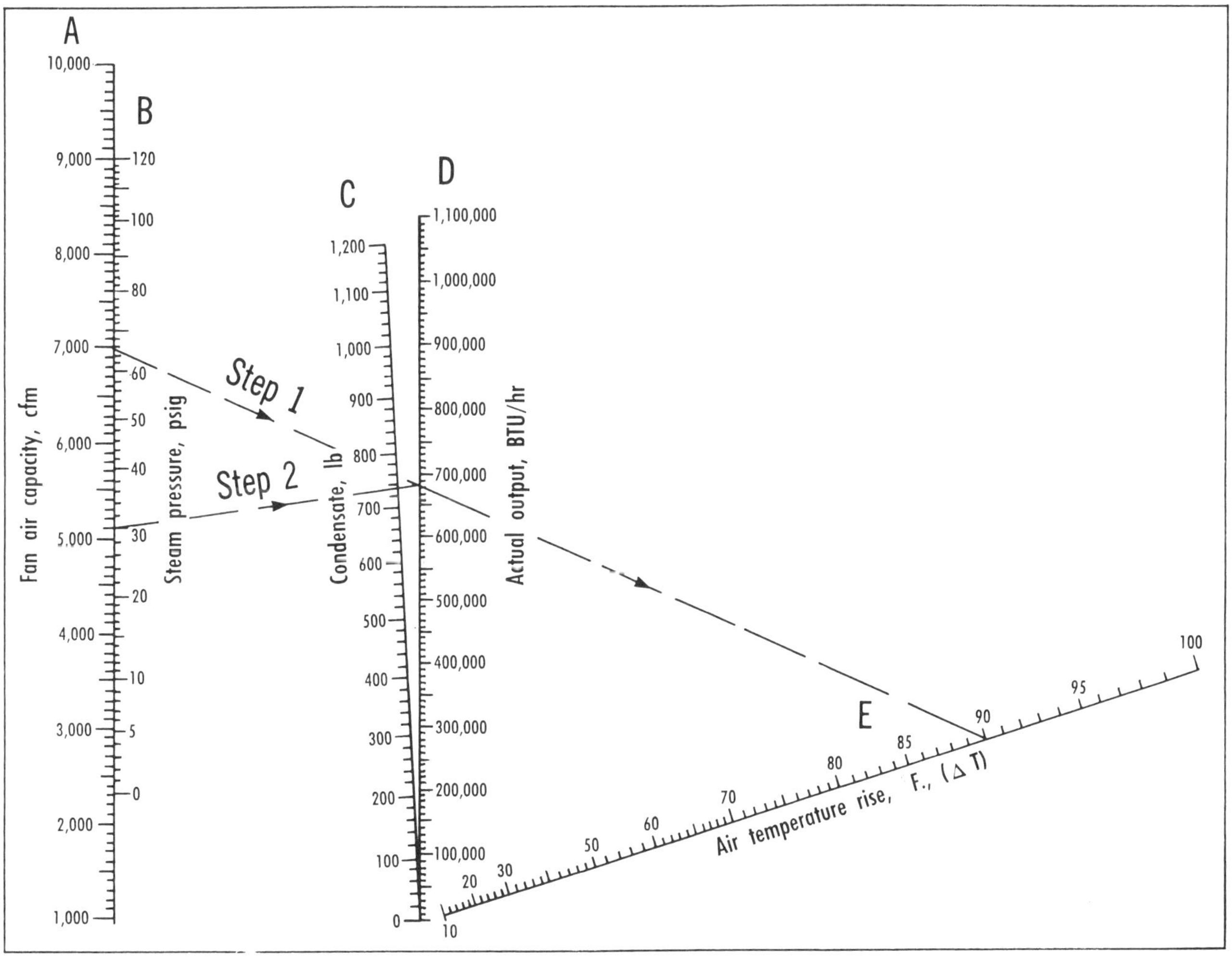

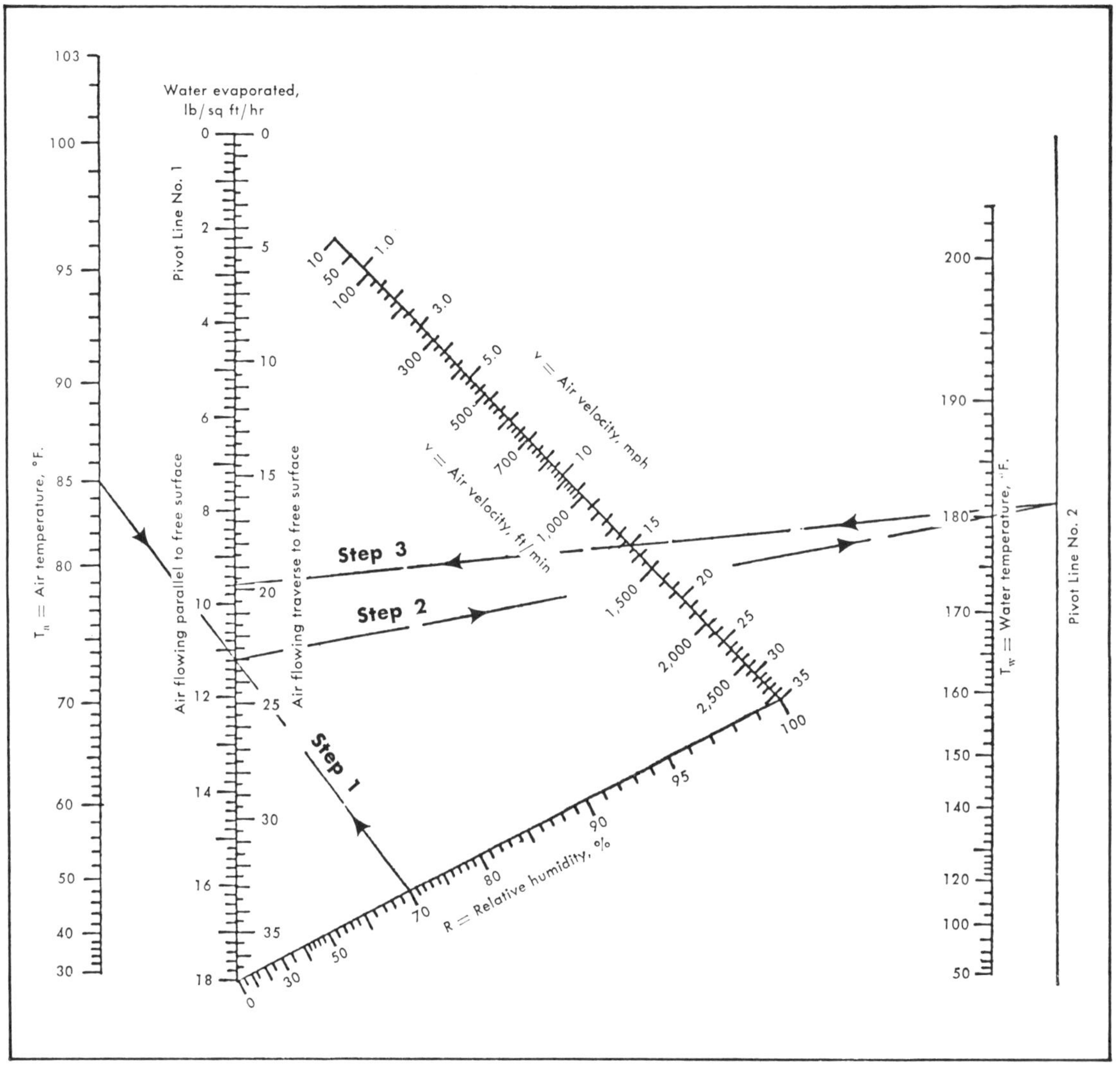

Nomogram determines water evaporated

Bill Sisson
Superintendent of Operations
Nipak Inc.
Pryor, Okla.

THE effect of cooling on water exposed to air movement is proportional to the exposed surface of the water. The rate of evaporation from a free water surface is controlled by the vapor pressure of the water at its temperature, vapor pressure of water in the surrounding air at the dew-point temperature of the air, average effective velocity of the air over the free surface of the water, and the direction of air flow with the respect to the free surface. The rate of evaporation with air flowing parallel to the free surface may be found with the formula:

$$W = [(95+0.425v)\ (P_w - P_a)]/h$$

and with air flow traverse to the free surface by:

$$W = [(201+0.88v)\ (P_w - P_a)]/h$$

Where: W = water evaporated, lb/sq ft/hr; v = longitudinal velocity of air, fpm; P_w = vapor pressure of water at its temperature, in. Hg; P_a = vapor pressure of moisture in air at air temperature, in. Hg; and h = latent heat of evaporation.

The nomogram can be used to determine the water evaporated for air flowing either parallel or traverse to the free surface as shown by the following example.

Example. If air is at a temperature of 85° F. and 70% relative humidity, water temperature is 180° F., and air velocity is 15 mph, what is the rate of evaporation if air is flowing parallel to a free surface and if the air is flowing traverse to a free surface?

Solution. Line 85° F. (Step 1) on air-temperature scale with 70% on Relative Humidity scale, and mark where line crosses Pivot Line No. 1. From this point (Step 2), line with 180° F. on Water Temperature scale, and extend to Pivot Line No. 2. From point found on Pivot Line No. 2 (Step 3), connect with 15 mph (1,320 ft/min) on Velocity scale, and project to Water Evaporated scale, and read 9.58 lb/hr/sq ft free surface for air flowing parallel to free surface and 19.8 lb/hr/sq ft free surface for air flowing traverse to free surface. END

Evaporation from lakes, ponds, reservoirs

Bill Sisson
Superintendent of Operations
Nipak Inc.
Pryor, Okla.

DATA on evaporation from lakes and reservoirs are not extensive. But there are formulas by which it may be computed. One of these; by Fitzgerald, has the form, $E_h = (S-F)(1+v/2)/60$; where E_h = evaporation rate, in./hr; S = vapor pressure of water at water temperature, in. Hg; F = vapor pressure existing in the air; and v = wind velocity, mph. Wind velocities are at the water surface and may be taken at one-half those recorded at an elevated station such as the Weather Bureau stations. For larger reservoirs, however, Weather Bureau values give results in close agreement with available direct measurements.

An alternative and substantially equivalent formula is given by Fitzgerald in more usable terms. Somewhat simplified and transformed; it is: $E_h = 0.0002\,(T_a - T_{wb})\,(1 + v/2)$; where T_a and T_{wb} are in the air temperature and wet-bulb temperature, respectively. The nomogram is based on the second formula. It includes the relative humidity for convenience.

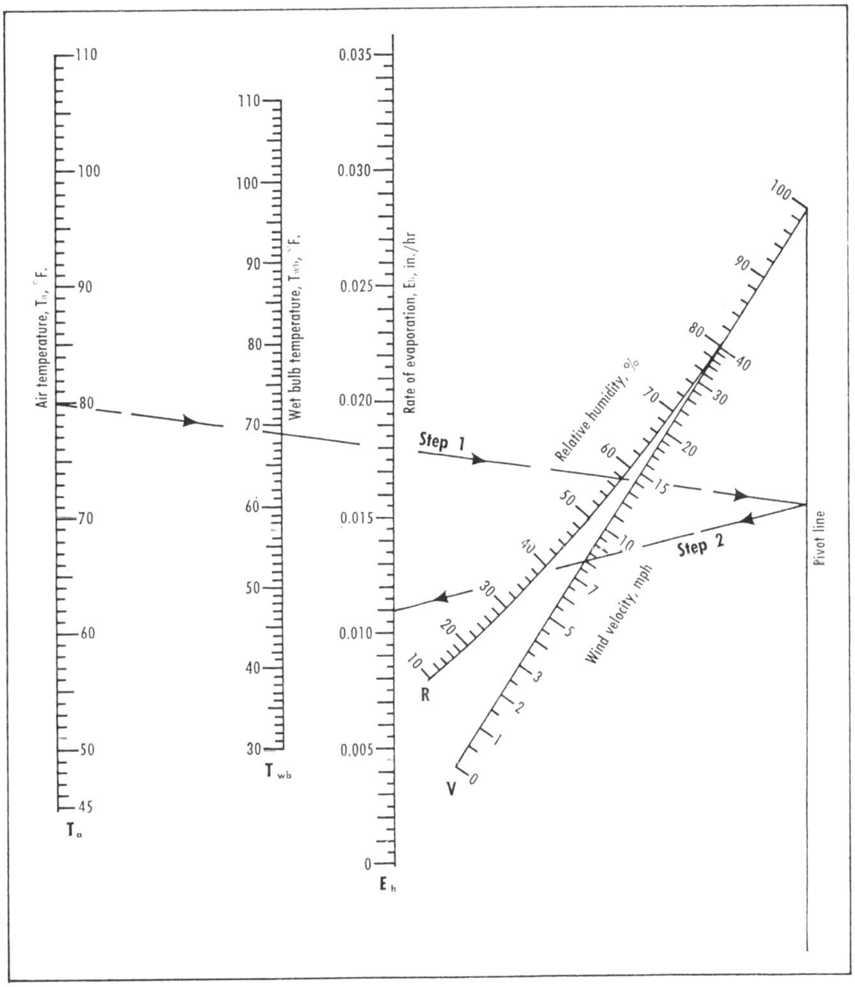

Example. Assume the "normal" or long-term monthly temperature, relative humidity, and wind velocity for a certain location are 80° F., 58%, and 8 mph; what is the "normal" wet-bulb temperature, and what is the evaporation rate per hour and per month of 31 days?

Solution. Step 1, line 80° F. on T_a scale with 58% on R scale, extend to Pivot line and mark. Also read wet-bulb temperature as 69° F. where line crossed T_{wb} scale. Step 2, from marked point on Pivot line, connect with 8 mph on V scale, extend to E_h scale, and read evaporation rate as 0.011 in./hr. The evaporation rate per month = $0.011 \times 24 \times 31 = 8.184$ in. END

Chart gives engine cooling-water needs

Bill Sisson
Superintendent of Operations
Nipak, Inc.
Pryor, Okla.

THIS nomogram can be used to quickly determine the water circulation required to cool diesel or gas engines. It can also be used to find the requirements for other related equipment including direct-driven compressors.

With a given engine brake horsepower and known heat-rejection rate, BTU/bhp/hr, you can use this nomogram to calculate the required water-circulation rate in gpm to give a desired water-temperature difference. Also, with a fixed circulation rate you can find the temperature difference this water flow will produce.

Note that the heat-rejection rate is normally included in the engine name-plate data. If not, it will be found in the manufacturer's catalog.

The chart is based on the equation,

$$q = Hu/500\ \Delta T$$

Where:

q = cooling water required, gpm
H = engine rating, bhp
u = heat to be dissipated to cooling water, BTU/bhp/hr
ΔT = temperature difference between inlet and outlet cooling water, °F.

Example. Determine the cooling-water circulation required for an engine rated at 1,000 bhp. Heat rejection rate is 4,000 BTU/bhp/hr, and water temperature difference is 20° F.

Align 1,000 bhp on the A scale with 4,000 BTU/bhp/hr, on the B scale, extend the line to the reference line and mark. From this marked point, align with 20° F. water temperature difference on the D scale, and read 400 gpm as the cooling-water requirement where the line crosses the C scale.

With a fixed water-flow rate, you can work the last step to determine the differential temperature produced. END

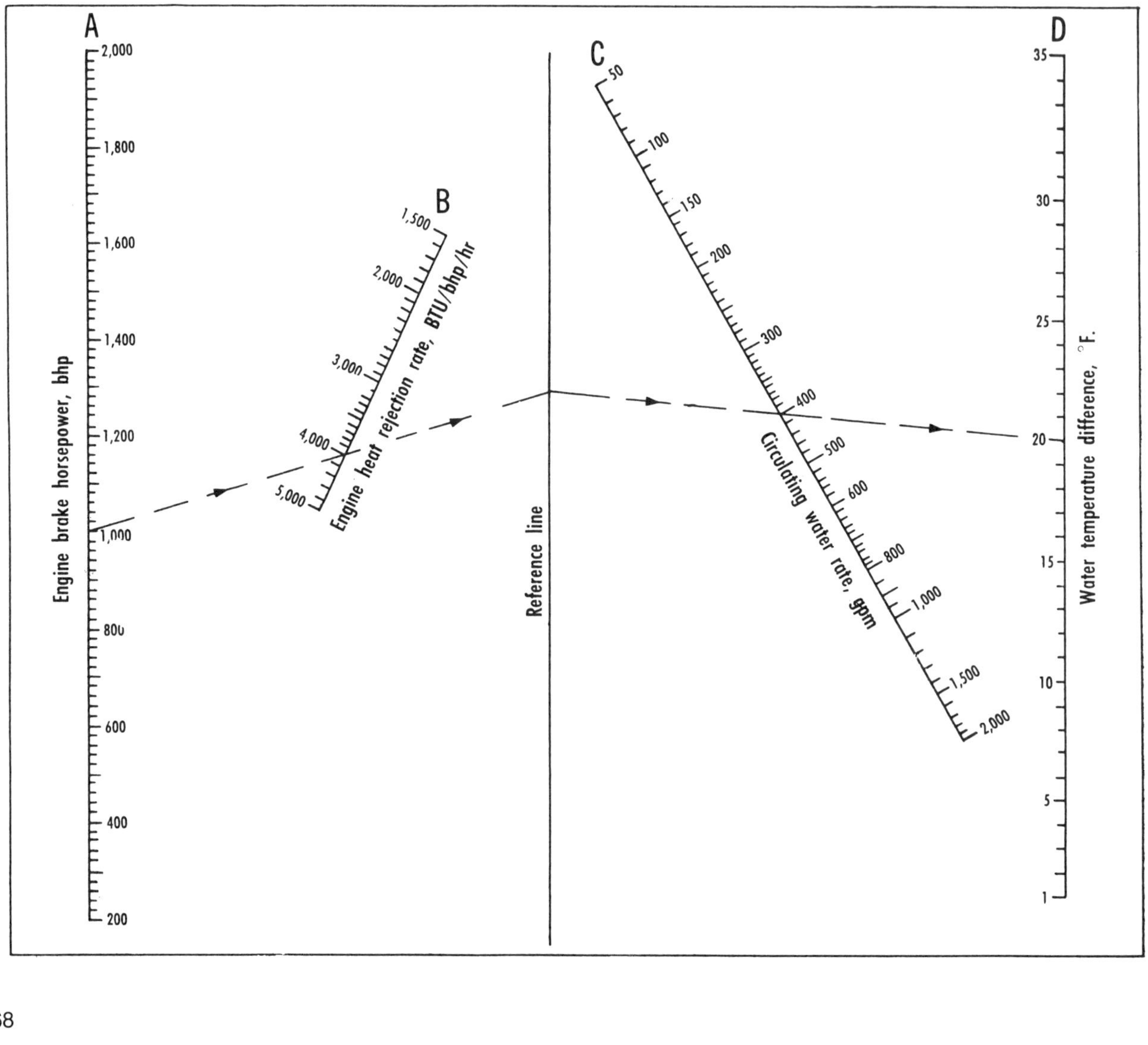

Use these nomograms to find the

Right size thermoelectric generator

Otmar Schreiber
Marketing Manager
Thermoelectric Div.
General Instrument Corp.
Hicksville, N.Y.

THESE nomograms are designed to help you select the required power rating for thermoelectric generators used for cathodic protection of pipelines and equipment. They are based on the assumption that the protection circuit resistance and required current are known.

Nomogram A helps you find the power required for series-connected thermoelectric generators that are electrically matched at 10 amp output current. Nomogram B helps you find the power required for two parallel banks of thermoelectric generators that are electrically matched at 20-amp output current (that is, two 10-amp generators in parallel). Use Nomogram A for current requirements from 3 to 15 amp and Nomogram B for current requirements from 15 to 30 amp.

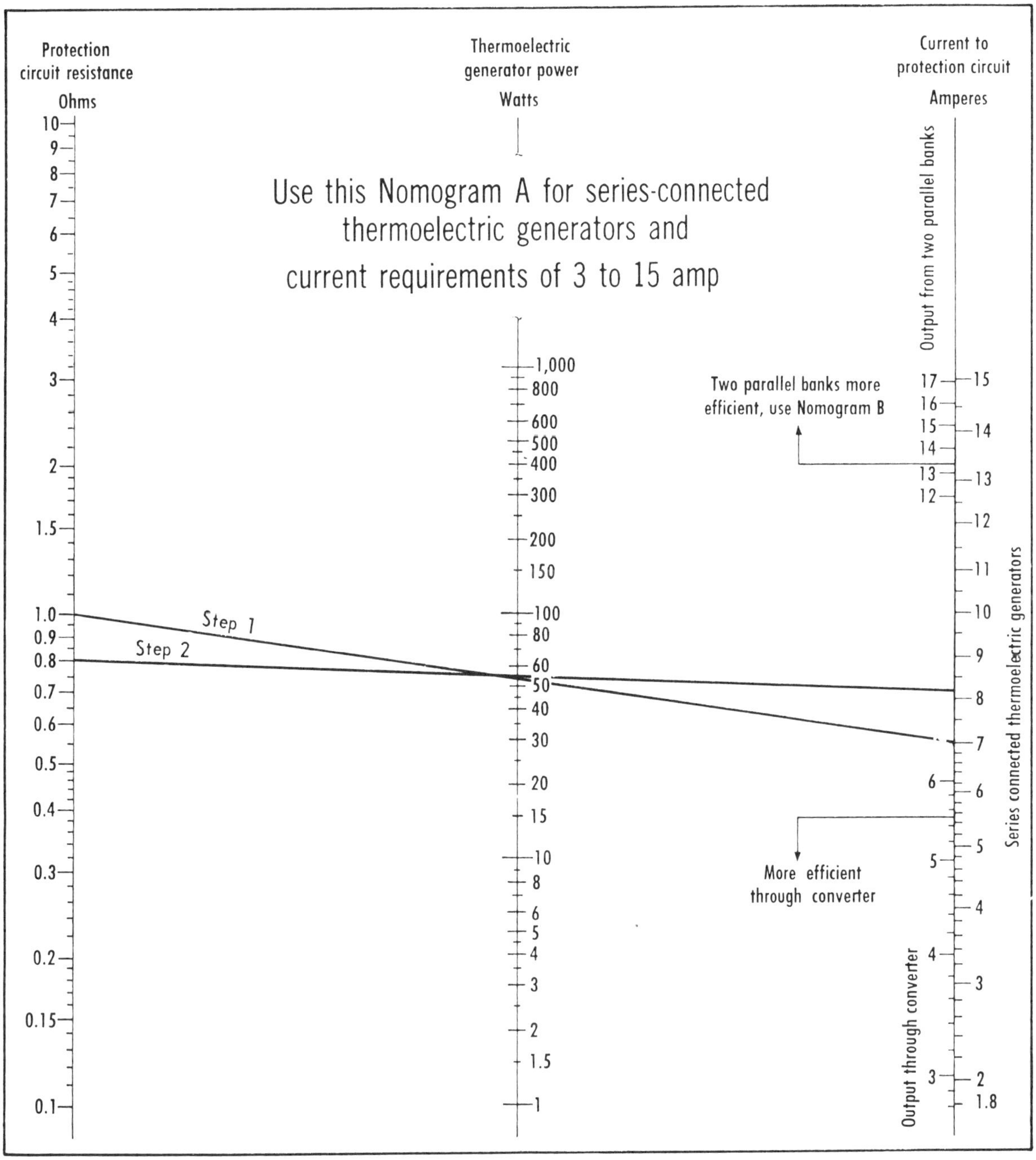

Noise level at various distances

Bill Sisson
Superintendent of Operations
Nipak Inc.
Pryor, Okla.

NOISE has become increasingly important. This is due to noise exposure standards of the Walsh-Healey Act and OSHA. This act set permissible daily noise-exposure levels that are to be measured in a certain way.

Sound is usually measured with a microphone and a sound-level meter that provides direct decibel readings on an indicating or recording instrument. There are three weighting networks (electrical circuits that react differently to sound, depending on its frequency) designated A, B and C. Readings are taken with the proper weighted network switched into the circuit. The unit of measurement is dbA if reading is taken on A scale, dbB if taken on B scale, and dbC if taken on C scale.

The A-weighted response corresponds approximately to the human ear's response at the threshold of audibility and provides a means of rank-order loudness or perceived noise levels of sound with similar characteristics. For this reason, A-weighted measurements, abbreviated dbA, have become the basis for most industrial noise regulations.

If the sound-pressure level at one distance is known, the sound pressure level at a second distance can be found from the formula:

$$SPL_2 = SPL_1 - 20 \times \log_{10} (D_2/D_1)$$

where SPL_2 = sound-pressure level at second distance, dbA; SPL_1 = sound-pressure level at distance where measurement is made, dbA; D_1 = distance where measurement is made, ft; and D_2 = second distance, ft. The nomograph shown will calculate the formula as shown by this example.

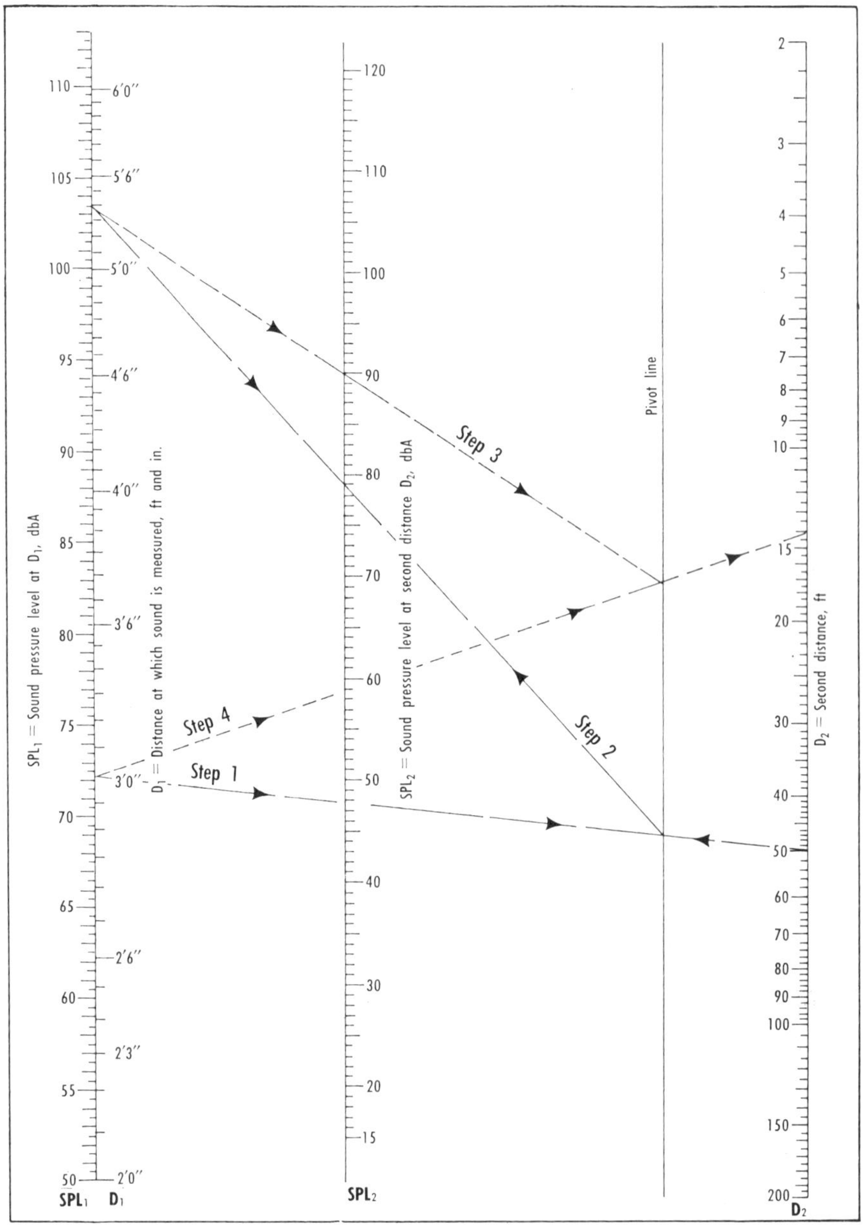

Example: If the sound-pressure level produced by a machine is measured as 103.5 dbA at 3 ft from the machine, what is the sound-pressure level at 50 ft? At what distance is the sound-pressure level 90 dbA?

Solution: Step 1, line 3 ft on D_1 scale with 50 ft on D_2 scale and mark where line crosses Pivot line. Step 2, from marked point on Pivot line, connect with 103.5 dbA on SPL_1 scale and read sound level at 50 ft as 79 dbA where line crosses SPL_2 scale. Step 3, to find distance at which sound-pressure level is 90 dbA, connect 103.5 on SPL_1 scale with 90 of SPL_2 scale and extend to Pivot line. Step 4, from point found on Pivot line, connect with 3 ft on D_1 scale, extend D_2 scale, and read 14 ft. END

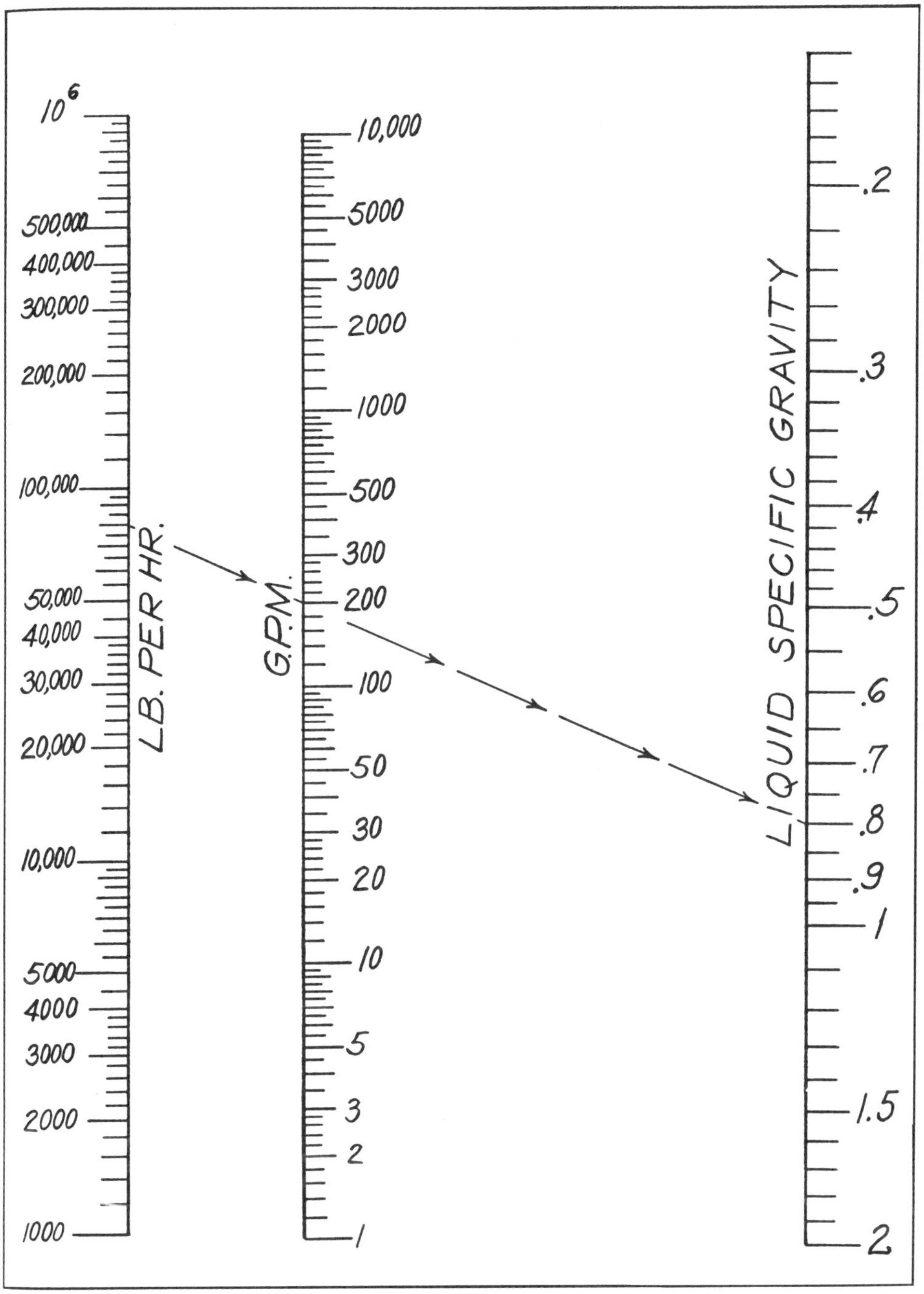

Convert flow units with this chart

IN MANY pumping calculations the flow rate may be expressed in either gallons per minute or pounds per hour. Frequently it becomes necessary to convert from one flow unit to another. This can be tedious, particularly if the specific gravity of the fluid is not 1.0. The accompanying chart makes these conversions easy.

Example: Oil having a specific gravity of 0.8 is flowing through a heater at the rate of 80,000 lb. per hour. What capacity pump would be required to handle this flow rate?

Solution: Enter the chart on the left at 80,000 lb. per hour and draw a straight line through the specific gravity, 0.8, on the right. At the intersection with the central scale read the flow rate as 200 g.p.m.

This chart can also be used to convert from gallons per minute to pounds per hour. Simply enter at the flow in gallons per minute and draw a straight line through the specific gravity. Extend this line to intersect the pounds per hour scale and read the result there.

The chart is based: Flow, gallons per minute = flow, pounds per hour/500S, where S = liquid specific gravity.

How Much Should Equipment Earn?

THE value of this chart is that it helps you determine, without involved calculations, how much money a plant or piece of equipment should bring in each year to pay back the investment plus interest. It's based on this equation:

$$A = (P-S)\left[\frac{i(1+i)^{n}}{(1+i)^{n-1}}\right] + Si$$

where:

A = annual earnings or annual cost of capital recovery
P = original investment
S = salvage value of equipment
i = interest rate or minimum attractive rate of return
n = economic life of equipment or investment, years

Example: How much should a $30,000 investment in equipment return each year to repay the investment plus interest? Assume the interest rate or desired rate of return is 10 per cent, the equipment life is 6 years, and its salvage value is zero. Further, what should be the annual return if the equipment has a salvage value of $5,000?

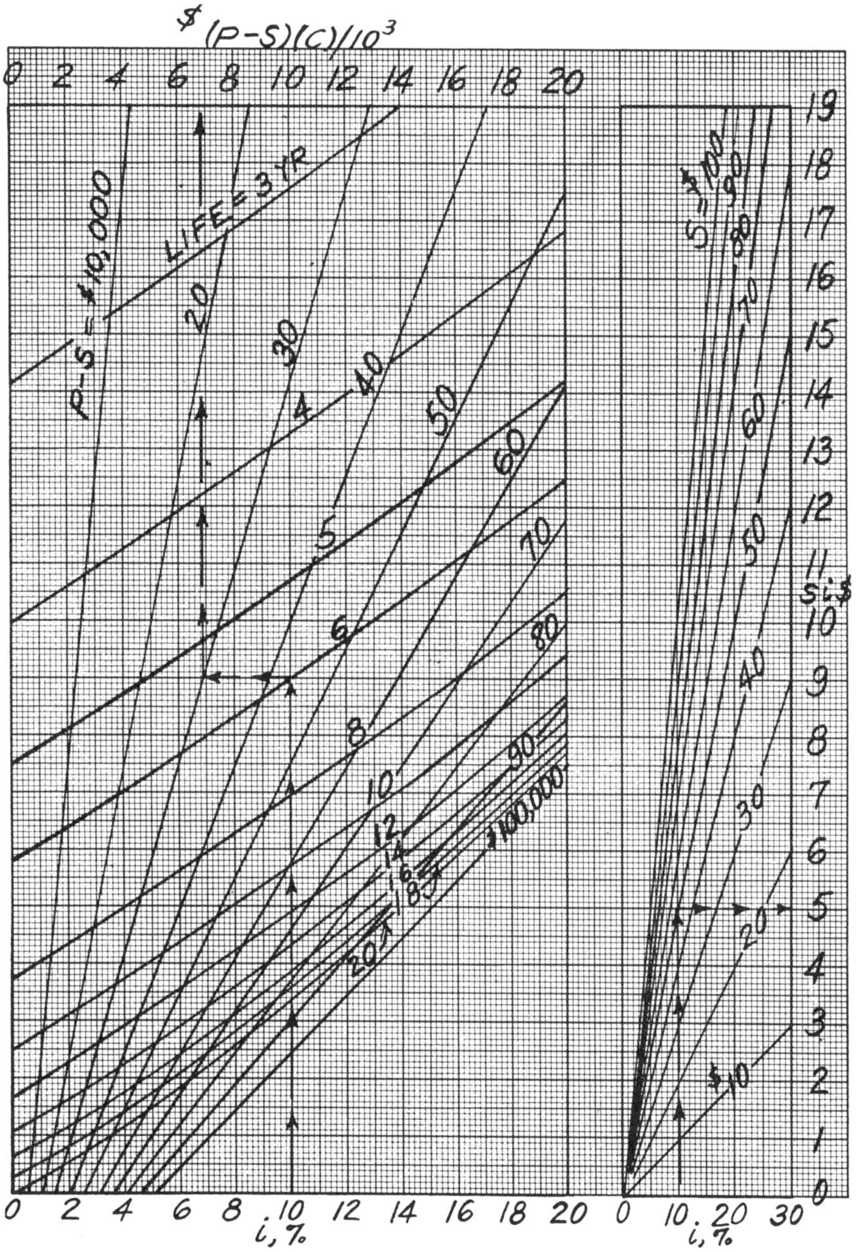

THESE GRAPHS can help you find how much money equipment should bring in annually to pay for investment and interest.

Solution: To solve the first part of this problem, enter the chart on the left at 10 per cent interest on the bottom. Then project vertically to the economic life of 6 years. From there project horizontally to the investment of $30,000. Read at the top of the chart an annual return of $6,900.

To solve the second part of the problem, enter the bottom of the chart on the right at the 10 per cent rate of return. Next project vertically to the curve marked 50. Read at the right the product of Si as $500. Enter the left chart and project as before, except that the value of P−S = $30,000 − $5,000 = $25,000. Read the annual return at the top of the chart as $5,750. Thus the annual return is now $5,750 + 500 = $6,250.

Extending the charts . . . Either chart can be used with numerical values of either P−S or S greater or less than plotted. Simply move the decimal point one or more places to the left or right in either value before entering the chart. Solve as before. Then move the decimal the same number of places to the left or right in the result. Thus, with a salvage value of $5,000, we moved the decimal point two places to the left and solved for $50. Then we moved the decimal point two places to the right in the result, $5, to obtain the correct value of $500. **End.**

Nomogram tells economic order

F. Caplan
Kaiser Engineers

THIS nomogram is designed to help you quickly solve the following formula which can be used to calculate the economic order quantity when you must carry a number of items in inventory:

$$Q = (2\ P\ Y / C\ I)^{1/2}$$

Where:
Q = economic order quantity
P = total cost of purchasing and receiving one order quantity, including clerical, accounting, inspection, and handling costs
Y = number of items used per year
C = cost per item, including freight
I = yearly inventory carrying cost, expressed as decimal fraction of a dollar, including cost of capital invested, insurance, space, heating, etc.

Example. What is the economic order quantity if P = \$6.00, Y = 200, C = \$1.50, and I = \$0.04? Step 1, align P = \$6.00 with Y = 200 and mark intersection with Pivot Line 1. Step 2, align C = \$1.50 with I = \$0.04 and mark intersection with Pivot Line 2. Step 3, align Pivot Line marks 1 and 2 and read Q = 200 units as the economic order quantity. END

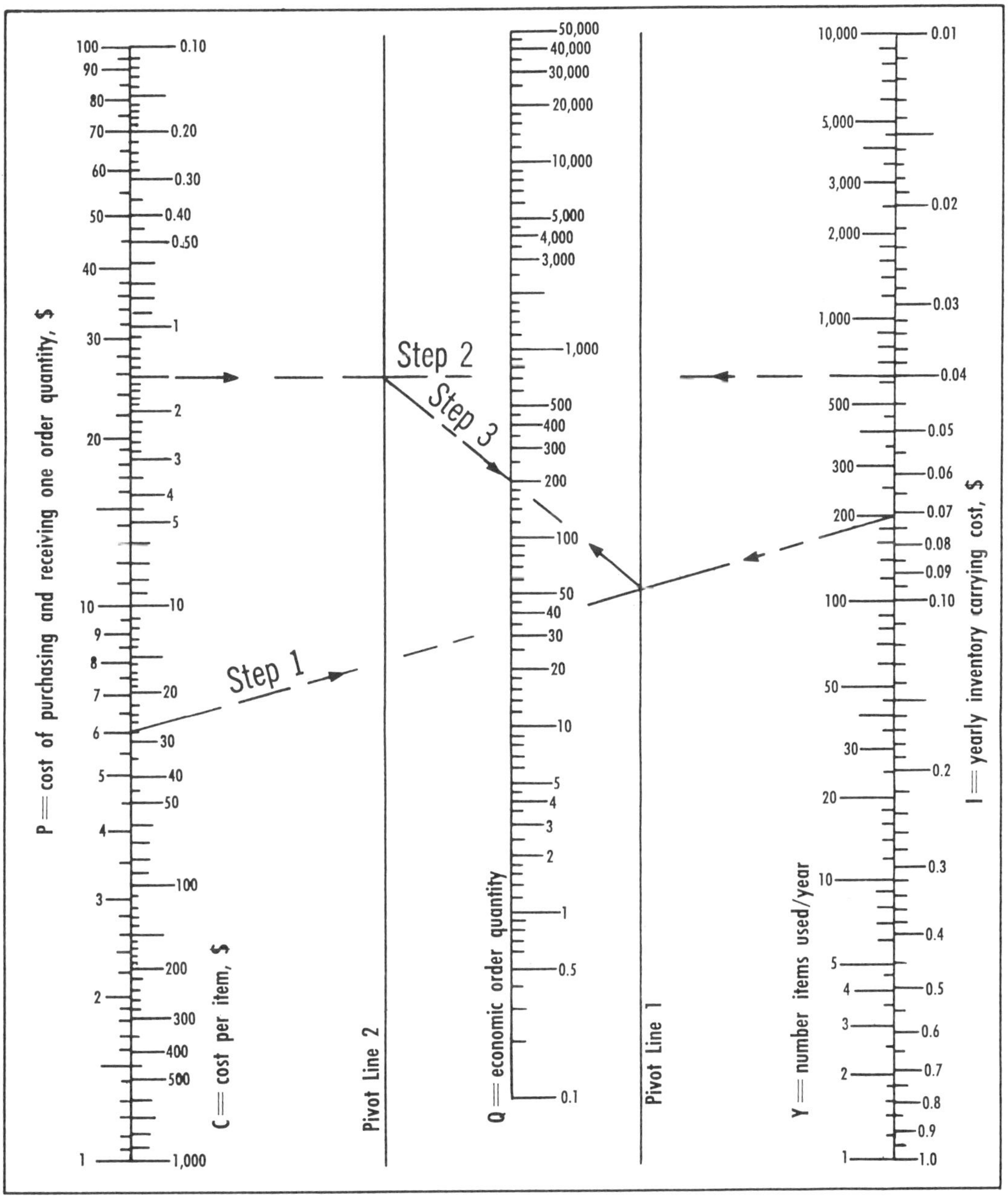

Nomogram finds day of week for any date

Bill Sisson
Superintendent of Operations
Nipak Inc.
Pryor, Okla.

EVERYONE, at some time or another, needs to know the day of the week for some specific past or future date. With many people, this happens rather frequently. When it does, they usually have to do considerable searching to find the necessary calendars. This is an annoying, time-consuming chore that all of us would like to avoid. How do we avoid this chore? Simple—use a nomogram. The nomogram shown here will find the day of the week of any date from 1960 through 1990.

Example. What day of the week is August 7, 1987?

Solution. Step 1, connect 1987 on Year scale with 7 on Date-in-month scale, and mark at intersection with Pivot line. Step 2, connect marked point with August reference point, and read 6 where line crosses day-of-week scale. Table 1 on nomogram shows 6 to be Friday. END

Day of week

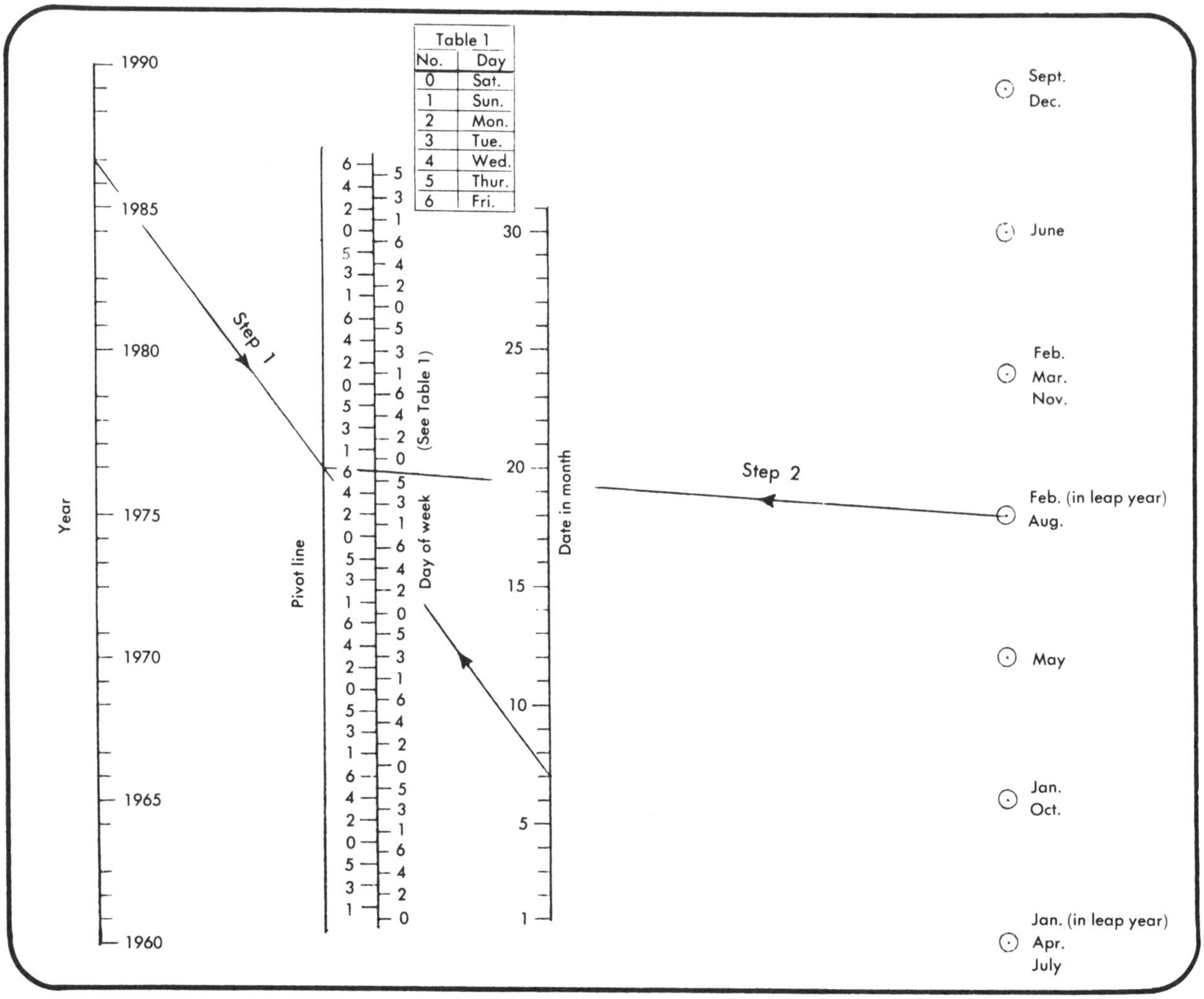

Table 1

No.	Day
0	Sat.
1	Sun.
2	Mon.
3	Tue.
4	Wed.
5	Thur.
6	Fri.

Nomogram tells powers and roots

F. Caplan
Kaiser Engineers

THIS nomogram can be kept handy to tell you the powers and roots of numbers fast. It is based on the equations,

$$W = Y^x$$

$$Y = W^{1/x}$$

Where:

W = Y raised to the x power
Y = root of W
x = power or root of number

Example. What is the square root of 200 (x = 0.5 or ½)? Step 1, align Y = 200 with x = 0.5 and read W = 14.1.

What is the value of 200 raised to the second power (squared)? Step 2, align W = 200 with x = 2 and read Y = 4 × 10⁴ = 40,000. END

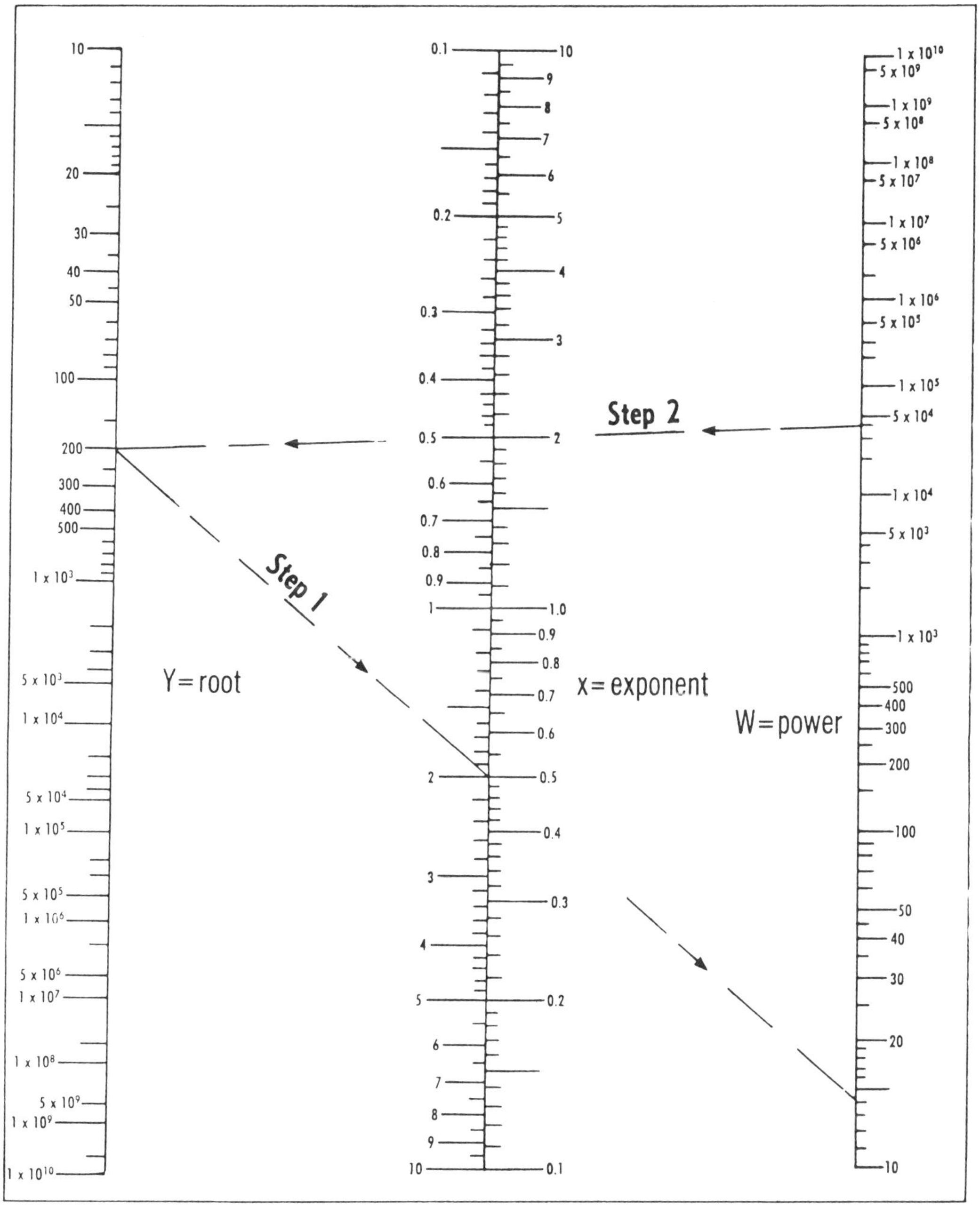

Nomogram gives volume of a cylinder, frustum of cone

Bill Sisson
Superintendent of Operations
Nipak Inc.
Pryor, Okla.

THE volume (V) of the frustum of a right circular cone can be calculated with the formula,

$V = \frac{1}{3} \pi H (R^2 + Rr + r^2)$, cu ft

The volume of a right circular cylinder is:

$V = \pi R^2 H$, cu ft

Where:

R=radius of large base, ft
r=radius of small base, ft
H=height of frustum, ft

This nomogram provides an easy method for computing these formulas as shown by the following example:

Example. What is the volume of the frustrum of a cone when the larger radius is 4 ft, smaller radius is 1.25 ft, and height is 16 ft?

Solution. Step 1, connect 4 ft on R scale with 1.25 on r scale, and mark at intersection with Pivot line. Step 2, line point found on Pivot line with 16 ft on H scale, extend to V scale, and read volume of frustrum of the cone as 378 cu ft. Step 3, to find the volume of the cylinder, connect 4 ft on R scale with 4 ft on r scale, and mark at Pivot line. Step 4, connect this point with 16 ft on H scale, extend to V scale, and read volume as 804 cu ft. **END**

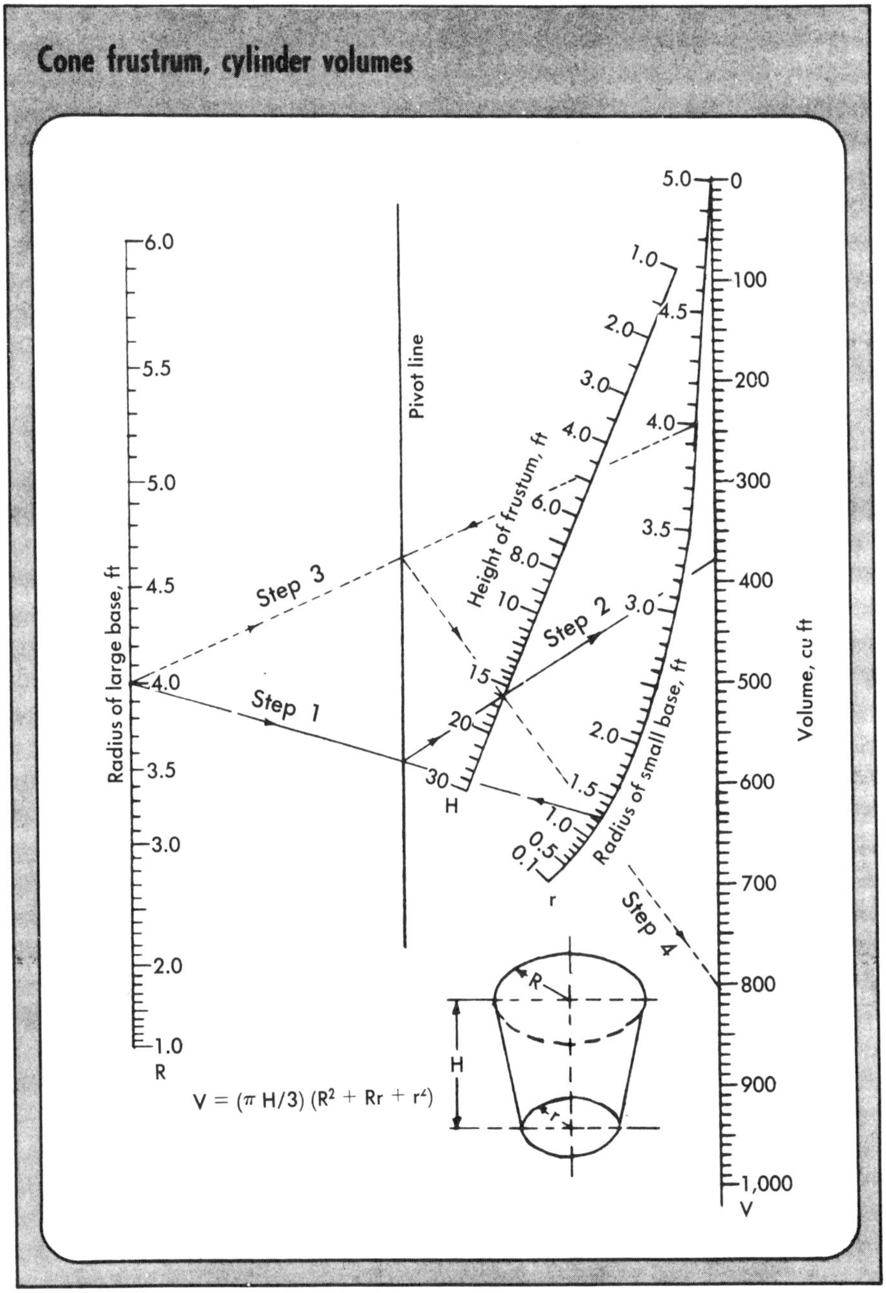